The twentieth century has seen biology come of age as a conceptual and quantitative science. Biochemistry, cytology, and genetics have been unified into a common framework at the molecular level. However, cellular activity and development are regulated not by the interplay of molecules alone, but by interactions of molecules organized in complex arrays, subunits, and organelles. Emphasis on organization is, therefore, of increasing importance.

So it is too, at the other end of the scale. Organismic and population biology are developing new rigor in such established and emerging disciplines as ecology, evolution, and ethology, but again the accent is on interactions between individuals, populations, and societies. Advances in comparative biochemistry and physiology have given new impetus to studies of animal and plant diversity. Microbiology has matured, with the world of viruses and procaryotes assuming a major position. New connections are being forged with other disciplines outside biology — chemistry, physics, mathematics, geology, anthropology, and psychology provide us with new theories and experimental tools while at the same time are themselves being enriched by the biologists' new insights into the world of life. The need to preserve a habitable environment for future generations should encourage increasing collaboration between diverse disciplines.

The purpose of the Modern Biology Series is to introduce the college biology student — as well as the gifted secondary student and all interested readers — both to the concepts unifying the fields within biology and to the diversity that makes each field unique.

Since the series is open-ended, it will provide a greater number and variety of topics than can be accommodated in many introductory courses. It remains the responsibility of the instructor to make his selection, to arrange it in a logical order, and to develop a framework into which the individual units can best be fitted.

New titles will be added to the present list as new fields emerge, existing fields advance, and new authors of ability and talent appear. Only thus, we feel, can we keep pace with the explosion of knowledge in Modern Biology.

James D. Ebert
Ariel G. Loewy
Richard S. Miller
Howard A. Schneiderman

Cells and Organelles

Alex B. Novikoff

Yeshiva University

Eric Holtzman

Columbia University

Holt, Rinehart and Winston, Inc.
New York   Chicago   San Francisco
Atlanta   Dallas   Montreal
Toronto   London   Sydney

To Professor Arthur W. Pollister

*from whom we both acquired much of our enthu-
siasm for cells. Between us we spanned 37 years of
his teaching histology and cytology at Columbia.*

Text and cover design by Margaret O. Tsao
Illustrations by George V. Kelvin • Science Graphics
Production supervised by Joseph Campanella

Preface

Cytologists study cells, all cells, and by many techniques. Because of their dependence on the microscope, they tend to analyze living systems in terms of visible structure, but in recent years cytologists have become increasingly concerned with biochemistry. *Cell biologists* start with a bias towards viewing cells in terms of molecules. Recently they have been much concerned with nucleic acids and proteins, the macromolecules which molecular biology has revealed to play primary roles in heredity. However, as studies progress, it is becoming more and more difficult to make a sharp differentiation among the various approaches to the cell—structural (classical cytology), physiological (biochemistry and biophysics), and molecular (cell biology). As distinctions between cell biology and cytology have become blurred, and perhaps out-moded, this book is about both cytology and cell biology.

The book is divided into five parts. The first part introduces the major features of cells and the methods by which they are currently studied. In the second, we consider each of the organelles in turn, presenting structural and functional information. Then, in the third part, we discuss the diversity of cell types constructed from the same organelles and macromolecules. The fourth part presents major mechanisms by which cells reproduce, develop, and evolve. The final part is a brief look at the progress and the future of cell study.

Many of the cells used to illustrate the principles we discuss are from higher animals. This is not only because the authors have had more direct experience with such cells. It also reflects the historical development of cytology; the cells of higher animals have been best analyzed from the viewpoint of correlated structure and function. However, increasing attention is being focused on protozoa, higher plants, and bacteria and related cells. Wherever possible we have referred to studies of such organisms.

We begin the book and each of the parts with introductions outlining the major themes of the chapters that follow and commenting upon some topics that do not fit conveniently into one chapter. In the chapters we combine descriptive and experimental information to provide key portions of the evidence on which contemporary concepts are based. The illustrations have been obtained from leading students of the cell. The figure legends and suggested reading lists will familiarize the reader with the names of some of those who have contributed to the progress of cytology and cell biology.

We hope that we convey some of the excitement felt today by students of the cell. In addition, we hope that not only past achievements, but important problems awaiting future solution become evident to the reader.

New York, N.Y.   *A.B.N.*
January 1970   *E.H.*

Contents

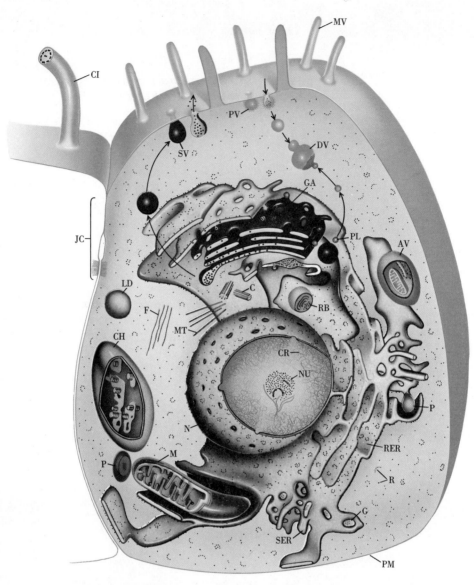

Schematic diagram of a cell and its organelles drawn to reveal their three-dimensional structure.

AV, autophagic vacuole; *C*, centriole; *CH*, chloroplast; *CI*, cilium; *CR* chromatin; *DV*, digestion vacuole; *F*, microfilaments; *G*, glycogen; *GA*, Golgi apparatus; *JC*, junctional complex; *LD*, lipid droplet; *M*, mitochondrion; *MT*, microtubules; *MV*, microvillus; *N*, nucleus; *NU*, nucleolus; *P*, peroxisome; *PL*, primary lysosome; *PM*, plasma membrane; *PV*, pinocytosis vesicle; *R*, ribosomes and polysomes; *RB*, residual body; *RER*, rough endoplasmic reticulum; *SER*, smooth endoplasmic reticulum; *SV*, secretion vacuole.

The organelles have been drawn only roughly to scale. Also, the sizes and relative amounts of different organelles can vary considerably from one cell type to another. For example, only plant cells show chloroplasts, and among animal cells only a few types show peroxisomes. A detailed enumeration of the organelle content of one cell type is presented in Chapter 2.12.

INTRODUCTION

The analogy between cells and atoms is a familiar one and like many familiar comparisons it is both useful and limited. Cells and atoms are units. Each is composed of simpler components which are integrated into a whole that exhibits special properties not found in any of the parts or in random mixtures of the parts. Both exhibit considerable variation in properties, based on different arrangements of components; the number of variations far exceeds the number of major components. Both serve as basic building blocks for more complex structures.

However, the analogy cannot be pressed too far; cells can reproduce themselves, whereas atoms cannot. The ability to utilize the nonliving environment to make living matter is probably the most fundamental property of life, and cells are the simplest self-duplicating units. Duplication is based on DNA (deoxyribonucleic acid) which can be *replicated* to form perfect copies of itself. Thus the genetic information encoded in DNA is perpetuated from one cell generation to the next, sometimes without significant variation over vast periods of time. DNA is unique among macromolecules in its replication. Only in certain viruses has another macromolecule (RNA, ribonucleic acid) been shown to replace DNA in its central hereditary role; the replication of RNA in these viruses is essentially like that of DNA.

Genetic information is expressed in cells by the mechanisms of *transcription* and *translation*. Transcription transfers the DNA-coded information to RNA molecules. Translation results in the formation of specific proteins whose properties are determined by the information carried by these RNA molecules. Among the proteins are *enzymes*, catalytic molecules that control most of the chemical reactions of cells. Enzymes differ in the kinds of molecules they affect (their *substrates*) and in the kinds of reactions they catalyze. Many are involved in the synthesis of the other cellular macromolecules, the nucleic acids (DNA and RNA), the lipids (fats and related compounds), and the polysaccharides (polymers made of many linked sugar molecules). Through this chain of transcription, translation, and enzymatic activities, DNA directs its own replication and controls as well the rest of *metabolism*, the sum total of all the chemical reactions that take place in cells. The chain is universal and thus all cells are made of the same classes of macromolecules (nucleic acids, proteins, lipids, polysaccharides) and smaller molecules such as water and salts. Duplication, and the presence in different cells of similar molecular and structural materials and mechanisms, are features of *cellular constancy*, one of the main themes of this book.

A second theme of the book is *cell diversity*. Cells may be classified into a large number of categories. Eucaryotic cells are distinguished from procaryotic cells, plant cells from animal cells, and muscle cells from gland cells. These distinctions derive from differences in morphology and metabolism. Eucaryotes differ from procaryotes in complexity of cellular organization. The unicellular protozoa, most algae, and the cells of multicellular plants and animals fall in the eucaryote category. In these cells, different specialized functions (such as respiration, photosynthesis, and DNA replication and transcription) are segregated into discrete cell regions which are often delimited from the rest of the cell by membranes. The cell's *organelles* reflect this segregation; they are subcellular structures of distinctive morphology and function. The most familiar of the organelles is the nucleus, which contains most of the DNA of the cell and enzymes involved in replication and transcription. The nucleus is separated by a surrounding membrane system from the rest of the cell, the cytoplasm. The cytoplasm contains many organelles including the *mitochondria*, the chief intracellular sites of respiratory enzymes; in plants, the cytoplasm contains *chloroplasts* in which are present the enzymes of *photosynthesis*, a metabolic process unique to plant cells. The mitochondria, chloroplasts, and a number of other cytoplasmic organelles are also delimited as discrete structures by surrounding membranes.

The procaryotes include the bacteria, the blue-green algae, and some other organisms. In contrast to the eucaryotes, they have relatively few membranes dividing the cell into separate compartments. This is not to say that all components are mixed together in a random fashion. The DNA, for example, does occupy a more or less separate nuclear region, but this is not delimited by a surrounding membrane. In fact, the traditional distinguishing feature of procaryotes is the absence of a membrane-enclosed nucleus (the suffix *caryote* refers to the nucleus). Respiratory and photosynthetic enzymes are not segregated into discrete mitochondria or chloroplasts, although, as will be seen, the enzymes are held in ordered arrangements within the cell.

The diversity of cell types owes its origin to evolution. By comparison with other macromolecules, DNA is remarkably stable. But *mutations* do occur at an appreciable, though low, frequency. Mutations alter the genetic information that is encoded in DNA and passed by a cell to its progeny; thus, they can produce inherited changes in metabolism. Some result in a *selective advantage*, roughly defined as an increase in the number of viable offspring produced per lifetime by an organism. In the evolving population, organisms carrying such advantageous mutations will slowly replace organisms without them. The pattern of the spread of a mutation in a population depends on reproduction and thus, ultimately, on mechanisms of division of cells.

Usually the daughter cells resulting from division of *unicellular organisms* are essentially similar to the parent cells; the daughters contain replicates of the parent cells' DNA and the DNA establishes the range of potential responses to the environment by specifying the available range of metabolic possibilities. Given a similar environment, there is little difference between parent and daughter. If the environment changes, parents and daughters will change in similar fashion and within genetically imposed limits. Diversity of cell types rests upon mutation.

In *multicellular organisms*; diversification of cell type without mutation is a regular feature of development. Most multicellular animals and plants start life as a single cell, a *zygote*, with a nucleus formed by the fusion of two parental nuclei. (Usually this results from fusion of sperm and egg or the equivalent.) The cell divides to produce daughter cells with identical DNA, but these *differentiate* into specialized cell types with different morphology and metabolism (for example, gland cells producing digestive enzymes or muscle cells rich in contractile proteins). Part of the mechanism for this is based on the fact that the immediate environment of a given cell is strongly influenced by the other cells of the organism. Cell interactions are of major importance in the development of multicellular organisms. They are key factors in determining what portion of its total genetic endowment will be expressed in a given cell. In different cell types, different portions of the DNA apparently are used in the transcription that underlies macromolecule synthesis; presumably, for a given cell type only a particular part of the total genetic information is responsible for the cell's characteristics. Thus constancy in DNA coexists with diversity in metabolism and morphology of the cells which carry that DNA. Differentiation implies that cells are not mere aggregates of independent molecules or structures each "doing its own thing," and that DNA molecules are not autonomous rulers of subservient collections of other molecules.

As in development, the normal functioning of adult multicellular organisms also depends upon the interaction of neighboring cells and upon long-range cell-to-cell interactions mediated, for example, by hormones or nerve impulses. Cells are integrated into tissues, tissues into organs, and organs into an organism. Similarly, cells are themselves highly organized; molecules are built into structures in which they function in a coordinated and interrelated manner, and often show properties not found in a collection of the same molecules free in solution. The products of one organelle may be essential to the operation of another. Cell functions depend upon mutual interaction of parts. *Cell organization* and the implications of organization for function are a third theme of the book.

Reproduction and constancy, evolutionary and developmental diversity, the integration of cellular components into a functional whole — all are subjects for investigation in cytology and cell biology. The fourth theme of the book is the dependence of major biological findings upon the development of *new methods of study* and upon the *choice of the best organism* for the problem at hand. We will illustrate the kinds of experiments and approaches currently used in cytology and cell biology. The central tool is the microscope, but (as outlined in the Preface) microscopy is increasingly supplemented by chemical and physical studies. Descriptive and experimental approaches supplement each other. The great diversity of cells and organisms presents opportunity for choice of cell types especially well suited for analyses of new problems. Investigations of pathological material and of cells experimentally subjected to abnormal conditions provide valuable clues to normal functioning.

Study of the cell is progressing rapidly and the solution of many problems presently unsolved may be anticipated with confidence. Some of the unsolved problems are of practical importance. As our understanding of cells increases, so does our ability to control and modify them. This ability is crucial for medicine and agriculture. It also raises important ethical and social questions.

part 1

CYTOLOGY
OF
TODAY

About forty years ago, the last edition of E. B. Wilson's great book, *The Cell in Development and Heredity,* was published. It was a summation and synthesis of a vast cytological literature. The work reflects the extraordinary ingenuity of early experimenters and the great excitement over what was then a recent appreciation of the roles of chromosomes in heredity. It is concerned mainly with eucaryotic cells. The nucleus had been extensively studied; some of the cytoplasmic organelles had been identified although clarification of their functions was only beginning. Biochemical analysis of cells was in its infancy.

Since that time, and especially in the last two decades, there has been a remarkable development of techniques applied to the study of cells. The electron microscope has extended the investigation of structure down to the level of macromolecules. Biochemists have separated and analyzed cell molecules and organelles and determined their metabolic functions. Cytology and biochemistry have been combined to the extent that modern cytology is often referred to as *biochemical cytology.*

To illustrate current views of *cell organization,* we will begin with the rat *hepatocyte,* the major cell type of the liver. This cell type has many important functions, ranging from the secretion of blood proteins and the storage of carbohydrates to the destruction of toxic material produced elsewhere in the body. This variety of physiological functions is one reason that hepatocytes are widely studied by biochemists. Biochemical study is facilitated by the relative homogeneity of the organ; hepatocytes constitute over 60 percent of the cells and 90 percent of the weight of the liver in the rat. (The remainder consists of cells of blood vessels, ducts, and supporting tissues and specialized *phagocytes*, cells that are believed to engulf dead red blood cells and remove them, along with other materials, from the blood.) Thus, constituents isolated from the liver come primarily from one cell type, the hepatocyte. Rats are readily available and they have large livers (almost 12 grams in an adult rat) from which relatively great quantities of cell constituents may be obtained. The hepatocytes are easily disrupted, to provide isolated organelles that can be studied by biochemical techniques. In addition, the liver is relatively easy to prepare for both light microscopy and electron microscopy.

To illustrate current views of *cellular metabolism*, we have chosen the metabolic pathway responsible for the formation of most of the ATP (*adenosine triphosphate*) of cells. ATP provides energy for virtually all cell functions: transport of substances into and out of the cell, chemical reactions within the cell, and integrated activities such as secretion, movement, and cell division.

c h a p t e r **1.1**

A PORTRAIT OF A CELL

The relations of hepatocytes to the architecture of the liver are shown schematically in Fig. I-1. Liver, like any other organ, must be nourished by molecules that enter the hepatocytes and other cells from the blood (for example, small molecules such as sugars, water, salts, and some macromolecules). The protein *albumin, lipoproteins* (complexes of lipid and protein), and other macromolecules are secreted by the hepatocyte into the blood, as are other materials including wastes (such as CO_2). Bile, which contains molecules formed in the breakdown of hemoglobin and of toxic substances, is excreted into small extracellular channels, the *bile canaliculi*, which lead to the bile duct; the latter in turn empties into the intestine where the bile facilitates fat digestion.

The flow of bile past the hepatocytes is in the opposite direction from the flow of blood. The hepatocytes are lined along the blood and bile spaces, and the arrangement of organelles within the cell reflects the polarization of functions; large areas of the cell surface are involved in the exchange with the blood and other smaller areas

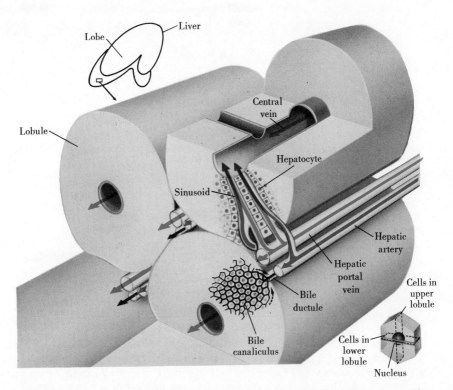

Fig. I-1 *Diagram of hepatic lobules (rat liver). The lobule at the upper right illustrates the relations of the liver cells (hepatocytes) to the blood; that at the lower right illustrates hepatocyte relations with the channels (bile canaliculi) into which the cells secrete bile. The two lobules diagram different views of the same cells as indicated at the lower right.*

in the secretion of bile. The key organelles are presented by the series of diagrams in Figs. I-2 through I-5. The diagrams are based upon light microscopy, electron microscopy (Fig. I-6), and biochemistry. To convey the three-dimensional structure of the organelles, a hypothetical "generalized" cell is also included (as the frontispiece), preceding the Preface of this text. Quantitative information on the number of different organelles is presented in Chapter 2.12. Two sets of definitions should be noted. *Intracellular* refers to things within the cell, *extracellular* to material outside the cell, and *intercellular* to components found between cells. *Vesicle* and *vacuole* are somewhat imprecise terms; both are used to designate a class of more or less spherical intracellular structures each delimited by a membrane. A vesicle is a small vacuole. The definition of *hyaloplasm* will be considered in Chapter 1.2B.

The diagrams indicate the complexity of cell structure and function. Each of the different organelles has a distinctive morphology,

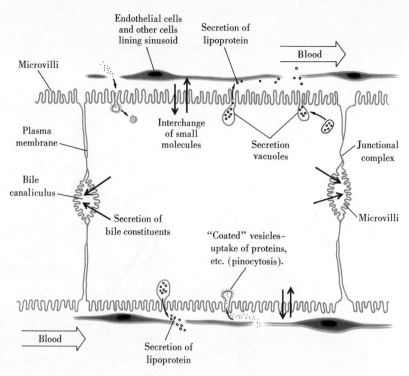

Fig. I-2 *Diagrammatic representation of a hepatocyte (from a rat) showing the* plasma membrane *that surrounds the cell. This membrane and structures associated with it control entry and exit of material. Arrows indicate interchanges of molecules between cells and blood. Macromolecules entering the cell are shown as small dots although generally they are invisible by current techniques. The lipoprotein particles secreted by the cell are seen as small spheres by electron microscopy. The* junctional complex *restricts movement of material between hepatocytes. For example, it prevents most molecules from moving directly from the blood sinusoids into bile canaliculi. The complex also probably helps in holding adjacent cells together.*

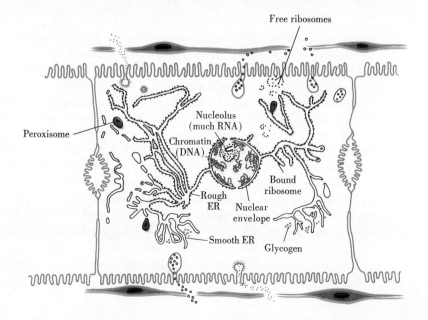

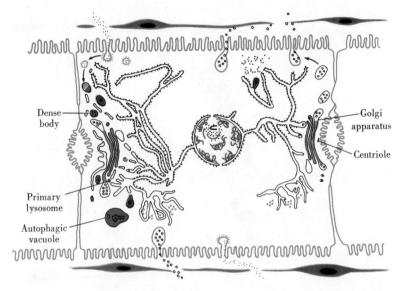

Fig. I-4 *The* Golgi apparatus, *which has been added to the structures seen in Fig. I-3, consists of flat saccules and vesicles of varying size. Its function includes concentrating of material produced in the ER and "packaging" of several classes of molecules into membrane–delimited vesicles. The large Golgi vesicles (often called vacuoles) are filled with secretion material such as lipoproteins (Fig. I-2). Among the small vesicles some probably are "primary"* lysosomes *involved in storage and transport of enzymes used in intracellular digestion. Other lyso-somes* shown in the diagram are dense bodies *and* autophagic vacuoles, *which are sites of intracellular digestion. A pair of* centrioles *is found close to the Golgi apparatus; these organelles function in cell division.*

◄ **Fig. I-3** *The* nucleus *and other organelles have been added to the structures seen in Fig. I-2. The* chromatin *of the nucleus contains most of the cell's DNA while the* nucleolus *is rich in RNA. The* nuclear envelope *is part of the endo-plasmic reticulum (ER), a complex interconnected system of membrane–delimited channels (large sacs and smaller tubules) in which certain types of macromolecules are transported within the cell.*

Ribosomes (granules composed of RNA and protein), usually complexed with information-carrying mRNA (messenger-RNA) to form polysomes, *are attached to part of the ER; other ("free") ribosomes are not attached to ER. ER with ribosomes attached is called* rough ER. *Proteins, synthesized on the polysomes of rough ER, in some unknown way enter the ER channels to be carried to other parts of the cell. The* smooth ER *lacks ribosomes; it is involved, for example in the synthesis of* steroids *and some other* lipids *(see Section 2.4.4).*

Stored carbohydrate, in the form of granules of glycogen *is found close to networks of smooth ER tubules.* Peroxisomes *are membrane–delimited structures containing enzymes that catalyze reactions many of which involve* hydrogen peroxide.

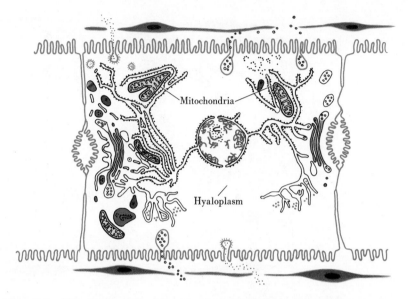

Fig. *I*-5 *The diagram now shows a few* mitochondria *(the average rat hepatocyte contains a thousand or more). The mitochondria are the chief sites of respiratory (oxygen-consuming) metabolism; they produce ATP (adenosine triphosphate, a key molecule in energy transfer and utilization) and many other important molecules. Mitochondria contain small amounts of DNA and RNA. The* hyaloplasm *is the "background" cytoplasm which, by current microscopic techniques, appears structureless.*

biochemical composition, and function. Detailed discussions of each organelle are found later in this section and particularly in Part 2. We now turn attention to the major techniques by which "portraits" of cells have become possible.

c h a p t e r **1.2**

METHODS OF BIOCHEMICAL CYTOLOGY

The microscopic study of cells is limited both by the microscope and by the manner of preparing the specimen for observation. In general, living cells and tissues are difficult to study directly with the ordinary light microscope. Multicellular tissues are usually too thick to permit penetration of light; single living cells are often transparent, with little visible internal detail. Thus, one line of development of techniques for cell study is centered around improvements in microscopes and in methods for preparing and observing cells.

A second line of development is concerned with coordinating

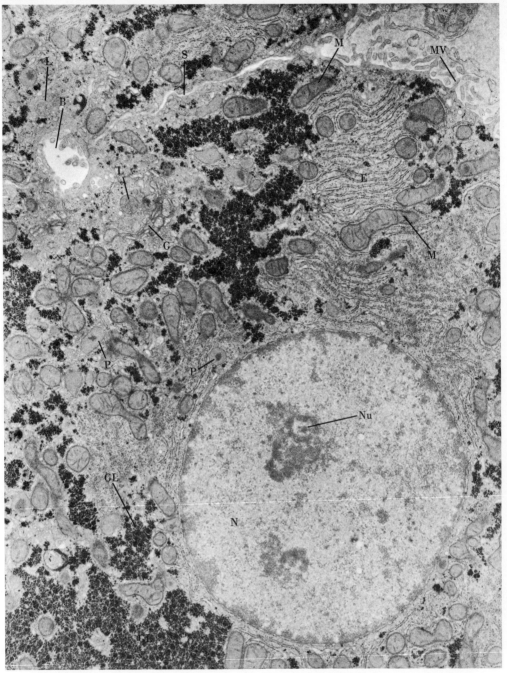

Fig. I-6 *Major organelles of a rat hepatocyte. This is a low magnification electron micrograph that illustrates the overall appearance of the cell. Subsequent micrographs at higher magnifications will show the detailed structures of organelles. (B) indicates a bile canaliculus and (MV) a microvillus at the sinusoid surface; many of the other microvilli near MV have been sectioned so as to appear unattached to the cell. The dense deposits at (GL) are masses of glycogen granules. (M) indicates mitochondria, (N) the nucleus, (NU) a nucleolus, (P) peroxisomes, (L) lysosomes, (G) the Golgi apparatus, and (S) the intercellular space separating this cell from an adjacent hepatocyte. (E) indicates endoplasmic reticulum. ×25,000. Courtesy of W.-Y. Shin.*

structural findings with biochemical information. Several methods have been devised which permit the direct use of the microscope in studying chemical and metabolic features of cells, and powerful, new nonmicroscopic techniques are being used in conjuction with them.

1.2A

MICROSCOPIC TECHNIQUES

During the latter part of the nineteenth and early part of the twentieth centuries, the light microscope approached the theoretical and practical limits of its performance as an optical instrument. Improvements in lenses and design resulted, by the beginning of this century, in instruments fundamentally similar to those in use today. During this period, the basic techniques of preparation of material for microscopy were also developed:

(1) *Fixing* of cells or tissue with agents that serve to kill and to stabilize structure with lifelike appearance and prevent post-mortem disruptive changes known as *autolysis*.

(2) *Embedding* in hard materials that provide support of the tissue for *sectioning*, the preparation of thin slices. Sectioning tissue permits study of complex structures by providing an unimpeded view of deep layers.

(3) *Staining* of cells with dyes that color only certain organelles, thus providing contrast, for example, between nucleus and cytoplasm or between mitochondria and other cytoplasmic structures.

At first glance, it might seem improbable that tissue put through such elaborate procedures would bear any resemblance to living material. Certainly there is reason for caution in interpretation. However, several generations of investigators have provided increasingly trustworthy methods for tissue preparation. When comparison is possible, the results often compare remarkably well with direct observation on living cells and with information obtained by various indirect means. A variety of preparative methods are available for study of specific cell features.

In the present century, development of cytological techniques has proceeded chiefly in four directions: (1) invention of microscopes based on newly understood physical principles, notably the *electron microscope* which permits use of much higher magnifications; (2) development of new optical devices, such as the *phase contrast* microscope, and perfection of others, such as the *polarizing* microscope, that facilitate detailed study of living cells; (3) evolution of *cytochemical methods* for obtaining chemical information about microscopic preparations; and (4) development of techniques for the disruption of cells and the selective *isolation* of organelles and other components for biochemical study.

1.2.1   *ELECTRON MICROSCOPY* The structures of interest to cytologists range widely in size. In terms of the units used for microscopic measurement (Fig. I-7), the diameter of a mitochondrion may be about 0.5 μ, the length of many bacteria roughly 1 μ, and the diameters of most mammalian cells in the range of 5–50 μ. All microscopes are characterized by limits of *resolution* which, in turn, determine the limits of useful magnification. These limits are inherent in the interaction of light with the specimen and with the optical system. No matter how perfect the microscope, an image can never be a perfect representation of the object (Fig. I-8). As a result, objects lying close to one another cannot be distinguished (*resolved*) as separate objects if they lie closer than approximately one-half the wavelength of the light being used. The wavelengths of visible light determine color. Blue light has a wavelength of 475 mμ, and red light of 650 mμ; the average wavelength of white light (for example, sunlight), which is a mixture of all colors, is about 550 mμ. Thus, for light microscopes, the limit of resolution is about 0.25 μ. If spaced apart by less than this limit, adjacent but separate objects appear in the light microscope as one object. This effect imposes an absolute limit on the details that can be distinguished with the light microscope irrespective of magnification. *Magnifica-*

Unit	Symbol		Chief use in Cytological Measurement
Centimeter	cm	= 0.4 inch	Macroscopic realm (naked eye) Giant egg cells
Millimeter	mm	= 0.1 cm	Macroscopic realm (naked eye) Very large cells
Micron	μ	= 0.001 mm	Light microscopy Most cells and larger organelles
Millimicron	mμ	= 0.001 μ	Electron Microscopy Smaller organelles, largest macromolecules
Angstrom unit	Å	= 0.1 mμ	Electron Microscopy, X-ray methods Molecules and atoms

Fig. I-7   *Units used in measuring the dimensions of cells and organelles.*

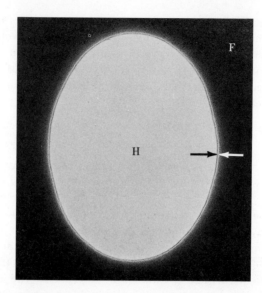

Fig. I-8   *An electron micrograph that illustrates one factor limiting resolution. A hole (H) in an electron-opaque film (F) was photographed under optical conditions that exaggerate the effect of* diffraction, *a phenomenon that occurs whenever light or electrons pass edges. The hole actually has a simple edge but the image of the edge shows a "diffraction fringe". The fringe consists of concentric bands, one dark and one bright (arrows). Under usual conditions of microscopy, the distorting effects of diffraction and other comparable optical phenomena are less severe but they cannot be totally eliminated.* × 120,000. *Courtesy of L. Biempica.*

tion refers to the apparent size of objects; *resolution* refers to the clarity of the image. Microscope images can be enlarged almost indefinitely by optical and photographic means, but beyond a point the image appears increasingly blurred and nothing is gained by further magnification. Thus, light microscopes are rarely used at magnifications greater than 1000–1500 ×.

The development of the electron microscope as a practical instrument in the 1940s and 1950s made possible useful magnifications of 100,000 × or greater. Electrons may be regarded as an extremely short-wavelength form of radiation. In practice, they can provide resolution of details separated by 1–5 Å, a thousand fold better than the light microscope. In the electron microscope, either magnetic or electrostatic lenses focus electrons in a manner analogous to the way glass lenses of ordinary microscopes focus light. The magnified image is viewed on a fluorescent screen like a TV screen and is also recorded on photographic plates. Most biological macromolecules have dimensions on the order of 10–100 Å. The diameter of a DNA molecule (not the length, which is very much greater) is 20 Å; many protein molecules have diameters of 50–100 Å. Thus, macromolecules cannot be distinguished in the light microscope, but are well within the range of resolution of the electron microscope.

Fixation for light microscopy and for electron microscopy usually depends on the use of agents that precipitate macromolecules and keep them insoluble. *Formaldehyde* and related compounds such as *glutaraldehyde* are widely used; they react with protein molecules to produce linking of previously separate molecules and other changes promoting insolubility. *Osmium tetroxide* ("osmic acid"), commonly used in electron microscopy, probably also can link separate molecules; it appears to react with lipids and other components but its reactions are incompletely understood (see Section 2.1.2).

For light microscopy, fixed tissue usually is embedded in paraffin wax. The tissue is dehydrated and soaked in molten wax which penetrates the tissues and then is cooled to form a hard matrix. Embedded tissue is sectioned (on mechanical slicers called *microtomes*) with sharp steel knives to produce sections 2–10 μ or more in thickness. The sections are mounted on glass slides for study in the microscope. On occasion, tissue is not embedded, but is instead frozen solid and sectioned while frozen. Freezing makes tissue hard enough for cutting reasonably thin sections, but also tends to damage cells. The technique is used for quick preparation (during surgical operations, for example, when it is desirable to examine tissue samples for diagnostic purposes) and for other applications such as retention of certain cell lipids and enzymes. Several of the many staining procedures used for light microscopy are outlined in Sections 1.2.3 and 2.2.2.

For electron microscopy, these techniques are inadequate. Electrons can pass only through very thin sections and are blocked entirely by glass. Thus, to embed tissue, it is soaked in unpolymerized plastic which is hardened by the addition of appropriate catalysts that promote polymerization. The plastics provide support for cutting sections 500–1000 Å or less in thickness. To accomplish this, sharp glass or diamond edges are used, as steel does not readily attain or keep an edge that is sharp enough. In place of glass slides, sections are mounted on a screenlike metal mesh and observed through the empty spaces of the screen.

The stains used in electron microscopy are not the colored dyes of light microscopy. Rather, they contain heavy metal atoms, usually lead or uranium, in forms that combine more or less selectively with chemical groups characteristic of specific types of structures in the cell. Since the presence of such atoms permits fewer electrons to pass through the specimen and produce an image, "stained" structures appear darker than their surroundings. Conventionally, a structure that appears dark in the electron microscope is referred to as *electron-dense* or *electron-opaque*.

Sectioning imposes difficulties of interpretation. If the object being studied is, for example, a cylinder 1.0 μ long, ten to twenty sections 500–1000 Å thick can be cut in planes perpendicular to the cylinder axis. Each section cut from the cylinder will appear then as a flat disc (Fig. I-9). To ascertain the three-dimensional structure, it is best to cut *serial sections*, a series of sections made at the same angle. These can be used to reconstruct the shape of the object in three dimensions. In our example, piling a series of discs one on top of another (mentally or by using cut photographs) would reproduce the original cylinder. However, serial sectioning for electron microscopy is difficult. Thus, most microscopists take sections at random and compare the appearances at various angles to the axis (Fig. I-9).

A number of ingenious techniques have been devised to supplement sectioning in the electron microscopic study of cells. Two of these are briefly described in Figs. II-12 and II-37. In addition, new types of electron microscopes are being developed to aid in the study of three-dimensional structure; for example, *scanning electron microscopes* permit direct viewing of outer surfaces, even with specimens far too thick for conventional electron microscopy.

1.2.2 *STUDY OF LIVING CELLS*   At the present time, living material cannot be examined by the electron microscope. Electrons are stopped by air, and the high vacuum consequently needed for the operation of the microscope kills cells.

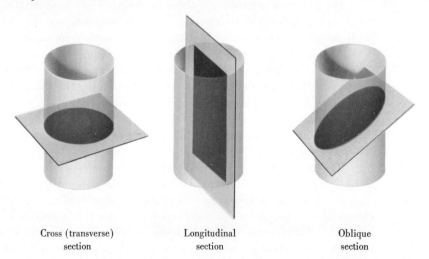

| Cross (transverse) section | Longitudinal section | Oblique section |

Fig. I-9 *Sectioning a cylinder along different axes can produce the appearance of a circle, a rectangle or an ellipse.*

For study of living cells, early light microscopists relied on a few large cell types (egg cells, some plant cells, and protozoa) and developed microsurgical techniques to introduce material into single living cells and to alter living cell structures. Aside from these efforts, cytology has been based predominantly on the study of fixed, sectioned, and stained material. The constant improvement of methods has enabled cytologists to obtain increasingly reliable information from fixed preparations, although it remains true that preparation of cells for microscopy may introduce *artifacts* (distorted and misleading appearances).

Living cells are dynamic entities, continually moving and changing; fixed and stained cells are stable and unchanging. Images from fixed material are at best only "snapshots" of dynamic cells. Often, a sequence of events must be reconstructed by careful study of a large number of such "snapshots" taken at intervals.

Fortunately, ways are now available to the light microscopist for examining cells in action. The behavior and speed of light as it passes through a portion of a cell depend on the concentration, nature, and organization of the cellular constituents. For example, cell regions differ in *refractive index*, which can be thought of as a measure of the speed of light through matter. Refractive index depends in large part on the density of the region through which the light passes, that is, on the concentration of matter per unit volume. The higher the density, the higher the refractive index, and the slower the rate at which light passes. Thus, light traveling through the nucleus will usually be affected differently from light traveling through the cytoplasm because of differences in the concentration of material. Differences in light paths due to factors of this type are not easily detectable in the ordinary microscope. The *phase contrast* and *interference* microscopes are light microscopes modified to translate refractive index differences into visible contrast, either in brightness or in color, by taking advantage of the phenomenon of *interference*. For example, under appropriate conditions two identical light waves can be made to cancel each other, causing darkness rather than brightness when they arrive together at a given point. This may occur when one wave passes through a medium of refractive index different from that encountered by the other wave. By using such effects to manipulate the light passing through the cell rather than manipulating the cell itself, an "optical staining" is achieved so that various organelles now stand out in sharp contrast to their surroundings (Fig. I-10). In addition, because of the influence of concentration on the optical properties of cells, the phase contrast and interference microscopes can also be used to estimate the amounts of material present in different cells or cell regions; an example of such quantitative use is outlined in Section 2.11.3.

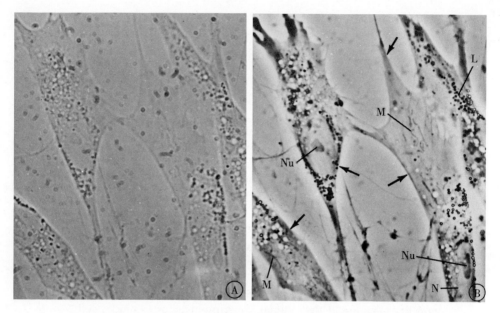

Fig. I-10   *The same living cells (chick fibroblasts grown in culture) photographed: **A**, through an ordinary light microscope and **B**, through a phase-contrast microscope. The phase-contrast image clearly shows cell borders (arrows), nuclei (N), nucleoli (NU), the elongate mitochondria (M), spherical lysosomes (L) and other details that are barely detectable in the ordinary microscope. × 800.*

The *polarizing microscope* can detect regions in cells where constituents are disposed in highly ordered array. A *polarizer* built into the microscope produces *polarized light* which is passed through the specimen. Ordinary light can be pictured as a beam in which wave-like vibrations are occurring in all possible directions perpendicular to the direction in which the beam is proceeding. In polarized light, the vibrations are confined to a single set of directions (for example, left and right, or up and down). Such light represents a sort of oriented field of electromagnetic energy that can interact in a distinctive fashion with specimen areas in which macromolecules or subunits are also arranged in a regular and oriented manner (for example, like a stack of coins or in parallel rows). Built into the microscope is an *analyzer*, an optical device that translates the behavior of polarized light that has passed through a specimen into visible contrast in the image. Ordered areas can be made to appear darker or brighter than areas of more random arrangement (Fig. IV-14); such areas are referred to as *birefringent*. The important point is that *order* can be detected in living cells, even when the oriented array is not visible by ordinary light microscopy.

The *ultraviolet microscope* makes use of the fact that certain substances in the cell, for example nucleic acids, strongly absorb ultraviolet light (nucleic acids absorb light with wavelength 260 mμ). Pictures taken in ultraviolet light show the locations of high concentrations of nucleic acids as regions darker than the rest of the cell.

In addition to improvements in optical techniques, the development of methods for growing cells in artificial conditions or *culture* has opened broad possibilities for studying living cells, both unicellular organisms and cells separated from tissues. An increasing variety of cell types can be maintained in artificial, well-defined growth media. Some properties of cultured cells will be discussed in Chapter 3.11.

Some activities of organelles in living cells particularly in unicellular organisms and in cells in culture, can be vividly demonstrated by taking motion pictures through a phase-contrast, polarizing, or interference microscope. Such *microcinematography* also *records* these activities and facilitates their detailed study; processes thus recorded may be speeded up photographically or may be viewed in slow motion. In Section 2.6.4, we refer to the study of mitochondrial movement, shape changes, and division. Another striking example is the "degranulation" (merger of lysosomes and other granules with phagocytic vacuoles) of white blood cells when bacteria are engulfed by these cells (see Section 2.8.2).

1.2.3   ***MICROSCOPE***   The procedures of microscope cytochem-
   CYTOCHEMISTRY   istry aim at producing a color or special
   contrast, such as enhanced brightness or darkness, at the sites of specific constituents within cells. Thus, in the *Feulgen procedure*, sections are treated with dilute hydrochloric acid. The DNA, because of the unique sugar it contains (deoxyribose), is the only constituent changed in the right way (aldehyde groups form) to react with the *Schiff reagent*, a colorless form of pararosaniline dye. The modified DNA reacts with this reagent to restore the bright red color of the dye. This Feulgen reaction generally stains only the nuclei, where almost all cellular DNA is found. Figure I-11 shows a Feulgen-stained nucleus; the DNA is seen to be localized in the *chromatin*, of which chromosomes are composed. The small amount of DNA present in some cytoplasmic organelles usually does not show, presumably because its concentration is below the limit needed to produce visible color.

If tissue sections are exposed to dyes that are soluble only in specific types of lipids, the cells will absorb the stain only in the areas that contain these lipids. Stains for RNA are also available. They show, for example, the localization of much RNA within a discrete intra-

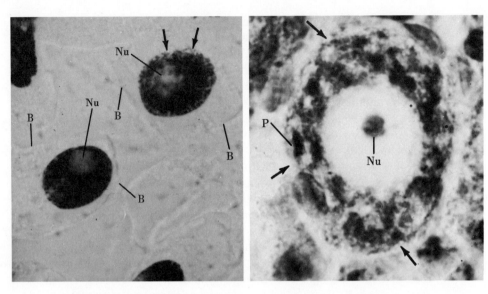

Fig. I-11 *and* **I-12**   *Cells stained by specific procedures for localizing nucleic acids. Fig. I-11 shows two cells from a root tip of* Tulbaghia, *stained with the Feulgen method for DNA (see Section 1.2.3). The cytoplasm is unstained and therefore the cell borders (B) are barely visible. Most of the nuclear volume is occupied by intensely stained chromatin and the arrows indicate points at which the tangled, thread-like character of the chromatin can be seen. Nucleoli (NU) are unstained. × 1,000. Fig. I-12 shows a neuron (nerve cell) from a rat ganglion stained with the dye, pyronine, which stains basophilic structures such as those rich in RNA (see Section 2.2.2). The red color of the dye shows as black when green light is used for photography, as it was in this figure. In the nucleus, the nucleolus (NU) stains intensely and in the cytoplasm, basophilic patches (P) are stained. Arrows indicate the cell's borders. × 1,600.*

nuclear body, the *nucleolus* (Fig. I-12). Stain also appears in areas of the cytoplasm seen,when viewed in the electron microscope, to contain many small RNA-protein granules, the *ribosomes.*

Such staining techniques may give quantitative information. For instance, the amount of dye taken up may be directly proportional to the amount of stained component, and methods (*microspectrophotometry*) have been devised to measure through the microscope the amount of dye present at a given site within a cell. This quantitative approach using the Feulgen stain was useful in establishing that the DNA content of cell nuclei was characteristic of the given species, a historical step in establishing the role of DNA.

As mentioned in the Introduction, enzymes vary in the compounds they affect (their substrates) and in the nature of the reactions they catalyze. Under specific conditions the actions of some enzymes can generate insoluble products visible in the microscope. This permits direct microscopic study of enzyme distributions in cells. For example, *phosphatases*, enzymes which split the phosphate group from specific substrates, are localized by incubating the section with the appropriate substrate in the presence of a metal ion such as lead. The lead ions combine with the phosphates liberated at the sites of the enzyme. Lead phosphate is insoluble and precipitates as it forms. The resulting deposits at the enzyme sites are visible by electron microscopy. For light microscopy, the lead phosphate is converted to lead sulfide which has a deep brown color. In Figs. I-13 through I-16, the plasma membrane, endoplasmic reticulum, Golgi apparatus, and lysosomes are shown to contain specific phosphatases which split different substrates. Figures I-17 and I-18 show two other organelles, mitochondria and peroxisomes stained through effects of enzymes within the organelles. Thus far, very few of the many enzymes known to biochemists have proved suitable for cytochemistry—these survive fixation and other steps in tissue preparation and are capable of yielding insoluble visible reaction products. Nevertheless, enzyme stains are useful in selectively staining organelles, as Figs. I-13 through I-18 demonstrate.

1.2.4   AUTORADIO-   GRAPHY   A valuable technique for tracing events in cells is *autoradiography*. Small *precursor* (metabolic forerunner) molecules that the cell uses as building blocks in the synthesis of macromolecules are made radioactive by incorporation of radioactive isotopes. The most widely used isotope is tritium (H^3), a radioactive form of hydrogen. Carbon^{14}, phosphorus^{32}, and sulfur^{35} are also employed. Probably the most frequently used radioactive precursor is *tritiated thymidine* (H^3-thymidine); this is thymidine (used by the cell to synthesize DNA) made radioactive by the substitution of tritium atoms for some of its hydrogen atoms. Typically, radioactive precursors are presented to cells which are then fixed at successive intervals and subsequently sectioned. The sections are coated with photographic emulsion similar to ordinary camera film (Fig. I-19). Exposure of camera film to light followed by photographic development leads to reduction of exposed silver salts in the film and to production of metallic silver grains. These grains form the image of the negative. Similarly, exposure to radioactivity and subsequent development produces grains in the autoradiographic emulsion. By examining the sections microscopically, both the underlying structure and the small grains in the emulsion are seen (Figs. I-20 and II-32). For example, if radioactive precursors of RNA are given to cells that are fixed only a few minutes later, almost all the grains are seen over the nuclei, suggesting that it is here that RNA is made from precursor molecules. If the interval between the exposure

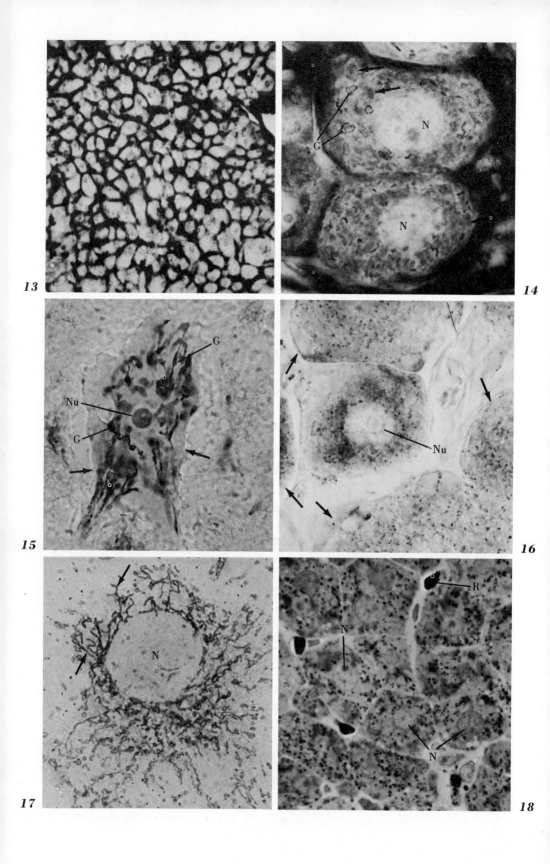

Figs. I-13 to I-18 *are light micrographs of preparations incubated to reveal sites of six different enzymes. In each case, "staining" is due to the deposition of a reaction product resulting from the action of the enzyme on a specific substrate. In Figs. I-13 to I-16 the enzymes are* phosphatases *hydrolyzing different phosphate-containing substrates.*

Fig. I-13 *Staining due to splitting of ATP by a* plasma membrane *enzyme (or enzymes) of cells in a liver tumor of a rat. Each light polygonal area is a cell outlined by the dark reaction product on its surface.* × 500.

Fig. I-14 *Staining due to splitting of the* diphosphate of inosine *(IDP) by enzymes at two sites in rat neurons (nerve cells). One site is the* Golgi apparatus *(G; see Fig. I-15). The other is the membrane systems of the* endoplasmic reticulum, *this results in staining of patches (arrows) in the cytoplasm comparable to the patches in Fig. I-12 where it is the RNA of the ribosomes bound to the reticulum that is stained. (N) indicates nuclei.* × 1,000.

Fig. I-15 *Staining of the* Golgi apparatus *in a rat neuron due to the activity of an enzyme that splits* thiamine pyrophosphate *(TPP) and therefore is called* thiamine pyrophosphatase *(TPPase). The Golgi apparatus is the network indicated by (G). Cell borders are seen at the arrows. (NU) indicates a nucleolus.* × 880.

Fig. I-16 *Staining of the* lysosomes *of rat neurons due to the presence of the enzyme,* acid phosphatase. *Portions of several neurons are seen in this micrograph. (NU) indicates the nucleolus within the nucleus of the one in the center of the field and the arrows indicate the edges of several others. The lysosomes are the dark granules scattered in the cytoplasm of the neurons.* × 800.

Fig. I-17 *Staining of the many* mitochondria *(two are indicated by arrows) in a mouse connective tissue cell grown in tissue culture. The enzyme responsible for staining transfers hydrogens of NADH (p. 32) to a soluble "tetrazolium" dye and thus converts the tetrazolium to an insoluble blue "formazan". The nucleus (N) is unstained.* × 1,000. *Compare this figure with the photograph of mitochondria in a living cell in Fig. I-10.*

Fig. I-18 *Staining of* peroxisomes *in rat hepatocytes due to a reaction in which a soluble colorless compound,* diaminobenzidine *is converted to an insoluble brown reaction product. The reaction is dependent on an enzyme within peroxisomes which are the dark cytoplasmic granules surrounding the unstained nuclei (N). Hemoglobin can also catalyze the reaction with the result that the red blood cells (R) in the hepatic circulation also stain.* × 800.

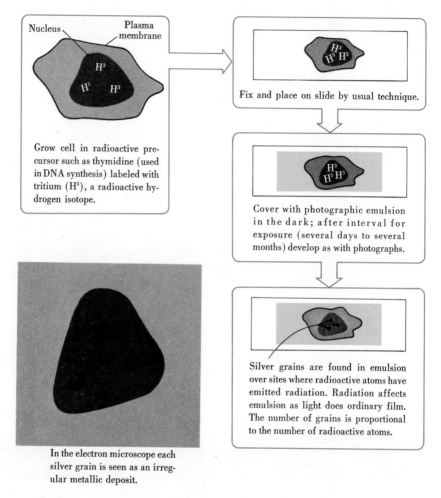

Fig. I-19   *Procedures of autoradiography.*

to radioactive precursors and the fixation is lengthened to an hour or two, fewer and fewer grains will appear over the nuclei and more and more over the cytoplasm (Figs. I-20, I-21). These observations suggest that once RNA has been made in the nucleus it moves out into the cytoplasm. Such use of "snapshots" taken at successive time intervals illustrates one of the most commonly used methods of studying dynamic biochemical events in cells.

The grains produced in the emulsion are similar, irrespective of the precursor used ; that is, radioactive RNA, protein, or DNA all "look" alike. Interpretation of autoradiograms depends on knowledge of biochemical pathways; precursors are carefully chosen so that they

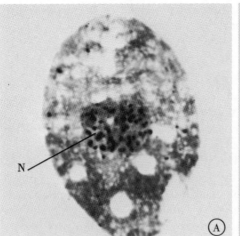 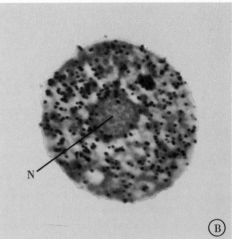

Fig. I-20 *Light microscope autoradiograms of* Tetrahymena, *a ciliated protozoan. The cells in* **A** *and* **B** *were both grown for approximately 10 minutes in tritiated cytidine (H³-C), a radioactive precursor of RNA. The cell in* **A** *was fixed almost immediately thereafter. The cell in* **B** *was allowed to grow in non-radioactive solutions for 90 minutes after exposure to H³-C. In* **A** *almost all grains are seen over the nucleus (N) whereas in* **B** *most are over the cytoplasm. Experiments of this type are known as* pulse-chase *experiments. A component is labeled during a brief exposure to radioactive precursor (a pulse) and the subsequent behavior of the radioactive molecules is followed in the* chase, *the period of growth in non-radioactive medium. In the present case, RNA molecules are labeled in the nucleus during the pulse. During the chase they migrate to the cytoplasm and are replaced in the nucleus by newly made (unlabeled) molecules. × 1,000 (approx.). Courtesy of D. Prescott, G. Stone and I. Cameron.*

are used by the cell to build only one kind of molecule. Further, "control" preparations are employed to verify interpretation. For example, after exposure to RNA precursors and incorporation of radioactivity, tissue sections can be treated with a solution of purified RNAase. This enzyme specifically breaks down RNA and removes it from tissues. If such treatment removes the radioactivity from the material

| | Percentage of grains over | | |
	Nucleolus	Rest of nucleus	Cytoplasm
30 minutes	54	41	5
30 minutes + 4 hours	14	19	67

Fig. I-21 *Mouse cells (L-strain fibroblasts) grown in culture were exposed for 30 minutes to radioactive RNA precursor (H³-Cytidine) then either fixed immediately or grown for an additional 4 hours in non-radioactive medium before fixation and autoradiographic study. "Rest of nucleus" refers to the non-nucleolar region of the nucleus. From the work of R. P. Perry.*

being studied (so that no grains are produced in the emulsion) it is reasonable to conclude that the precursor had been incorporated into RNA.

1.2B
CELL FRACTIONATION

Cell fractionation is an important and versatile technique for studying cell chemistry. The different subcellular structures are separated by centrifugation of homogenates. The latter are prepared by disrupting cells of liver or other organs in media that preserve the organelles. Solutions of sucrose or other sugars are generally used because they can be readily adjusted to maintain the integrity of organelles and to counteract the tendency of organelles freed from the cell to clump together. Some organelles, such as mitochondria and plastids, remain essentially intact. On the other hand, the endoplasmic reticulum, which in the living cell is an interconnected network of membrane-bounded cavities, is broken into separate pieces that form rounded vesicles (Fig. I-22). Similarly, the plasma membrane is fragmented into pieces of variable size.

The behavior of a particle in a centrifugal field depends chiefly upon its weight and upon the resistance it encounters in moving through the suspension medium. Thus, size, density, and shape influence the movement of a particle in a centrifugal field. The most widely used method for separating cell organelles is called *differential centrifugation*. It is based on differences in the speed with which structures sediment to the bottom of a centrifuge tube. At a given centrifugal force, structures that are relatively large, dense, and heavy sediment most rapidly. Thus, using comparatively low forces (slow speed of centrifuge), nuclei, large fragments of plasma membrane, and some mitochondria may be sedimented as a pellet (Fig. I-22), while the remaining cell structures remain in suspension. With somewhat higher centrifugal force, a second fraction containing most of the mitochondria

Fig. I-22 *Cell fractionation. Cells can be disrupted* (homogenized) *by a* ▶ *variety of procedures. The one illustrated involves a homogenizer employing movement of a close-fitting glass or plastic pestle within a tube. The space between pestle and tube wall can be varied to achieve breakage of cells and organelles with minimal damage. Inevitably, the plasma membrane and endoplasmic reticulum are fragmented.*

Modern ultracentrifuges are capable of rotating at 50–100,000 revolutions per minute. This imposes forces in excess of 100,000 times the force of gravity and permits sedimentation of the smallest cell structures and of many macromolecules.

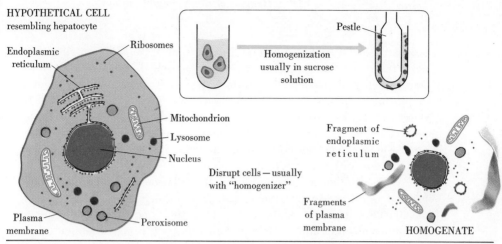

HYPOTHETICAL CELL
resembling hepatocyte

Endoplasmic reticulum

Ribosomes

Homogenization usually in sucrose solution

Pestle

Mitochondrion

Lysosome

Nucleus

Disrupt cells — usually with "homogenizer"

Fragment of endoplasmic reticulum

Fragments of plasma membrane

HOMOGENATE

Plasma membrane

Peroxisome

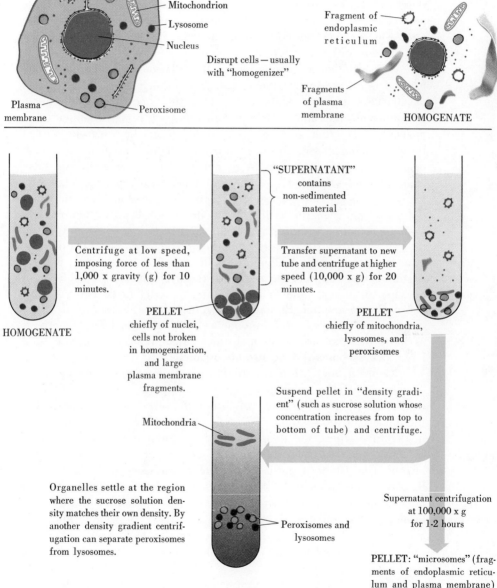

HOMOGENATE

Centrifuge at low speed, imposing force of less than 1,000 x gravity (g) for 10 minutes.

"SUPERNATANT" contains non-sedimented material

PELLET
chiefly of nuclei, cells not broken in homogenization, and large plasma membrane fragments.

Transfer supernatant to new tube and centrifuge at higher speed (10,000 x g) for 20 minutes.

PELLET
chiefly of mitochondria, lysosomes, and peroxisomes

Suspend pellet in "density gradient" (such as sucrose solution whose concentration increases from top to bottom of tube) and centrifuge.

Mitochondria

Organelles settle at the region where the sucrose solution density matches their own density. By another density gradient centrifugation can separate peroxisomes from lysosomes.

Peroxisomes and lysosomes

Supernatant centrifugation at 100,000 x g for 1-2 hours

PELLET: "microsomes" (fragments of endoplasmic reticulum and plasma membrane)

SUPERNATANT: ribosomes not bound to membranes and soluble molecules.

is separated. Most lysosomes and peroxisomes are sedimented in the next fraction. The so-called microsomes, which consist mainly of fragmented endoplasmic reticulum but also of some small fragments of plasma membrane, come down with still greater force (note that there are no "microsomes" *as such* in the cell). Free ribosomes (those not bound to membranes) often remain unsedimented, together with other small structures and soluble molecules. The ribosomes can be brought down by prolonged use of high centrifugal forces.

The solution remaining after the ribosomes have been sedimented is referred to as the *soluble fraction* or *cytosol*. This fraction usually contains some material lost from the organelles in the course of homogenization and centrifugation. However, it also contains other molecules, including some enzymes, that are not known to be part of the intracellular structures identifiable in the microscope. Such molecules are usually thought to derive from the *hyaloplasm* (also called ground substance, cell matrix, or cell sap) which is defined by microscopists as the apparently structureless medium that occupies the space between the visible organelles. As will be evident in subsequent chapters, with the improvement of techniques less and less of the cell appears structureless; thus the number of components thought of as truly part of the hyaloplasm is diminishing. However, at present it still remains appropriate to regard the hyaloplasm as a solution, or perhaps an almost fluid gel, containing material in transit from one cell region to another. Probably it also includes a number of dissolved or suspended enzymes that function as individual molecules or perhaps as parts of small groups of molecules rather than as components of large, microscopically recognizable organelles.

Centrifugation through gradients of increasing concentration of sucrose or other substances can yield separations on the basis of density; the higher the concentration of dissolved sucrose, the greater the density of the solution. Organelles situate themselves at the levels in the tube where the densities of the solution match the densities of the organelles. Density gradient centrifugation (Fig. I-22) often enables separation of particles of roughly the same size but different density. Thus peroxisomes may be separated from lysosomes, or microsomes may be separated into those fractions in which the membranes are studded with ribosomes and those in which ribosomes are absent. Plasma membrane fragments also have been isolated in this manner. Modified procedures of gradient centrifugation can be used to achieve separations of macromolecules, such as RNA and DNA.

Isolated fractions are examined by electron microscopy to determine the integrity of the organelles and the purity of the fractions. This step is of importance since fractions thought to contain only one organelle may prove to be contaminated with others. For example, some lysosomes and peroxisomes may sediment with the mitochondria and some mitochondria sediment with the lysosomes and peroxisomes. Fractions that consist exclusively of one organelle are difficult to obtain so that most work is done with fractions that are less than pure.

Fractions can be disrupted by chemical or physical means and subfractions can then be separated by additional centrifugation. Thus, fragments of rough endoplasmic reticulum (isolated as microsomes) can be treated with detergents such as deoxycholate that solubilize the membranes and free the ribosomes.

Isolated fractions can be subjected to the full range of biochemical analysis to learn the chemical composition, enzyme activities, and metabolic capabilities of subcellular structures. With the use of radioactive precursors and by isolating fractions at different time intervals, insights may be gained into temporal aspects of metabolic events. If cells are exposed, for instance, to radioactive amino acids, and fractions are isolated early after exposure, radioactive proteins are found only in the fractions that contain ribosomes, suggesting that ribosomes are the sites of protein synthesis. As the interval between exposure to radioactive amino acids and isolation of fractions is lengthened, other cell structures are found to contain radioactive protein. Thus it is possible to trace the transport of the protein, initially synthesized on the ribosomes, through its subsequent transport in the cell.

Biochemical techniques give information about the *average* properties of the material in a fraction. Often much is gained when they are supplemented by autoradiographic studies of intact cells or of isolated material; autoradiography can focus on single organelles and thus reveal the distribution of properties in the individuals underlying the average. Similarly, autoradiographic work is greatly strengthened by parallel biochemical investigations.

c h a p t e r **1.3**

A MAJOR METABOLIC PATHWAY

Figure I-23 outlines major metabolic sequences responsible for the breakdown of glucose. Full details are given in *Cell Structure and Function* by Ariel Loewy and Philip Siekevitz and in other references listed at the end of this part. Emphasis here will concern only certain features especially relevant to cell organization and useful for subsequent discussions.

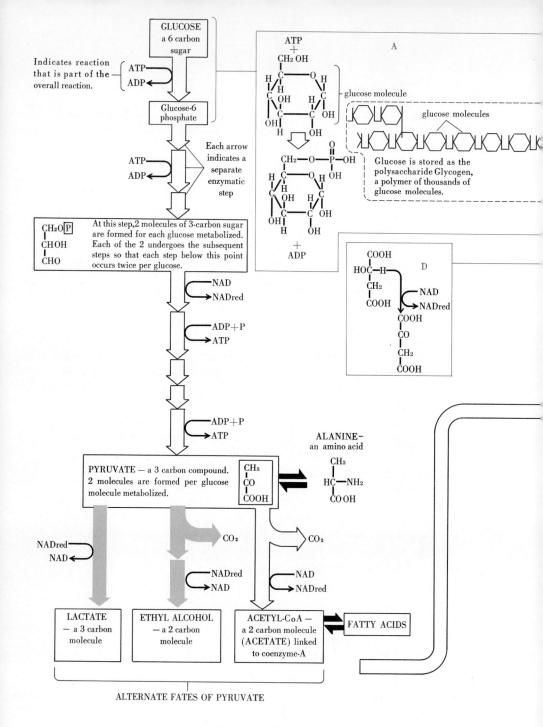

Fig. I-23 *Some features of the reaction sequences in glucose metabolism. The chemical structures of several key molecules are indicated. **A–D** illustrate important biochemical reactions: **A** shows the* phosphorylation *of glucose,* **B**, *the* condensation *of two molecules,* **C**, *a* decarboxylation *(the removal of a carboxyl group (COOH) by its conversion to CO_2) and **C** and **D**, the* transfer of hydrogen atoms to coenzymes *(the* dehydrogenation *or oxidation of a substrate molecule coupled with the* reduction *of a coenzyme molecule). Also illustrated are some of the many pathways that intersect with the glucose pathway.*

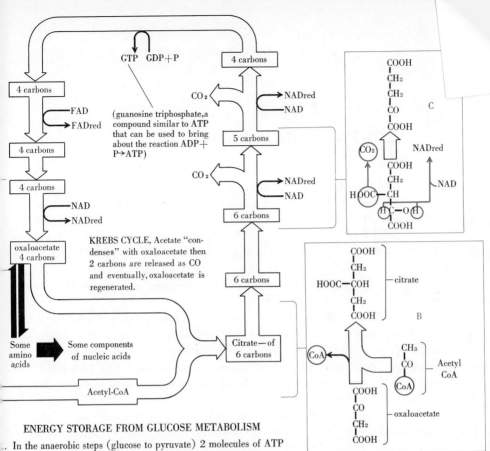

ENERGY STORAGE FROM GLUCOSE METABOLISM

1. In the anaerobic steps (glucose to pyruvate) 2 molecules of ATP are consumed and 4 ATPs are formed per glucose broken down.

2. In the Krebs cycle starting with pyruvate, the equivalent of one ATP (a GTP) is formed per pyruvate, or two per glucose. In addition, reduced coenzymes (NADred, FADred) are formed. Via electron transport and oxidative phosphorylation the reduced coenzymes can give rise to a total of 14 ATP's per pyruvate (28 per glucose).

3. Two NADred molecules per glucose are formed in the glucose to pyruvate steps. Their metabolic fate may differ somewhat from the Krebs cycle coenzymes but they represent the equivalent of 6 ATP's that could be formed by electron transport and oxidative phosphorylation.

4. TOTALS: Overall, 38 molecules of ATP formed per glucose.
   "Aerobic" steps: 30 ATP's from Krebs cycle and associated
                   pathways.
                 6 ATP's from electron transport based on
                   coenzymes reduced in glucose to
                   pyruvate steps.

   "Anaerobic" steps: 2 ATP's from glucose to pyruvate steps. (In anaerobic metabolism, the reduced coenzymes from the glucose to pyruvate steps can be used in formation of *ethyl alcohol* or *lactate* rather than entering electron transport).

> Terminology: *NAD was once known as* diphosphopyridine nucleotide *and thus is often referred to as DPN. Sometimes, for convenience, the glucose-to-pyruvate steps are also referred to as* glycolysis *although strictly speaking, the term designates the path from glucose to lactate. "Aerobic" metabolism requires oxygen, "anaerobic" metabolism does not.*

1.3A

METABOLIC BREAKDOWN OF GLUCOSE

1.3.1   AEROBIC AND ANAEROBIC PATHWAYS; SOME DEFINITIONS

In hepatocytes, glucose (a 6-carbon sugar) is broken down first to two molecules of the 3-carbon compound, *pyruvic acid* (pyruvate). As seen in Fig. I-23, this breakdown occurs in a series of steps, each step catalyzed by a specific enzyme. The product of each step serves as the substrate for the next enzyme; the products are referred to as *intermediates*.

The further metabolism of pyruvate to form carbon dioxide and water depends on a second series of reactions known as the *Krebs cycle*. Each pyruvate molecule is first converted to a 2-carbon *acetate* molecule. The acetate undergoes the reactions shown in Fig. I-23. Each molecule is combined with a 4-carbon molecule (oxaloacetate) to produce a 6-carbon product. This 6-carbon molecule is broken down by a series of steps that ultimately regenerate an oxaloacetate molecule, which can then go through another turn of the cycle. At two steps in the series of reactions, CO_2 is released; these CO_2 molecules plus the one CO_2 released in the pyruvate to acetate conversion account for all three carbons of the pyruvate molecule.

In five of the Krebs cycle reactions, hydrogen atoms are transferred to *coenzymes*, either NAD (nicotinamide adenine dinucleotide) or FAD (flavin adenine dinucleotide). Coenzymes constitute a class of small nonprotein molecules that function, together with enzymes, in many metabolic reactions. In the reactions we are considering, the coenzymes accept hydrogen atoms enzymatically released from substrates and are thus *reduced*. From *reduced* NAD (often *NADH_2* or *NADH^+* is used as a shorthand designation) and *reduced FAD* (*FADH_2*), these hydrogen atoms are transferred eventually to oxygen.

The steps from glucose to pyruvate are *anaerobic*—they require no oxygen. Virtually all organisms are capable of degrading sugars by similar anaerobic pathways. In some microorganisms, the pyruvate molecules are further metabolized (anaerobically) to produce ethyl alcohol molecules; such production of alcohol from sugar is known as *fermentation*. Other cells instead convert pyruvate molecules anaerobically to lactic acid (lactate) molecules; the path from glucose to lactate is called *glycolysis*. Some bacteria, yeasts, and other microorganisms metabolize exclusively by fermentation, glycolysis, and other anaerobic pathways. Cells of most organisms have enzymes both for anaerobic pathways (glycolysis and others) and the Krebs cycle and related aerobic (oxygen-dependent) pathways. As Fig. I-23 makes evident, the aerobic metabolism of glucose via the Krebs cycle makes use of the same glucose-to-pyruvate steps as glycolysis.

1.3.2   ***GLUCOSE***
METABOLISM
AND ATP

Generally, metabolism of glucose or related sugars and other carbohydrates is the major source of cellular ATP. ATP is usually formed by the addition of a phosphate group to ADP (adenosine diphosphate). The chemical bond formed in this reaction is called a high-energy bond; its formation requires an input of energy derived from metabolism. When such a bond is broken, energy is made available. ATP is the major molecular form in which energy is stored and transferred to sites of use in cell functions.

Reduced coenzymes also are carriers of metabolic energy. In glucose metabolism, coenzymes are reduced in several reactions. These reduced coenzymes may be used in a special sequence of reactions that generate ATP. The ATP-generating reactions will be discussed in greater detail in Part 2. In brief they are as follows:

(1) The electrons from H atoms carried by the coenzymes enter a sequence of reactions known as *electron transport*. This involves a set of enzymes known as the *respiratory chain*.

(2) The terminal reaction of the series results in the transfer of the electrons plus the protons (H^+ ions) from hydrogen atoms to oxygen. This produces H_2O.

(3) Much ATP is generated in the course of electron transport. The process of formation of ATP is referred to as *oxidative phosphorylation*; it links oxygen consumption to the addition of phosphate to ADP.

Two conclusions from Fig. I-23 should be noted. First, the aerobic pathways generate much more ATP than the anaerobic pathways; this reflects chiefly the production of reduced coenzymes in the Krebs cycle. In some anaerobic organisms mechanisms have evolved which supplement sugar-based ATP production with other anaerobic pathways that store energy. This facilitates survival of these organisms despite the absence of aerobic pathways. Second, the linking of glucose metabolism to ATP production by aerobic cells involves three more or less readily distinguishable sequences of reactions: the formation of pyruvate; the degradation of pyruvate molecules in the Krebs cycle; electron transport and oxidative phosphorylation. The relations of this conceptual separation of glucose metabolism to actual cell organization will be discussed shortly.

*1.3*B
METABOLIC INTERRELATIONS

1.3.3   ***CARBOHY-*** As with many other metabolic sequences,
            DRATES, the glucose pathways intersect with other
         PROTEINS pathways (Fig. I-23). For example, pyruvate
          AND FATS can be formed from the amino acid *alanine*
                by a reaction that is reversible, so that (if
appropriate nitrogen sources are present) alanine can also be formed from pyruvate. By such reactions, carbohydrates can be used to synthesize amino acids which are combined to form proteins, or amino acids can be metabolized via the Krebs cycle to generate ATP. Fatty acids, components of fats, can be converted to acetate and degraded by the Krebs cycle, or fatty acids can be built up from acetate units derived from carbohydrates.

Reduced NAD is used in a variety of metabolic pathways as a reducing agent (a source of hydrogens or electrons). ATP participates in a very large number of reactions.

The controls that determine which sequence within the complex network of possible pathways an individual molecule will follow are largely unknown. The overall metabolic pattern responds to the availability in the cells of ATP, NAD, oxygen, various intermediates, and such agents as hormones. For example, in very active muscle, much glucose is rapidly broken down to pyruvate, but since the oxygen supply is inadequate for immediate breakdown of all the pyruvate to CO_2 and water, some pyruvate is converted anaerobically to lactate. This conversion is reversible, and the lactate is converted back to pyruvate either in the muscle itself during periods of rest, or in other tissues which the lactate reaches via the blood stream. More will be said later about metabolic controls (see, for example, Section 3.2.5).

1.3.4   ***METABOLIC*** If hepatocytes or other eucaryotic cells are
    ORGANIZATION homogenized and cell fractions prepared by
     AND INTERAC- centrifugation the following distribution of
           TIONS OF enzymes is observed:
     ORGANELLES IN
        EUCARYOTIC   (1) Enzymes of the glycolytic pathway
              CELLS are present chiefly in the unsedimented
                supernatant fluid.
(2) Krebs   cycle,   oxidative   phosphorylation,   and   respiratory

chain enzymes are largely confined to the mitochondria.

(3) If the mitochondria are disrupted (for example, by exposure to *ultrasonic vibration,* very high frequency sound waves) most Krebs cycle enzymes are released as soluble material. Centrifugation of a preparation of disrupted mitochondria results in a pellet containing the membranous remnants of mitochondrial structure to which are attached electron transport and oxidative phosphorylation enzymes. Most Krebs cycle enzymes remain in the supernatant.

The generally accepted conclusion from these observations is that the glycolytic enzymes are not firmly bound to any of the visible cellular organelles. Probably they are in the hyaloplasm, the supposedly structureless medium in which the organelles are suspended (see Chapter 1.2B); however, conceivably they are so loosely bound to in-

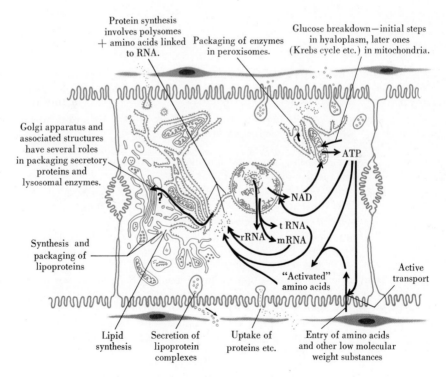

Fig. I-24   *A few examples of the probable interrelations of organelle functions in important metabolic sequences — in this diagram the synthesis, transport and "packaging" of proteins and lipoproteins. The roles in protein synthesis of the several classes of RNA are outlined in Fig. II-14; the synthesis takes place on polysomes (complexes of mRNA with ribosomes) and those molecules synthesized on polysomes bound to the ER enter the ER and are transported to other cell regions.*

tracellular structures that they are readily freed during homogenization. In contrast, the enzymes of the Krebs cycle, electron transport,
and oxidative phosphorylation are located in the mitochondria. By
centrifugating disrupted mitochondria, further compartmentalization
is demonstrated; the electron transport and oxidative phosphorylation
enzymes are firmly attached to the mitochondrial membranes, whereas
most Krebs cycle enzymes are not.

Such observations form the basis of Fig. I-24, a final diagram
of the hepatocyte. In many metabolic pathways, there is interaction
among several different organelles. Even our discussion of the glucose
pathway could easily be extended to include other organelles. For
example, in hepatocytes, the *synthesis* of NAD molecules (as distinct
from the addition of H atoms to NAD) depends on an enzyme apparently
localized in the nucleus. Figure I-24 illustrates only a tiny fraction of
the known metabolic interrelations.

FURTHER READING

Grimstone, A. V., *The Electron Microscope in Biology*. New York: St. Martin's,
    1968. 54 pp. An elementary description of basic theory and practice
    of microscopy.

Hughes, A., *A History of Cytology*. New York: Abelard-Schuman, 1959.
    158 pp. A concise history of the field.

Lehninger, A. L., *Bioenergetics*. New York: W. A. Benjamin, Inc., 1965. 258 pp.
    Discusses key aspects of metabolism, especially as related to the forma-
    tion and use of ATP.

Loewy, A. G. and Siekevitz P., *Cell Structure and Function*, 2d ed. New York:
    Holt, Rinehart and Winston, 1969. 512 pp. A wide-ranging discussion of
    cells, stressing their biochemistry.

The Encyclopedia of Biochemistry. Williams, R. J. and E. M. Lansford (eds.),
    New York: Reinhold, 1967. 876 pp. Contains concise introductory mate-
    rial on organelles and macromolecules.

Many articles of interest appear in the monthly, *Scientific American*. Of these,
some earlier articles of current interest have been collected as a book, *The
Living Cell*, edited by D. F. Kennedy (San Francisco: Freeman, 1965. 296 pp).

The Bobbs-Merrill reprint series (Indianapolis, Howard W. Sams and Co.)
includes a collection of reprints of research articles on cell structure and function (Biology, Category C).

ADVANCED GENERAL READINGS

Many professional journals contain articles of interest to cytologists
and cell biologists. These include, among many others, *The Journal*

of Cell Biology, The Journal of Cell Science, The Journal of Ultra-structure Research, The Journal of Molecular Biology, The Journal of Biological Chemistry, International Review of Cytology, Proceedings of the National Academy of Sciences of the United States and *The Journal of Histochemistry and Cytochemistry.*

Brachet, J. and Mirsky, A. E. (eds.), *The Cell,* Volumes I–VI. New York: Academic Press, Inc., 1959–63. A collection of articles on many aspects of cells; some are out of date but many provide useful background.

DuPraw, E. J., *Cell and Molecular Biology.* New York: Academic Press, Inc., 1968, 739 pp. A recent book with especially interesting discussions of the nucleus and chromosomes.

Grey, P. *Handbook of Basic Microtechnique,* 3rd. ed. New York: McGraw-Hill, 1964. 302 pp. A practical manual on using the light microscope.

Lima-de-Faria, A. (ed.), *Handbook of Molecular Cytology.* (See p. 160 for description.)

Methods in Cell Physiology, 3 Volumes. Prescott, D. M. (ed.), New York: Academic Press, Inc., 1964–68. A collection of short articles describing methods of cell study.

Physical Techniques in Biological Research, 2d ed., Volumes 1–3. Oster, G., and A. W. Pollister (eds.), New York: Academic Press, Inc., 1966. A collection of articles detailing the theory and aspects of the practice of many techniques.

Sjostrand, F. S., *Electron Microscopy of Cells and Tissues,* Volume 1. New York: Academic Press, Inc., 1967. 462 pp. Discusses the practice of electron microscopy.

Stern, H. and Nanney, D. L., *The Biology of Cells.* New York: Wiley, 1965. 548 pp. An introductory text.

Wilson, E. B., *The Cell in Heredity and Development,* 3d ed. New York: Macmillan, 1925. 123 pp. Outstanding summary of cytology of its period; problems discussed in manner still of importance to students of cytology and cell biology.

p a r t **2**

CELL ORGANELLES

This part discusses major features of the organelles of eucaryotic cells; Part 3 will consider the procaryotes and will include additional information on eucaryote organelles as specific cell types are discussed. Almost all of Part 4 is devoted to further consideration of the nucleus, and Part 5 contains a summary of some recent information about nucleoli.

The membranes of eucaryotic cells divide the cell into compartments that are distinctive in both morphology and metabolism. Membranes surround the cell as a whole and the *membrane-bounded* or *membrane-delimited* organelles: the nucleus, mitochondria, chloroplasts, lysosomes, peroxisomes, and the cavities of the endoplasmic reticulum and Golgi apparatus. The membranes are not simple, mechanical barriers. They are highly ordered arrays of molecules, chiefly lipid and proteins, in which enzymes are integrated; they participate in diverse activities. Membranes provide selective barriers that control the amount and nature of substances that can pass between the cell and its environment and between intracellular compartments. Within several of the organelles, notably mitochondria and chloroplasts, the arrangement of enzymes on membranes provides remarkable efficiency of function, since the different enzymes involved in sequential reactions of metabolic pathways are held in close relationship with one another.

But must a structure be membrane-bounded to qualify as an organelle? As techniques improve, less and less of the cell appears unstructured, but only some of the organization involves membranes. Nucleoli, chromosomes, ribosomes, centrioles, and microtubules all are distinctively structured and have specialized roles in the cell, but no membrane surrounds them. The non-membrane-bounded organelles grade down in size and complexity to protein filaments composed of a few hundred molecules. At the lower end of the size spectrum, the distinctions between organelle and macromolecule become difficult to define and perhaps meaningless. Should a multienzyme complex, in which a few or a few dozen enzymatically active protein molecules are complexed as a functional unit (Section 3.2.4), be called an organelle or a molecular aggregate? Are nucleoli to be considered organelles despite their being contained in other organelles (nuclei)? The decision appears to be a matter of arbitrary definition.

Organelles are not static units of unchanging size and shape, with fixed relations to other organelles and with inflexible function. All show movement within most cells. Some can grow and produce duplicates. Most exhibit evidence of *turnover*. For example, if the mitochondria in a hepatocyte are labeled by the incorporation of radioactive amino acids into their proteins, by eight to ten days later, half of the label has been lost from the mitochondria. In number and mass, the population of mitochondria has remained the same, but half the organelles have "turned over." Does this mean that new mitochondria are continually arising as others are destroyed? Or does each mitochondrion add new material to its structure while simultaneously losing an equivalent amount of older constituents? Or do both processes take place? The answers are not yet known. The amount of a given component present in a cell at a given time depends on the rate of formation and on the rate of degradation. Both formation and degradation must be subject to precise cellular controls, and the relevant mechanisms are being sought. Macromolecules also turn over; a notable exception, however, is DNA, in all but a very few special situations.

The information that will be reviewed in this part suggests that cytologists and cell biologists may, before long, be able to specify the particular molecular arrangements in each organelle and the detailed chemical reactions that occur within it. The analysis of several organelles has passed from the stage of mere identification of the structure and of the molecular constituents to the point where the normal modes of formation and duplication of the organelles are being studied in detail, and functional parts of organelles are being reconstituted from simpler components in the test tube.

chapter **2.1**
THE PLASMA MEMBRANE

A plasma membrane is present at the surface of all cells. Although some other closely associated structures play important roles, this membrane is the primary barrier that determines what can enter or leave the cell. Its properties strongly influence the formation of multicellular aggregates (such as tissues), as well as the passage of material between closely associated cells. Specialized plasma membrane regions are often present. In Figure I-2 the *microvilli* were mentioned; these are tubular projections that increase the surface area of hepatocytes and of many other cell types. These and other specializations of the plasma membrane will be discussed in Part 3. There we will also consider the extracellular *cell wall* surrounding plant cells. At this point our concern is with some general properties of plasma membrane structure and function.

2.1.1   ***MICROSCOPIC***   Light microscopy can not reveal the presence of the plasma membrane directly, since the thickness of the membrane is well below the resolving power of the light microscope. However, before the advent of electron microscopy, a great deal of indirect information about the membrane was accumulated from physiological experiments. Artificial systems were devised, in which two solutions containing different concentrations of given substances dissolved in water were separated by a membrane that resembled the plasma membrane in being semipermeable (permitting only some types of molecules to pass). If a given component can pass readily through the membrane, a net movement or *diffusion* of the component will take place from the solution of higher concentration to that of lower concentration ("down the concentration gradient"), until the concentrations on both sides of the membrane are equal. If the membrane is impermeable or only slightly permeable to the dissolved components, but if, like the plasma membrane, it is moderately permeable to water, then water will pass rapidly into the more concentrated solution. This net movement of water in the direction that tends to equalize the concentrations of dissolved materials on each side of the membrane, is referred to as *osmosis*. In early experiments, cells were suspended in solutions containing different molecules dissolved in water, and the volume changes of cells were measured. The details of behavior of a cell in a given solution will depend on the total concentrations of all molecules dissolved inside the cell and in the surrounding solution as well as on the relative rate of passage through the membrane of dissolved molecules as compared with water; to a first approximation, at equilibrium, the total number of molecules of all dissolved material per unit volume inside and outside will be equal as will the concentration of each individual component that can diffuse through the membrane. Under some experimental conditions, extensive volume changes can be taken to indicate relative impermeability to the particular dissolved molecules and consequent influx or efflux of water; cells may shrink drastically or swell to bursting depending on the concentration of the solutions. Studies of this type, and more sophisticated experiments in which concentrations of substances in cells are measured directly, have made it clear that a plasma membrane exists and exhibits great selectivity as to what can pass through. Gases move across the membrane with little difficulty. Water and other small molecules can pass through more readily than larger molecules with comparable chemical properties. Material that is soluble in lipids generally enters the cell more rapidly than nonlipid-soluble substances. Some substances pass through by passive diffusion, others, only with the expenditure of energy by the cell.

This selectivity and other indirect observations suggest that the membrane is a highly organized and structurally complex entity. However, the electron microscope usually shows a relatively simple and uniform structure, and the structural basis of complex membrane functions is still unknown. In most electron microscope preparations, the plasma membrane appears to be a series of three layers with a total thickness of 75–100 Å (Fig. II-1). This three-layered structure is referred to as a *unit membrane*. As will be noted in subsequent chap-

Fig. II-1 *Portion of the surface of a human red blood cell (C). The arrows indicate the three-layered (dense-light-dense) "unit membrane" structure of the plasma membrane.* × *250,000. Courtesy of J. D. Robertson.*

ters, many intracellular membranes also may show unit membrane structure with the overall thickness of the membrane varying in different organelles. At present, there is considerable controversy about the significance of the unit membrane structure. Some believe that it represents an accurate image of the structure of most membranes of the living cell; others consider it misleading and argue that membranes are far more complicated. Discussion of this problem will continue below.

2.1.2 **A MOLECULAR MODEL OF MEMBRANES; LIPIDS** From a variety of correlated chemical and structural studies, a general model thought to apply to the plasma membrane and to other cellular membranes has been proposed as a conceptual framework for further investigation. Although it is now evident that this model is oversimplified, it has stimulated a great deal of invaluable experimentation and discussion.

The model is based on the fact that cellular membranes contain lipids. This has been concluded both from direct chemical analysis and from the permeability characteristics of the plasma membrane which provide indirect evidence for the presence of lipids. Because proteins are present as the other major membrane constituent, the membranes are referred to as *lipoprotein membranes;* plasma membranes of different cell types usually contain from one to four times as much protein as lipid, although a few cellular membranes contain more lipid than protein. This estimate is on a *weight* basis: Since lipid molecules are usually much smaller than protein molecules (lipids have molecular weights on the order of 1000 whereas proteins range from 10,000 to over 100,000) there are usually far more lipid than protein molecules in membranes. The ratio may approach 10 to 100 lipid molecules per protein molecule in some cases. In very rare cases, such as the membranes surrounding the gas vacuoles of certain procaryotes, membranous structures are found that seem to contain much protein but little lipid. These membranes have distinctive microscopic appearances, but are poorly understood.

Figure II-2 illustrates three of the major types of lipids found in nature: fats, phospholipids, and steroids. Fats consist of *fatty acids*, long hydrocarbon chains, linked to a "backbone" three carbons long. The backbone is formed from the compound *glycerol*. When all three glycerol carbons are attached to fatty acids, the resultant lipid is known as a *triglyceride*, the major component of most fats. Mono- and diglycerides are constructed, as their names suggest, with one or two fatty acids linked to glycerol. Most phospholipids have a structure similar

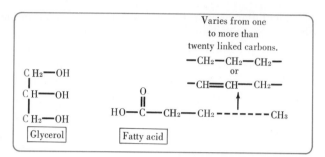

Varies from one
to more than
twenty linked carbons.

$-CH_2-CH_2-CH_2-$
or
$-CH=CH-CH_2-$

$C H_2-OH$
$C H-OH$
$C H_2-OH$

$HO-C-CH_2-CH_2-----CH_3$

Glycerol   Fatty acid

Glycerol
carbons   Fatty acid
$C H_2-O-C-CH_2-CH_2-CH_2---CH_3$
$C H-O-C-CH_2-CH=CH----CH_3$
$C H_2-O-C-CH_2-CH_2-CH_2---CH_3$

Triglyceride

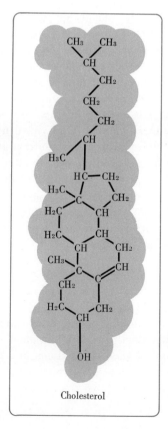

Cholesterol

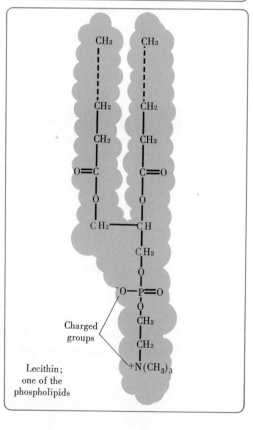

Charged
groups

Lecithin;
one of the
phospholipids

◄ *Fig. II-2* *Structures of lipid molecules. After J. B. Finean and others. Common natural fatty acids have 16–20 carbons in the region indicated by the dotted lines. The nitrogen-containing compound attached to the phosphate on the third glycerol-derived carbon in lecithin is* choline.

to triglycerides, except that in place of one of the fatty acids they have a more complex chain including phosphate and nitrogen-containing groups. Steroids have an architecture that is quite different; they are relatives of *cholesterol*, a molecule containing interconnected rings of carbon atoms. *Glycolipids* (lipids with sugar groups attached) also are found in cellular membranes.

The phospholipids and steroids are polar molecules; different ends of the molecules have different properties. One end is *hydrophobic*; that is, it tends to be insoluble in water. However, groups such as the hydroxyl (OH) group found at one end of the cholesterol molecule, or the charged phosphate and nitrogen groups at one end of a phospholipid molecule, confer *hydrophilic* properties on the regions of the molecules in which they are present. The result is that these regions have a great affinity for water.

Plasma membranes and other cellular membranes are rich in polar lipids, and the model being discussed makes use of this fact. The model suggests that the central region of membranes consists of two layers of lipid molecules, largely phospholipids and steroids (Fig. II-3).

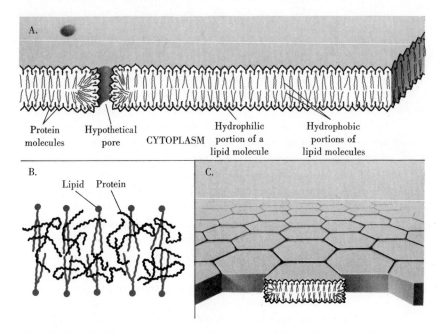

Fig. II-3 *Models of the plasma membrane. After H. A. Davson, J. F. Danielli, L. Rothfield, A. Finkelstein and others.*

Each layer is one molecule thick. The layers are arranged with the hydrophobic ends of the molecules in one layer associated with the hydrophobic ends of the molecules in the other layer. As will be outlined later (Section 4.1.1), hydrophobic regions of molecules have considerable affinity for one another and, as a result, lipids will often associate spontaneously in ordered groups (including layered arrays). The hydrophilic portions of the membrane lipids are presumed to face outward towards the surfaces of the "bimolecular layer" of lipid. At these surfaces the lipids are associated with proteins (which are known to contain many hydrophilic groups and to associate readily with water). The protein layers on the inner surface of the membrane (the surface facing the cell cytoplasm) probably differ, in specific molecular composition, from the protein layer that faces the external environment.

This model is attractive since it agrees with the usual electron microscope image; presumably the central layer of the unit membrane would correspond to the hydrophobic portions of the lipids, while the two dense lines would represent the proteins and the hydrophilic regions of the lipids. Unfortunately, this correspondance by itself is only conjectural. Interpretation of the electron microscope image of a membrane in chemical terms is difficult. Electron density (Section 1.2.1) in most preparations depends largely on the location of osmium atoms from the osmium tetroxide used in fixation and on the locations of other heavy metal atoms used as stains. On chemical grounds, osmium tetroxide should react strongly with some of the hydrophobic portions of the membrane lipids (specifically, with double bonds, $C{=}C$, in the fatty acids); if the model is correct, this would be expected to produce a dense central layer rather than the two outer dense layers of the unit membrane. On the other hand, when artificial mixtures with known arrangements of lipids or lipids and proteins are fixed with osmium tetroxide or stained with heavy metal stains, most electron density is produced in the regions where hydrophilic groups are expected; this empirical finding has been used to argue that the electron dense lines in the unit membrane are in the position that the model predicts they should be.

These considerations and others that are comparably contradictory are the basis for a major current debate in cell biology. For several years, the above model was almost unchallenged as a plausible picture of molecular arrangements in cellular membranes. It is still widely accepted and has not been disproved; but a growing body of evidence has suggested the need for revision or for serious consideration of alternatives. Membrane chemistry has been studied in very few cell types. Most attention has been devoted to the membranes surrounding red blood cells and to the special layered arrangements of plasma membranes in the myelin sheath surrounding some nerve cells (Section 3.7.4). These membranes are relatively easy to study and the model was derived largely from work on them. Mature mammalian red blood cells, for example, have no internal organelles. The plasma membrane surrounds a cytoplasm consisting chiefly of the protein, hemoglobin; the absence of a nucleus and cytoplasmic membranes facilitates plasma membrane isolation and purification. Analysis of red blood cell membranes indicates the presence in each cell of roughly the amount of lipid needed to make a layer two molecules thick, and this has been taken as support for the model under discussion. However, both the red blood cells and the myelin sheath are so specialized in function that there is no assurance that their membranes are typical. The limited information available from other cells tends to indicate that the plasma membranes of different cell types differ in the specific proteins, lipids, and other components present; in the ratios in which different molecules are found; and in the enzymatic and other activities of the membrane. Structure might vary with composition. And even the similarity of overall visible structure among different plasma membranes may reflect only a rough similarity in functional organization. In addition, it is likely that different portions of the plasma membrane of a single cell differ from one another and that a given region varies in composition and in activity at different times and under different conditions. (Only the barest hints of such dynamic variation are obtained with current techniques.)

These considerations are even more important when one thinks not only of plasma membranes but also of the variety of different cellular membranes with very different roles. It may be unrealistic to expect that all have comparable structures.

Several microscopists have reported that some cellular membranes perhaps including the plasma membrane, show a more complex structure than the unit membrane if special methods are used to prepare the cell for microscopy. They propose that a membrane (or areas of it) is a pavement of small globular or polygonal subunits rather than a continuous layer (Fig. II-3). Other microscopists have found that unit membrane structures can be observed even in cellular membranes from which almost all lipid has been experimentally removed; they argue that this may mean that the microscopically visible unit membrane need not reflect a molecular arrangement of the type proposed by the model. Currently there is disagreement over the significance of these observations. For example, some claim that subunits really exist, others that they are artifacts, and still others that the same membrane can exist either as a unit membrane or in subunit form depending on conditions inside and outside the cell. Resolution of this controversy probably will require newer techniques for cell preparation, but much important insight will come also from the analysis of membrane function, particularly from studies of transport.

2.1.3   ***TRANSPORT***   The first consideration involves the way
              ACROSS THE   in which molecules in aqueous solution
              PLASMA   (i.e., dissolved in water) outside the cell pass
              MEMBRANE   through the lipoprotein plasma membrane
                                 and gain access to the aqueous solution
inside the cell. The model of the plasma membrane as an uninter-
rupted bimolecular layer of lipid with proteins at the surface accounts
for the ability of lipid-soluble material to enter the cell; such material
could pass down concentration gradients from the "outside" solution
into the lipid layers and then into the "inside" solution. However, it
is not obvious that water, inorganic ions [for example, sodium ions
(Na^+) or potassium ions (K^+)] or a variety of other substances that
are not readily soluble in lipids but enter the cell could easily pass a
barrier of continuous lipid layers. Some molecules require enzyme
activity to pass the membrane, and the presence of only a few molecules of the appropriate specific enzyme per square micron of membrane would suffice to explain their ability to enter the cell. Many
others, however, appear to pass through without enzymatic interven-
tion. To explain the passive diffusion of some molecules into the cell,
and also to explain some restrictions on the sizes of molecules that
can readily enter, it is often hypothesized that special channels pass
through the membrane, interrupting the lipid layers. Usually these
channels are referred to as *pores*, although one need not think literally
in terms of holes. Rather, the channels are presumed to be arrangements of membrane molecules or of portions of molecules, such that
there extends from one surface of the membrane to the other a continuous zone of hydrophilic groups (for example, portions of proteins or
charged groups of lipids) and perhaps even of water molecules loosely
bound to membrane components. Such a hypothetical pore is diagrammed in Fig. II-3. Probably, a variety of nonlipid-soluble molecules
could diffuse through such channels far more rapidly than through continuous layers of hydrophobic groups. For example, inorganic ions associate readily with groups bearing charges opposite to their own and with
water; in solution they are surrounded by a loose "shell" of associated
water molecules. The pores are usually thought of as having diameters
of a few Ångstroms so that passage through them would be restricted to
water or material such as inorganic ions and small molecules with diameters less than that of the pore.

Several alternative types of channels have been proposed. The
molecules in the membrane are in constant motion; such movement
might result in a transient realignment of appropriate portions of
molecules that creates a temporary channel, a *dynamic pore*. Or perhaps more permanent channels exist, but their small size, irregular
structure, or poor preservation during fixation prohibit identification

by current microscopic techniques. Another suggestion is that the regions between membrane subunits, if such subunits exist, would be equivalent to pores. Also, some recent speculation suggests that the membrane is actually a network of protein with lipids associated as diagrammed in Fig. II-3; such a membrane could readily permit the passage of nonlipid-soluble material. This protein network model is under intensive investigation as an interesting possible alternative to the model discussed above; along with their hydrophilic groups, proteins have hydrophobic regions that could associate with the hydrophobic portions of lipids.

A number of enzymes, including some called *permeases*, aid the passage of various molecules across the plasma membrane. For example, glucose enters some animal cells by "facilitated" or "mediated" transport mechanisms in which membrane-associated enzymes take part. The existence of such enzymes is indicated in part by the fact that sugars only slightly different in chemical structure may enter cells at very different rates, even if they are of similar sizes, lipid solubilities, and other features that govern passive movement through membranes. This argues for highly selective sites, presumably on proteins, that bind the sugar molecules as part of the transport mechanism. Membrane enzymes are involved also in energy-dependent transport, termed *active transport*. Interruption of energy production by cells (for example, by exposure to low temperature or by treatment with metabolic poisons such as iodoacetic acid which interrupts glycolysis) can strongly affect the movement in or out of the cell of several important types of molecules, including inorganic ions.

Energy-dependent transport can operate against concentration gradients; that is, it can result in concentrations within the cell that are considerably higher or lower than concentrations outside the cell, for materials that would tend to be at equal concentration inside and out were active transport not operating. One important transport of this kind involves sodium ions (Na^+). Intracellular sodium concentration in most cells of higher organisms is kept much lower than extracellular concentration. When radioactive sodium ions are introduced into a solution containing nonradioactive cells, some of the radioactive ions rapidly enter the cells and replace nonradioactive sodium ions that leave. This "tracer" experiment indicates that sodium normally continually enters and leaves the cell and that the low intracellular concentration is not due to an inability of the ions to cross the membrane. Rather, an active extrusion mechanism (a sodium "pump") balances the entry of sodium ions and maintains the low intracellular level. Potassium ions (K^+) have the reverse distribution; they are more concentrated within the cell than outside. The relevance of these ion distributions to nerve transmission will be discussed in Section 3.7.3.

For the moment, it should be noted that a potential difference, comparable to the differences between poles of a battery (or, more accurately, between surfaces of a capacitor) exists across the membrane; the inside of the cell is normally electrically negative in comparison to the outside. Physical chemical theory indicates that the existence of this potential reflects the asymmetries in ion distribution and the fact that some ions penetrate the membrane more readily than others.

Energy-dependent transport is usually attributed chiefly to the presence of hypothetical "carrier" molecules in the membrane, although other mechanisms (such as transient "pore" formation) have also been postulated. Carriers are presumed to attach to molecules at one surface of the membrane, carry them through the membrane, and release them at the other surface. One enzyme, an ATPase that splits ATP at a much accelerated rate if Na^+ and K^+ are present in the medium, has been found in many plasma membranes, and it is considered to participate in Na^+ transport. The usual suggestion is that the enzyme picks up a sodium ion from inside the cell and exchanges it for a potassium ion outside; the K^+ is then carried to the inside and exchanged, in turn, for another Na^+. That the ATPase is involved in such ion transport is indicated: (1) by the required presence of both Na^+ and K^+ for rapid activity in splitting ATP; and (2) by the fact that the drug *ouabain* inhibits both the splitting of ATP by the enzyme and the transport of Na^+ out of the cell. How the carrier molecule transports ions is not established but molecular mechanisms are known that provide suggestive models. Perhaps the carrier remains fixed in position in the membrane but changes its three-dimensional shape, so that portions of the molecule move and therefore carry components linked to them. Or, perhaps a complex between the molecule being transported and the entire carrier molecule moves across the membrane.

Although much remains to be learned about the ways in which molecules cross the plasma membrane, new information is coming from many directions. Methods for isolating and purifying the plasma membrane are being improved. Another approach of importance is the study of artificial membranes made from purified phospholipids and other components. For example, membranes have been constructed that are believed to have the simple lipid bilayer structure proposed in the model discussed in Section 2.1.2; surprisingly, water can pass through these membranes almost as rapidly as it can enter the cell. Of interest is the fact that the permeability to water of artificial membranes varies inversely with the proportion of cholesterol added to the phospholipids, a finding that may have important significance in explaining differences among natural membranes. Some features of the electrical behavior of cell surfaces (Section 3.7.3) are mimicked if a specific mixture of proteins is added to the artificial membrane. Inorganic ions (Na^+, K^+, and so forth) usually pass through the artificial membranes far more slowly than they do through natural membranes. However, if an antibiotic known as *valinomycin* is present there is a great increase in the rate at which Na^+ and K^+ can diffuse through the artificial membrane. Furthermore, with valinomycin K^+ passes through artificial membranes at a faster rate than Na^+, an important selectivity feature that is characteristic of the passive movement of these ions through many natural membranes. Valinomycin is a small molecule that can form a specific type of complex with appropriate inorganic ions; it is hypothesized that an ion-valinomycin complex is far more soluble in lipid than the ion alone, probably in part because valinomycin has substituted for the "hydration shell" of water molecules that normally surrounds the ion.

The significance of observations on artificial systems for understanding natural transport mechanisms is not clear, and compounds resembling valinomycin are not known to be present in natural membranes. However, study of the effect of such compounds on artificial membranes should provide insight into possible transport mechanisms and thus help orient future research with the more complex natural membranes. For example, the valinomycin effect may provide a model for natural "carrier" mechanisms. Other antibiotics are thought to form pores in artificial membranes and these may aid in the search for natural pores.

2.1.4 **BULK TRANSPORT: PINOCYTOSIS AND PHAGOCYTOSIS** Participation of the plasma membrane in another kind of transport can be studied in the microscope. *Phagocytosis* of bacteria or other large structures, by protozoans or by phagocytic cells of higher animals, involves the incorporation of these particles into intracellular vacuoles that originate by the folding of the plasma membrane around the material being engulfed. *Pinocytosis* is a similar phenomenon, except that here the particles taken up are of molecular or macromolecular dimensions. Proteins or other molecules are adsorbed to sites on the plasma membrane, and the membrane folds in to form small vacuoles or vesicles that move into the cell, carrying drops of the medium. This may be seen by the light microscope in living amebae, because the vacuoles are quite large. In most cells of multicellular organisms, the vesicles are too small for light microscopy but can readily be seen by electron microscopy. In many cases, the occurrence of pinocytosis can be demonstrated by introducing tracer molecules into the organism or the media in which cells are grown. One

such tracer is the iron-containing protein, *ferritin*, which is visible
in the electron microscope because the large number of iron atoms
in each molecule scatter electrons much more effectively than the cell
components do. Another tracer is the enzyme *peroxidase* isolated from
horseradish; many cells will take up this protein by pinocytosis. If
cells are then fixed and incubated with the enzyme's substrate, the sites
of peroxidase activity will be marked by a dense product (Fig. II-4).
During the incubation many molecules of reaction product are formed

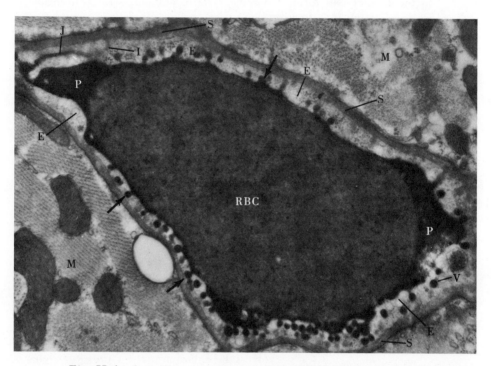

Fig. II-4   *A capillary (cut transversely) in the heart muscle of a rat that had
been injected intravenously with the enzyme peroxidase 5 minutes prior to
fixation. The tissue was treated as mentioned in Section 2.1.4 so that peroxidase sites are marked by dense reaction product. Most of the space within the
capillary is occupied by a red blood cell (RBC). Surrounding this is a peroxidase-
filled region, the plasma (P) suspending the blood cells. The capillary wall
consists of thin endothelial cells (E) outside of which a space (S) separates the
vessel from the muscle cells (M). Many pinocytosis vesicles (V) are present in
the endothelial cells; some (arrows) are seen still to be connected to the cell
surface (see also Figs. I-2 and III-31). Under the experimental conditions used,
peroxidase also can pass between the endothelial cells and gain access to the
space at S (see Section 3.5.2). This has begun to take place at (J). In preparations fixed at later times the remainder of the space between adjacent cells (I)
shows peroxidase as does the region at S. × 25,000. Courtesy of M. J. Karnovsky.*

by the action of each molecule of peroxidase taken up by the cell; thus, the method is very sensitive and sites of only a few peroxidase molecules can be detected.

When pinocytosis and phagocytosis proceed rapidly, large amounts of plasma membrane are removed from the surface and taken into the cell as the membranes of the vacuoles. Formation of new membrane replenishes the surface; the Golgi apparatus has been implicated as one probable source of new cell surface material.

As will be seen in Section 2.8.2, the macromolecules inside phagocytic and pinocytic vacuoles may be digested by enzymes acquired when the vacuoles fuse with lysosomes. An interesting feature of many pinocytosis vesicles (Figs. III-30, 31) is the "coating" they show on the membrane surface facing the cytoplasm. Similar coating is sometimes seen on the surface of vesicles not involved in pinocytosis (e.g., some vesicles produced by the Golgi apparatus or associated systems). While the nature of the coat is not known, a reasonable hypothesis is that it represents special organization of cytoplasm involved in the pinching off of the small membrane-enclosed vesicles from larger membrane structures. (See Fig. II-30).

2.1.5 **POLYSAC-CHARIDES AND PLASMA MEMBRANES** Polysaccharides, as a class, are polymers of carbohydrate molecules. *Glycogen*, composed of chains of glucose units, is the major intracellular storage form of sugar in animals. A similar polysaccharide, *starch,* is used in sugar storage by plants. Complex polysaccharides (for example, the *acid mucopolysaccharides*) contain groups such as sulfate groups and various small organic molecules linked to the sugars. Many polysaccharides are found complexed with proteins.

Animal cells often have a layer of protein and polysaccharide at their outer surface. Sometimes this forms a distinct coat or "fuzz" (Figs. II-5 and III-18). Even in compact tissues such as liver, the plasma membranes of adjacent cells are separated almost always by a space of 100–200 Å. (Only at special membrane regions called *tight junctions* do adjacent cells approach much closer than this (Section 3.5.2).) The space between adjacent cells is often occupied by a matrix of proteins and polysaccharides that helps to cement cells together. In addition, carbohydrates are built into some membrane lipids. For example, lipids known as *gangliosides* contain the sugar derivatives *sialic acid* or *neuraminic acid*. It is interesting that treatment of some living animal cells with the enzyme *neuraminidase*, which removes these

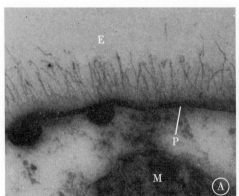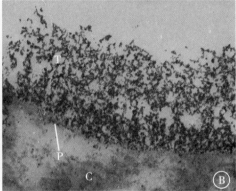

Fig. II-5   *A. Portion of the surface of an ameba. Numerous fine filaments are seen attached to the extracellular surface of the plasma membrane (P). The tip of a mitochondrion is present at (M). (E) indicates the space outside the cell. × 55,000.* **B.** *Portion of a similar cell that had been exposed to thorium dioxide particles in suspension. The electron-dense metal-containing particles are adsorbed to the filaments (T). Such adsorption is the first step in the uptake of material through pinocytosis. (C) indicates the cell cytoplasm. × 35,000. Courtesy of G. Pappas and P. Brandt.*

components, considerably alters the surface properties of the cells; for example, it reduces the net negative charge characteristic of the surfaces of many animal cells and removes some sites at which proteins and other material can be bound prior to uptake into the cell. Some cell proteins, known as *glycoproteins*, also have carbohydrate molecules attached to them.

A number of animal cells and tissues have extensive special extracellular polysaccharide material. *Chondroitin sulfate* is a major component of cartilage, and *hyaluronic acid* is present in sites such as joints, the lower layer of the skin, and also in the external coats that surround some egg cells. The cell walls of plant cells are made largely of the polysaccharide *cellulose*, as will be discussed in Section 3.4.3.

2.1.6   **MEMBRANE SPECIFICITY**   The specific polysaccharides, proteins, and lipids present at cell surfaces vary among different cell types. The plasma membrane apparently combines a relatively simple framework and a highly ordered arrangement of specific molecules. Cells have selective surface binding sites, including some that adsorb extracellular molecules which the cell then engulfs by pinocytosis (Fig. II-5). The factors responsible for blood type are identifiable at the surfaces of red blood cells. The responses of cells to complex stimuli, such as nerve impulses or the presence in the extracellular fluid of hormones, begin at the plasma membrane. There are specific receptor sites built into the membrane for the transmitters liberated by nerve endings (Section 3.7.5). A much-discussed hypothesis for the action of many hormones is based on the presence of a plasma membrane receptor system that includes the enzyme *adenyl cyclase* in cells responsive to a given hormone. A variety of hormones are thought to stimulate the action of this enzyme in various cell types, thus producing a rise in the intracellular level of the enzyme's product, *cyclic AMP* (adenosine monophosphate). The cyclic AMP apparently affects the metabolic pattern of the cell. It is likely that different cell types possess different batteries of surface-associated enzymes and receptor sites which respond to different extracellular agents.

One indication of the existence of a complex surface architecture is seen in surface specificity during cell association. Embryonic cells from different organs are separated from one another by appropriate gentle means (usually reduction of the calcium content of the medium is part of the procedure) and the cells of two organs are intermixed. The cells from each organ will "recognize" one another on contact. In time, cells of a given type separate from the mixture and reconstitute separate aggregates. These aggregates may even develop some features of the original organ. Such aggregation probably reflects differences in molecular arrangements at the surfaces of different cells.

c h a p t e r **2.2**
THE NUCLEUS

Cells without nuclei have limited futures. The only common animal cell type without a nucleus, the mammalian red blood cell, lives only a few months; aside from its role in oxygen transport, it is extremely limited in its metabolic activities. Egg cells from which nuclei have been experimentally removed may divide for a while, but the products of division never differentiate into specialized cell types and eventually they die. Fragments without a nucleus, cut from such large unicellular organisms as amebae or the alga *Acetabularia*, survive temporarily but ultimately they die unless nuclei from other cells are transplanted into them. Thus the nucleus is essential to long term continuation of metabolism and to the ability of cells to alter significantly their structure and function (as in differentiation). In part, this reflects the primary role of the nucleus in producing the RNA required for protein synthesis. When cells change, their new functions and structures require new proteins. Even cells that are constant in metabolism and structure show continual replacement (*turnover*) of macromolecules and probably of organelles, including portions of the cytoplasmic protein-synthesizing machinery.

The nucleus is of central importance in cell heredity; it determines key morphological and metabolic features of a cell. For example, if a fragment containing the nucleus is cut from an *Acetabularia* of one species, characterized by a given morphology, the fragment will regenerate a whole cell of that species. This regenerative ability permits experiments of the type illustrated in Fig. II-6, in which nuclei of one species are combined with cytoplasms from different species. The conclusion drawn from such experiments is that the nucleus produces some material, probably containing RNA, that enters the cytoplasm and participates in the control of cell growth and cell morphology. The crucial finding is that the morphology of the regenerated cells eventually becomes like that of the species from which the nucleus is taken. In the hybrid fragments with the nucleus from one species and most of the cytoplasm from the other species, old cytoplasmic material persists for a while and may influence cell form. Eventually, however, this is depleted and replaced by newly produced material from the nucleus.

2.2.1   ***DNA***   The approximate composition of rat hepatocyte nuclei is 10–15 percent DNA, 80 percent protein, 5 percent RNA, 3 percent lipid. (This is on a percent of dry weight basis; water is a primary constituent of all cells and organelles and accounts for 70 percent of hepatocyte weight.) An early cytochemical finding was the presence in nuclei of virtually all the DNA of cells; only later was the presence of small but significant amounts of DNA in some cytoplasmic organelles recognized. The DNA content of cells of different organisms varies greatly. However, quantitative cytochemical studies of cells in multicellular animal and plants established that the DNA content of most nuclei in a given organism is twice ($2 \times$) that of the sperm or egg cell nuclei (or, for some cells, a multiple of $2 \times$ based on continued doubling, such as $4 \times$, $8 \times$, or $16 \times$. This is as expected for the genetic material: each gamete contributes an equal amount of DNA to the zygote nucleus, which by duplicating, gives rise to all the nuclei in the cells of the organism. These considerations will be discussed further in Part 4.

DNA molecules are composed of two strands coiled together in a double helix (Fig. II-7). Each strand is a chain of nucleotides (a *polynucleotide* strand); all DNA nucleotides consist of a 5-carbon sugar (deoxyribose) with a phosphate group attached at one end and a nitro-

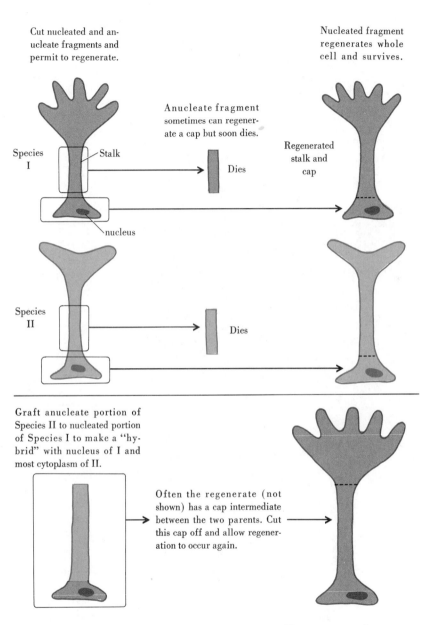

Cut nucleated and an-
ucleate fragments and
permit to regenerate.

Nucleated fragment
regenerates whole
cell and survives.

Anucleate fragment
sometimes can regener-
ate a cap but soon dies.

Species
I

Stalk

Dies

Regenerated
stalk and
cap

nucleus

Species
II

Dies

Graft anucleate portion of
Species II to nucleated portion
of Species I to make a "hybrid" with nucleus of I and
most cytoplasm of II.

Often the regenerate (not
shown) has a cap intermediate
between the two parents. Cut
this cap off and allow regeneration to occur again.

New regenerate has mor-
phology of Species I. This
characteristic remains stable.

Fig. II-6 *Experiments with the large single-celled alga,* Acetabularia. *After
the work of Gibor, Hammerling and others.*
    *The intermediate caps sometimes formed in the initial "hybrids" probably
reflect the fact that some time elapses before "old" material in the cytoplasm
is depleted and replaced by new material from the nucleus.*

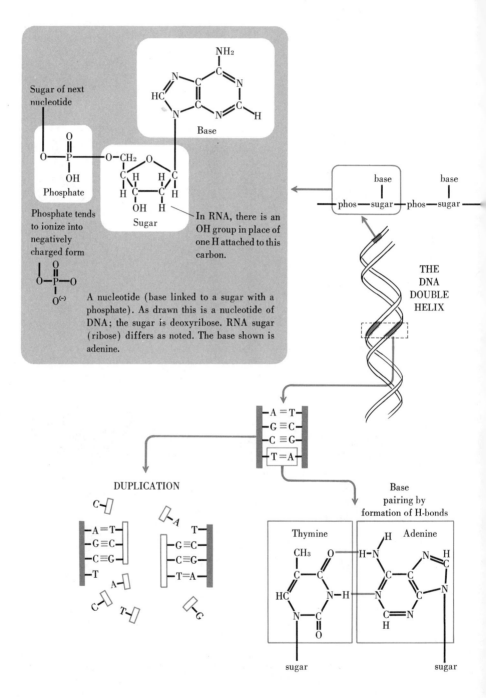

Sugar of next nucleotide

O
‖
O—P—O—CH₂
|
OH
Phosphate

Base

NH₂

Sugar

In RNA, there is an OH group in place of one H attached to this carbon.

Phosphate tends to ionize into negatively charged form

O
|
O—P—O
|
O⁽⁻⁾

A nucleotide (base linked to a sugar with a phosphate). As drawn this is a nucleotide of DNA; the sugar is deoxyribose. RNA sugar (ribose) differs as noted. The base shown is adenine.

base   base
| |
—phos—sugar—phos—sugar—

THE DNA DOUBLE HELIX

—A = T—
—G ≡ C—
—C ≡ G—
—T = A—

DUPLICATION

—A = T—
—G ≡ C—
—C ≡ G—
—T

—T—
—G ≡ C—
—C ≡ G—
—T = A—

Base pairing by formation of H-bonds

Thymine   Adenine

sugar   sugar

Fig. II-7 *Some features of nucleic acids.*

gen-containing ring compound (known as a *base*) at the other. Four different bases are present on different nucleotides: adenine (A), guanine (G) cytosine (C), and thymine (T). The two strands of nucleotides are aligned according to simple pairing rules: Every A on one strand pairs with a T on the other, while every G pairs with a C. The paired bases are held together by molecular interactions, chiefly *hydrogen bonds* in which a hydrogen atom that is part of one base is also attracted to and loosely held by the second base. The strands are referred to as *complementary*, since the sequence of bases on one strand exactly dictates that of the other. It is the base sequences that carry the hereditary information. The replication of DNA results in the duplication of the base sequences (Fig. II-7) and this underlies the role of DNA in the transmission of hereditary information, as will be discussed in Part 4. (See also p. 323.)

2.2.2   **PROTEINS**   In the nucleus, DNA is complexed with proteins among which are the *histones*. The phosphate groups of the DNA are negatively charged. When cells are stained with basic dyes (those dyes that are positively charged), the DNA- and RNA-containing regions stain because their negatively charged phosphate groups attract the oppositely charged dye molecules. Such regions are referred to as *basophilic*. The histones are basic proteins; that is, they have a high content of the basic amino acids lysine, arginine, and histidine which contain derivatives of amino (NH_2) groups; amino groups can assume a positively charged (NH_3^+) form. Histones are *acidophilic*; their positively charged amino group derivatives attract acid (negatively charged) dyes. In nuclei, however, histones do not stain strongly with acid dyes unless the DNA is first removed by chemical or enzymatic means. This suggests that many of the DNA phosphate groups bind to the oppositely charged histone amino groups, thus neutralizing the histones' charge and minimizing the attraction for acid dye molecules. When DNA is removed, the histone groups are free to bind dye. DNA and histone are usually present in nuclei in approximately equal amounts, and the number of amino groups in the histones is roughly the same as the number of phosphate groups in the DNA.

In addition to the histones, nuclei usually contain even larger amounts of other, nonbasic, proteins. These include enzymes and nonhistone proteins involved in maintaining intranuclear structure. The contribution of the nucleus to total cell mass may vary from approximately 5 percent (some muscle cells), to 5–10 percent in hepatocytes, and up to 50 percent or more in cells of the thymus gland

and in other rapidly dividing cells such as plant root-tip cells and cancer cells. Variations in the ratio of nuclear volume to cell volume fall within the same range. Measurements on the nuclei of different cell types of a given organism indicate that the DNA per nucleus usually is essentially constant from cell to cell or varies as a multiple of the $2 \times$ amount; the histone content often parallels the DNA. In contrast, the content of nonbasic proteins and of RNA varies considerably. Sperm cell nuclei, which are metabolically inactive, have virtually no RNA or nonbasic protein; DNA may account for one-third to one-half of the nuclear dry weight. In cells with metabolically active nuclei, such as hepatocytes, DNA accounts for only 10–20 percent of the nuclear dry weight, and much nonbasic protein and RNA are present.

2.2.3   ***RNA***   A most important aspect of DNA function is the production of RNA. Autoradiographic evidence outlined earlier (Section 1.2.4) indicates that most of the RNA of cells is produced in the nucleus in close association with DNA. DNA and RNA are similar in that both are polynucleotide chains, but RNA differs from DNA in several ways. RNA is single-stranded, except in certain viruses where, like DNA, it is double-stranded. The RNA 5-carbon sugar is *ribose* instead of deoxyribose. Finally, in RNA the base *uracil* replaces the thymine of DNA. (See Fig. II-7.)

The base sequences of RNA molecules are determined by DNA. RNA nucleotides align by base pairing along one DNA strand. (What determines which of the two strands is copied remains unclear.) The A base of RNA pairs with T of DNA, G of RNA with C of DNA, U of RNA with A of DNA, and G of RNA with C of DNA. A transcription enzyme, RNA *polymerase*, joins the nucleotides to form an RNA strand that is complementary to a DNA strand. By this mechanism, DNA acts as a *template* for RNA synthesis.

The complementary relationship of RNA with DNA is the basis of an extremely valuable technique called *molecular hybridization*, as shown in Fig. II-8. If test-tube "hybrids" can form between particular DNA's and RNA's then the two probably contain many complementary base sequences. Several different classes of RNA are present in cells; they differ in size and in role, as will be seen in the next chapter. Using hybridization, it can be shown that all major types of RNA are complementary with DNA from the nucleus (with the probable exception of the nucleic acids associated with mitochondria and plastids). This strongly suggests that all are synthesized in the nucleus.

Once the RNA is made in the nucleus, much of it moves rapidly into the cytoplasm. In rat hepatocytes, less than 10 percent of the total

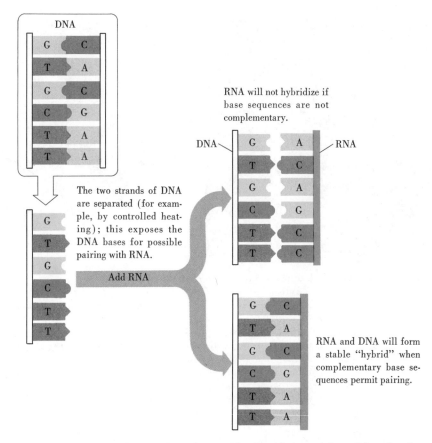

Fig. II-8 *Hybridization of nucleic acids. Purified nucleic acid molecules are mixed in the test tube. The formation of hybrids can be detected in several ways. For example, hybrids between radioactive single-stranded RNA and non-radioactive DNA will be double-stranded (one RNA strand, one DNA strand) and radioactive. Their radioactivity distinguishes them from the original double-stranded DNA while their double-strandedness produces differences (for example, in density) which permit them to be separated (by centrifugation or other methods) from the original single-stranded RNA molecules.*

cellular RNA is present in the nucleus at any given moment; much of this is probably in process of synthesis or in transit to the cytoplasm. The transfer of much RNA from nucleus to cytoplasm can be nicely demonstrated in nuclear transplantation experiments using amebae; Fig. II-9 shows the results of one such experiment.

It has been found that some RNA made in the nucleus does not pass into the cytoplasm; eventually it breaks down within the nucleus. The existence of this type of RNA has only recently been generally accepted and its significance is now being studied.

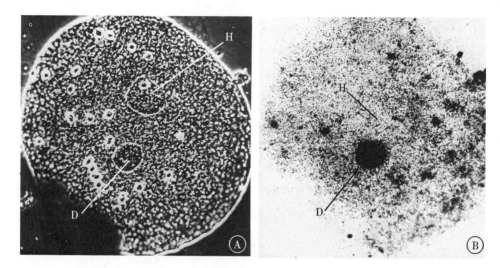

Fig. II-9 *A*. Phase-contrast micrograph *of an ameba grown in a non-radio-active medium into which a nucleus (D) was transplanted from an ameba in which the RNA had been made radioactive by growth in P³²-phosphate medium. (H) indicates the nucleus of the host cell.* **B**. *Autoradiogram of the same cell. The site of the donor nucleus appears black due to the accumulation of many closely packed grains in the autoradiographic emulsion (see Fig. I-19). Grains are also numerous, though less closely packed, over the cytoplasm. There is little if any radioactivity in the host cell nucleus; the grains seen over the region of this nucleus probably result from radioactive molecules in the cytoplasm immediately surrounding the nucleus. × 150. Courtesy of L. Goldstein and W. Plaut.*

2.2.4   ***NUCLEAR***   Nuclei have enzymes (polymerases) thought
              ENZYMES   to participate in the production of DNA and
                                RNA. Other enzymes are present as well,
but these have been studied only to a limited extent. Previously (p. 36), allusion was made to one such enzyme involved in the synthesis of NAD in hepatocytes. Nuclei of some cells may generate ATP by a pathway still not fully elucidated. Whether nuclei synthesize their own proteins or receive them from the cytoplasm is of current interest. A growing body of information suggests that at least some nuclear proteins originate in the cytoplasm and migrate to the nucleus. However, several laboratories report that preparations containing isolated nuclei, at least from some cell types, can synthesize their own proteins. Although it is sometimes difficult to be sure that the isolated nuclei have been completely freed from cytoplasm, there is a possibility that different nuclear proteins may originate in different parts of the cell or that cell types vary in the ability of their nuclei to synthesize proteins.

Even these fragmentary findings make it evident that nuclei cannot be regarded as mere libraries of DNA sequences that periodically extrude information in the form of RNA. However, most of the details of nuclear metabolism remain to be unravelled by future studies.

2.2.5   ***STRUCTURE***   As seen in the light microscope, the nucleus is bounded by a nuclear membrane. Within the nucleus there is a heterogeneous collection of fibrils and of dense areas that include *euchromation, heterochromation,* and *nucleoli* (Figs. I-10–I-12, II-10, and II-11). Chromatin refers to the DNA-containing structures in the nucleus. The finest structures seen in the electron microscope are fibrils 20–40 Å in diameter, corresponding to DNA molecules complexed with proteins. These fibrils are parts of the chromosomes. During cell division, each chromosome coils into a compact configuration and is readily identifiable as an individual unit. (This will be discussed at greater length in Section 4.2.5.) In cells not in division, chromosomes exist in a relatively uncoiled state: in euchromatin, this uncoiling is maximal; in heterochromatin, the

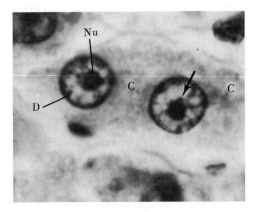

Fig. II-10   *Two hepatocytes in a section of rat liver prepared by widely used routine procedures: fixation in formaldehyde, embedding in paraffin, preparation of sections 5–10 microns thick and staining with hematoxylin (colors chromatin blue) and eosin (gives the cytoplasm (C) a contrasting pink color). Prominent nucleoli (NU) are present within the nuclei. Much of the chromatin is dispersed and difficult to see in a black and white photograph; the rest is in accumulations close to the nucleolus (arrow) or the nuclear envelope (D). While fixation can clump dispersed chromatin producing "artifacts" (Section 1.2.2), dense accumulations such as these are often seen in living cells. The term heterochromatin is used for the condensed form of chromatin (see also Fig. II-11). × 1,500.*

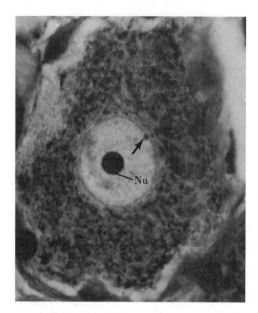

Fig. II-11 *Within the nucleus of the nerve cell of a cat, a prominent nucleolus is visible (NU). Most of the chromatin is in the form of fine threads dispersed throughout the nucleus. The arrow indicates the* sex heterochromatin *or "Barr body", a characteristic accumulation of dense chromatin found in several cell types in females of many mammals. ×1,000. Courtesy of M. L. Barr.*

chromosomes or parts of them are in a more compact coiled array. Some nuclei show a stable, clearly nonrandom, pattern of heterochromatin. For example, in many cells of some female mammals, a small specific heterochromatic body is present. This is an X chromosome, one of the chromosomes that determines the sex of the animal (Fig. II-11). This heterochromatic region is absent in males, which have only one X chromosome as opposed to two in the cells of the females. Cells from males and females can be distinguished on the basis of this cytological characteristic. Evidence outlined later (Section 4.4.6.). supports the tentative conclusion that a given region of chromatin is active when in the diffuse euchromatic state and inactive when converted to the condensed heterochromatic state.

Nucleoli are especially rich in RNA and proteins. Although the pattern varies from organism to organism, often there are from one to four nucleoli per nucleus; generally they take the form of large spherical or ovoid bodies. Nucleoli are prominent in RNA metabolism (Section 2.3.4).

It is believed that intranuclear structures are surrounded by a more or less fluid component generally called *nuclear sap.* Among the substances dissolved or suspended in it are probably those in transit from nucleus to cytoplasm. Structures visible in the nuclear sap include small ribonucleoprotein-like granules, other granules, and fine fibrils of unknown nature.

2.2.6 **THE NUCLEAR ENVELOPE** The structure of the "nuclear membrane" is of great interest for nuclear-cytoplasmic interactions. The electron microscope has revealed it to be a flattened and expanded part of the endoplasmic reticulum that surrounds the nuclear material. The term *nuclear envelope* is used to convey the fact that the nucleus is surrounded by a flattened sac rather than a simple membrane. The outer surface of the sac, the one exposed to the cytoplasm has RNA-containing granules attached to it; probably these are ribosomes similar to the ones in the rest of the cytoplasm. The inner surface, exposed to the nucleus, lacks ribosomes. Often, much chromatin is aggregated along the inner surface of the envelope. In addition, in some cell types, the inner membrane surface is coated with a layer of electron dense material (the *inner dense lamella* or *fibrous lamina*) which may be a few hundred Å thick.

Nuclear "*pores*" occur at regular intervals (Fig. II-12). These are circular or polygonal areas where inner and outer surfaces of the sac appear fused and where the sac is apparently interrupted by openings. In some cells, the pores seem to be occupied, at least in part, by a thin diaphragm or by some sort of plug; these appearances are not well understood but it may be an oversimplification to regard the pores simply as holes in the envelope. However, material of similar appearance in the electron microscope is sometimes seen adjacent to nuclear and cytoplasmic surfaces of the nuclear envelope and even inside the pores (Fig. II-13). Such images probably represent material in transit through the pores from nucleus to cytoplasm. Experiments with marker substances indicate that material can also move through the pores in the other direction. When particles of gold up to 100 Å in diameter are injected into the cytoplasm of amebae, some of the particles gain access to the nucleus; the electron opacity of the gold permits direct visualization of the particles in the electron microscope. It is difficult to conceive how particles this large could enter the nucleus if they did not pass through the pores; they are far too large to pass directly through the membranes. Overall, the pores, each several hundred Å across, may occupy as much as 10–25% of the nuclear surface in some cell types. While this provides an appreciable potential channel system for nuclear-cytoplasmic exchanges, the mechanisms of movement remain unknown. In some cases even small molecules and inorganic ions pass across the envelope at rates slower than would be expected were the pores substantially open paths for diffusion. A few observations have been made that point toward some type of active or enzyme-facilitated transport through the envelope. At present, much attention is being focused on the ring-like region (*annulus*) of envelope

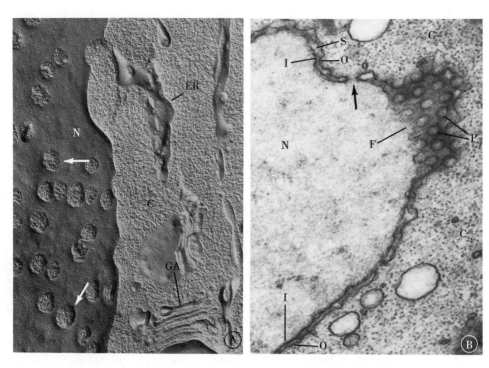

Fig. II-12 *The* nuclear envelope *as seen with different preparative procedures for electron microscopy.* **A.** *Portion of an onion root tip cell prepared by* freeze-etching. *This method provides views of surfaces within cells without the need for fixation or sectioning. Tissue is rapidly frozen and then fragmented; fracturing often occurs along preexisting intracellular surfaces such as membranes. Replicas (special molds) of the surfaces exposed by the freezing-fragmentation procedure are examined in the electron microscope. At (N), a portion of the nuclear surface is seen; the arrows indicate two of the* pores *that appear as circular depressions or interruptions of the surface. In the cytoplasm (C) portions of the Golgi apparatus (GA) and endoplasmic reticulum (ER) are visible.* × *30,000. Courtesy of D. Branton.* **B.** *Portion of an oöcyte of the toad,* Xenopus, *prepared by conventional methods of fixation, embedding and sectioning. (N) indicates the nucleus and (C), the cytoplasm. In cross section, the nuclear envelope appears as a membrane-enclosed sac (S). The membrane at the surface of the sac facing the cytoplasm (O) and that at the surface facing the nucleus (I) are continuous at the edges of the pores; thus in cross section, the pores appear (arrow) as gaps in the envelope. At (F), the envelope is twisted and the section provides a face view showing the pores (P) as circular openings.* × *40,000. Courtesy of J. Wiener, D. Spiro, and W. R. Loewenstein.*

membrane and associated electron-dense material that borders each pore and there are hints from microscopy that this region is specially organized. Also, limited evidence suggests that pore properties may vary somewhat among different cell types and perhaps within a given cell under varying conditions.

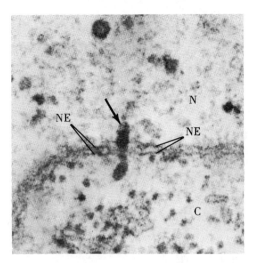

Fig. II-13 *Small portion of a cell from the salivary gland of the insect,* Chironomus. *The arrow indicates a structure that seems to be passing through a pore in the nuclear envelope (NE). The structure is thought to contain RNA in passage from nucleus (N) to cytoplasm (C). × 100,000. Courtesy of B. J. Stevens and H. Swift.*

In some metabolically active cells, the nuclear envelope shows many infoldings which increase the surface available for nucleo-cytoplasmic exchange. In other cells "annulate lamellae" are seen near the envelope. They are stacks of parallel, flattened sacs with "pores" similar to the nuclear pores. It has been tentatively proposed that these lamellae derive from the nuclear envelope of these cells and that they give rise to endoplasmic reticulum.

c h a p t e r **2.3**

NUCLEOLI AND RIBOSOMES

Ribosomes, the intracellular sites of protein synthesis, are present in virtually all cells. The few exceptions (for example, mature mammalian red blood cells and mature sperm cells have few, if any, ribosomes) show no protein synthesis. These exceptional cells have restricted functions and relatively short life spans.

Ribosomes possess RNA of distinctive size and base composition known as *ribosomal* RNA (rRNA). Production of this RNA is a major function of nuclei. Hybridization experiments on a few organisms indicate that well over a hundred (sometimes many hundreds) of apparently identical rRNA molecules can bind simultaneously to the DNA from one nucleus. This indicates that many duplicate (*redundant*) copies of DNA sequences responsible for rRNA production are probably present in the chromosomes. A visible, intranuclear organelle, the *nucleolus*, is prominent in most cells of eucaryotes; it is the site of rRNA synthesis.

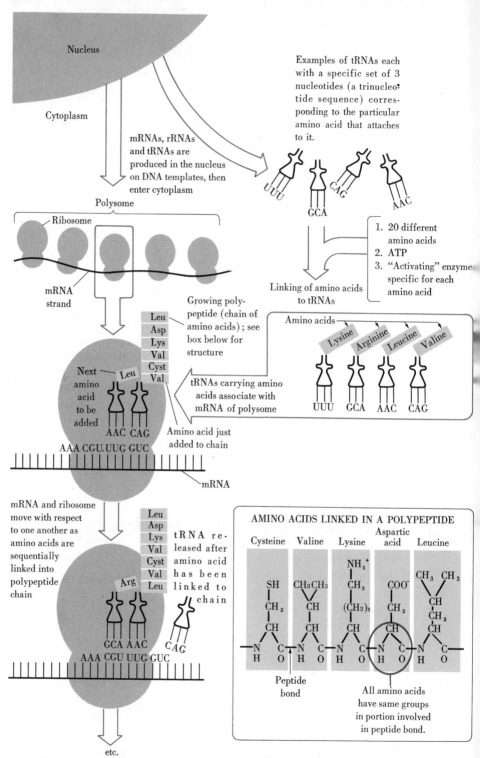

Fig. II-14 *The current concept of protein synthesis. The diagram shows only a few of the many different tRNAs and of the 20 amino acids. The trinucleotide sequences shown for the tRNAs are hypothetical but they are plausible from knowledge of the genetic code and tRNA structure; the actual sequences may include modified bases in addition to the usual four.*

2.3.1   ***PROTEIN***   As this process is currently pictured (Fig.
                    SYNTHESIS   II-14), specific "activated" amino acids are
                                    attached in the cytoplasm to specific *trans-
fer RNA's* (tRNA), relatively small molecules approximately 75–90
nucleotides long and with folded compact configurations. Some feature
of tRNA's, perhaps particular sequences of bases, permits a specific
enzyme to recognize a tRNA molecule and attach the proper amino
acid to it. The amino acid-tRNA complexes become aligned on a molecule of *messenger RNA* (mRNA) which has derived its base sequence
from a specific region of DNA, the gene reponsible for the synthesis
of a particular protein. This mRNA base sequence can be pictured as
a series of sets of three nucleotide bases, each of which is complementary to a three-nucleotide sequence of a tRNA. By base pairing, the
nucleotide sequence of the mRNA specifies the alignment of the specific tRNA sequence. Since each tRNA carries a specific amino acid,
amino acids are thus aligned according to the mRNA base sequence.
The amino acids are linked together sequentially to form a growing
polypeptide chain which, when complete, is released from the RNA
(Fig II-14). A protein molecule consists of one or several such chains
folded in a specific, three-dimensional form. The amino acid sequence
in a polypeptide chain results from an mRNA base sequence and thus,
ultimately, from a particular base sequence in DNA. It is the sequence
of amino acids that determines the specificity of the protein. As will
be outlined later (Section 4.1.2), the folding of polypeptide chains is
based upon their amino acid sequences. Different enzymes differ in
amino acid sequences. Thus they differ in their *active sites*, the region
of the molecule that binds to and acts upon the substrate; the active
site of a particular enzyme consists of specific amino acids arranged
in specific geometrical array.

The interaction of tRNA and mRNA, and protein synthesis, take
place on a ribosome-mRNA complex. A given mRNA molecule is
simultaneously complexed at different points with a number of ribo-

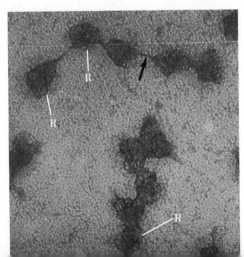

Fig. II-15 *Two* polysomes
isolated from human reticulocytes *(progenitors of red
blood cells). Each appears to
be composed of 5 ribosomes
(R). The polysome at the top of
the micrograph was stretched
during preparation and the
thin strand (arrow) running
between its ribosomes may be
a strand of mRNA.* × 350,000.
*Courtesy of A. Rich, J. R.
Warner and C. E. Hall. See
also Fig. II-30.*

somes to form a polyribosome or *polysome* (Fig. II-15). Present theory suggests that the ribosomes and mRNA molecule move with respect to one another as successive portions of the mRNA molecule are "read" or "translated" into amino acid sequence.

Each ribosome of a polysome apparently synthesizes a polypeptide chain so that several chains are probably being synthesized simultaneously along a given polysome. The number of ribosomes per polysome varies for different cells and proteins; on occasion polysomes are reported with as many as 10–20 ribosomes or, rarely, many more.

Even a relatively small bacterium may contain 5000–10,000 ribosomes; a larger cell may contain from 100,000 to several million. Thus, many proteins can be made simultaneously by one cell. On the average, synthesis of a complete polypeptide chain may take from 10 or 20 seconds to a minute or two, depending on the cell type and on the length of the chain. For example, in mammalian cells producing the blood protein *hemoglobin*, about one minute is required for completion of a polypeptide containing 150 amino acids. (The polysomes on which this occurs are thought to contain five ribosomes.) Bacterial cells may synthesize their proteins at a somewhat faster rate than this.

2.3.2　　　　**RIBOSOMES** Ribosomes have been well characterized by a variety of techniques. They appear in the electron microscope as spherical to ellipsoidal bodies, roughly 150–250 Å in diameter and consisting of two distinct subunits (Fig. II-21). The subunits can be reversibly separated by various treatments, for example, by lowering the Mg^{2+} concentration of the medium. The two subunits differ in size. Generally, ribosome sizes are spoken of in terms of the speed with which they sediment in a centrifugal field. The *Svedberg unit*, designated as S, is the unit used in measuring this speed; it is a rough measure of particle or molecule size, but it is influenced also by the density and shape of the particles and by the medium in which they are suspended. Intact ribosomes of eucaryotic cells sediment at 80 S and have a molecular weight of about 5 million. These figures are derived from mammalian cells; other eucaryotes may differ slightly, and bacterial cell ribosomes are distinctly smaller (Section 3.2.3). The two ribosomal subunits sediment respectively at about 60 S for the larger and 40 S for the smaller; the larger subunit accounts for 60–70 percent of the weight of the ribosome. The larger subunit contains an RNA molecule (28 S) the molecular weight of which is about 1.5 million, twice that of the RNA (18 S) in the smaller subunit; the number of nucleotides can be estimated from the fact that each nucleotide contributes a molecular weight of approximately

300. Protein, some of it resembling histone, is an integral part of each subunit. Each, then, is a *ribonucleoprotein* particle, with rRNA and protein present in roughly equivalent weights. Studies have suggested that in the functional polysome, the mRNA is associated with the small subunit of the ribosome, and the newly synthesized protein, as it forms, with the large subunit.

Ribosomal RNA molecules are large, and they may account for a substantial proportion of the total cellular RNA (often 75 percent; in some bacteria 90 percent). But, as yet, the details of their roles are unknown. In comparison to ribosomal RNA's, the tRNA's are small; they have molecular weights of 25,000. Messenger RNA's vary considerably, as would be expected; each contains at least the three nucleotides per amino acid required to synthesize a polypeptide chain, and the lengths of these chains very greatly in different proteins. Insulin contains about 50 amino acids, the pancreatic digestive enzyme precursor, *chymotrypsinogen*, about 250; other single polypeptide chains may have many more amino acids (500–1000 or more have been reported for some bacterial and viral proteins). The proteins coded for by one mRNA molecule may consist of two or more different chains. The length of the particular mRNA molecule present is presumably a chief determinant of polysome size.

Experiments may prove that the RNA in ribosomes holds the components of the complex system of protein synthesis together in some specific manner; very likely, tRNA's have special base sequences for binding to the ribosomes and perhaps for binding to one another as well as to mRNA. It may be speculated that rRNA's stabilize the interaction of mRNA and tRNA by binding to both, thus holding them in fixed relation to one another; or perhaps that ribosomal proteins have enzymatic functions necessary for protein synthesis. However, experimental evidence is too scant to draw final conclusions, as evidenced by the fact that new components of the system are still being discovered. Only recently, a form of RNA called 5 S RNA (mol wt 40,000) was found to be present in ribosomes, probably part of the large subunit.

Because ribosomes play an essential role in protein synthesis, the mechanism by which they themselves are made is of great interest. Recent experimental analysis points toward the nucleolus as playing a central role in the formation of rRNA.

2.3.3   *NUCLEOLI*   Autoradiographic evidence shows nucleoli to be the sites of extensive RNA synthesis (see Fig. I-21). Cytochemical and cell fractionation studies indicate that at least 5–10 percent of the nucleolus is RNA; the rest is mainly protein. Preparations of isolated nucleoli also contain some DNA, usually in chromatin that is attached to the nucleoli. In the electron microscope, nucleoli show a complex architecture (Fig. II-16); large numbers of small ribonucleoprotein granules are present, as well as areas of fine fibrillar and amorphous material. Often portions of the nucleolus are organized in the form of twisted strands, made of aggregated granules or fibrils in an amorphous matrix; the name *nucleolonema* is sometimes used for such strands. There is no membrane at the nucleolar surface. The nucleolus is associated with a special region of chromatin, *nucleolus-associated chromatin*, which may penetrate deep into the substance of the nucleolus. When discrete chromosomes become visible in cell division, nucleoli are often seen associated with specific *nucleolar organizer* regions of specific chromosomes (Figs.

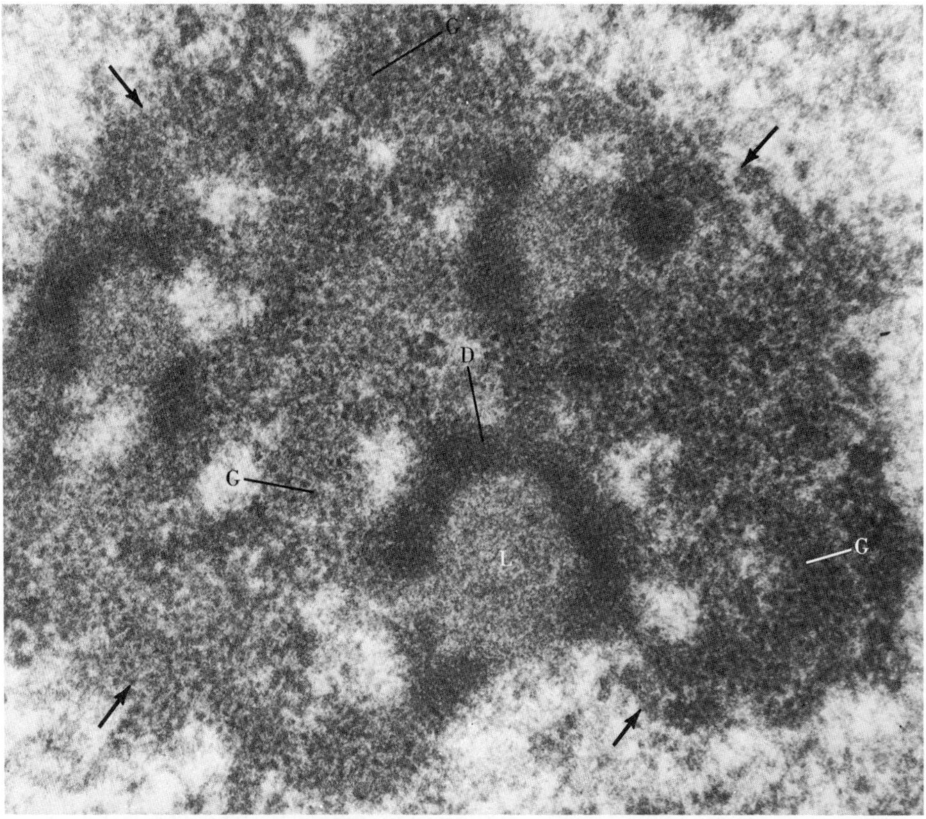

Fig. II-16 Nucleolus *from a cell in the vaginal lining (human). Arrows indicate the approximate edges of the nucleolus. The organelle contains numerous granule-like structures (G), dense fibrous regions (D) and amorphous, somewhat less dense regions (L).* × 60,000. *Courtesy of J. Terzakis.*

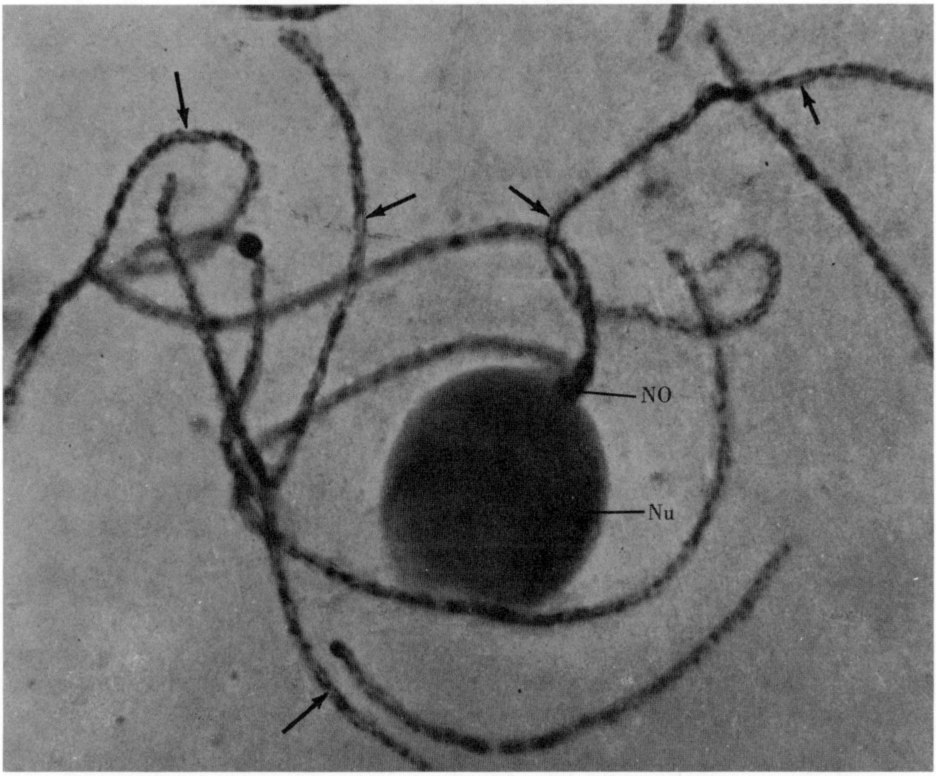

Fig. II-17 *Preparation of a cell in corn during male gamete formation. The cell has been squashed prior to fixation, thus separating organelles and permitting observation of some details. The thread-like structures are pairs (arrows) of closely associated chromosomes (pachytene stage of meiosis; see Chapter 4.3 and Figs. IV-20 and IV-21). The nucleolus is prominent (NU). It is associated with a specific chromosomal site, the nucleolar organizer (NO). × approx. 2,000. Courtesy of M. M. Rhoades and D. T. Morgan, Jr.*

II-17, IV-5). Presumably the nucleolar organizer regions are equivalent to the nucleolar-associated chromatin seen when cells are not in division. Some fibrillar or fibrous-appearing zones in nucleoli may contain chromatin-derived DNA along with RNA and proteins.

2.3.4   **NUCLEOLI AND RIBOSOME SYNTHESIS**   Cytologists noted long ago that nucleoli are particularly prominent in cells that have high rates of protein synthesis. Information outlined in the preceding paragraph suggests a role in RNA metabolism. A much more detailed analysis of nucleolar function is possible from several additional lines of evidence.

One important series of observations has been made on a mutant strain of the clawed toad *Xenopus*. It was initially noted that matings of certain *Xenopus* individuals produce offspring of which one-quarter die. (The explanation for this percentage of inviable offspring will be outlined in Section 4.3.2). For present purposes, the key fact is that cytological and genetic studies show that death can be attributed to an inheritable chromosomal defect affecting nucleolus formation. The nuclei of embryos carrying defective nucleolar chromosomes contain either no visible nucleoli or grossly abnormal ones. The embryos develop normally until shortly after hatching and then die. Death occurs at the stage when normal embryos show a great acceleration in ribosome synthesis; before this, both normal and abnormal embryos rely on ribosomes stored in the egg during its maturation preceding fertilization. The embryos that die cannot make ribosomes, and the lack of nucleoli is a visible manifestation of this inability. Ribosomal RNA purified from normal *Xenopus* cannot hybridize with DNA from abnormal embryos, indicating that the DNA base sequences for rRNA formation are missing in the mutant chromosome. These observations demonstrate a central role for nucleoli in ribosome formation.

Initially it was believed that the small granules normally present in the nucleolus represented newly formed ribosomes that had accumulated there. However, the evidence now available does not support this idea. Properly isolated nucleoli contain few complete ribosomes. They do contain much RNA metabolically related to rRNA, but it seems to be in the form of extremely large molecules (up to 45 S; molecular weight about 4 million) that must be fragmented and otherwise modified to become mature ribosomal constituents (see Fig. III-33). Some steps in this modification apparently occur in nucleoli. The sites and mechanisms of the final assembly of RNA and protein into ribosomes and polysomes are being sought.

The most obvious places to examine for the DNA template sequences for the synthesis of rRNA would be the chromatin associated with the nucleolus (Section 2.3.3). It is not yet possible to isolate specific portions of chromatin so that direct study of isolated nucleolus-associated chromatin or nucleolar organizers is not feasible (but see Fig. V-1 for a recent breakthrough in this regard). However, there is autoradiographic evidence in some cell types that the sites of initial incorporation of radioactive RNA precursors into nucleoli are in the associated chromatin and in the fibrillar regions of the nucleolus. Presumably, the RNA then is transformed into material of the granular regions of the nucleolus which, in turn, give rise to ribosome subunits.

In some strains of the fruit fly *Drosophila*, it is possible by selective breeding technique to control the numbers of nucleolar organizers on the chromosomes of the offspring. Experiments with such strains have shown a relation of the number of nucleolar organizers per nucleus to the amount of DNA with base sequences complementary to rRNA. As shown in Fig. II-18, DNA isolated from normal nuclei (two organizer regions) contains half as many sites that will hybridize with purified rRNA as DNA isolated from nuclei that have four such regions. The data indicate that the organizer regions contain the DNA templates for rRNA; the presence of twice as many organizers results in the presence of twice as many templates and, thus, twice as many sites with which added rRNA can bind in the hybridization experiments. Careful bookkeeping of the sizes and amounts of hybridizing molecules shows that a normal *Drosophila* nucleus with two organizer regions contains enough DNA sites to bind about 250 28 S rRNA molecules and the same number of 18 S rRNA molecules.

Consideration of the features of nucleolar organizer regions of chromosomes will arise at several points in subsequent chapters. A summary of the discussions is presented in Chapter 5.1, and a photograph of what is believed to be organizer DNA in process of synthesizing RNA is shown in Fig. V-1. The major point of concern at the moment is that the nucleolar organizers in several organisms have been shown to be sites at which are grouped as many as several hundred copies (apparently identical) of the genes responsible for the synthesis of the RNA molecules that give rise to rRNA.

Another role for the nucleolar organizer has also been proposed. During cell division, the nucleolus usually disappears as a discrete entity, reappearing only in the terminal stages of division. Studies on plant cells indicate that the first sign of nucleolar reappearance is the occurrence of many small nucleolus-like bodies that eventually fuse to form the one or two nucleoli of the usual interphase cell. Although the origin of the small bodies is not yet known, such observations have led to the hypothesis that the nucleolar organizing region collects either material synthesized at many points along the chromo-

Fig. II-18 *An experiment by F. M. Ritossa and S. Spiegelman on the relations between nucleolar organizers and rRNA in the fly,* Drosophila. *The DNA purified from nuclei of individuals with different numbers of nucleolar organizer regions in the chromosomes is tested for the number of sites that can hybridize with purified rRNA (graphed as the percent of the DNA per nucleus that hybridizes).*

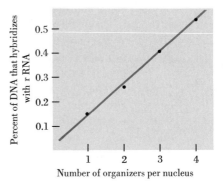

somes or perhaps material that is stored on the chromosomes when the nucleolus disappears early in cell division (Section 4.2.9).

Finally, there are some organisms whose cells form many small nucleoli either normally or under experimental conditions; presumably the cells have many separate nucleolar organizer regions in their chromosomes rather than only a few.

2.3.5   ***FREE AND*** Two distinct populations of ribosomes are
         BOUND recognized: those *bound* to membranes
   RIBOSOMES (endoplasmic reticulum); and those that are
         free, that is, not visibly bound to membranes. Both types form polysomes with mRNA and both play similar roles in protein synthesis, but the free ribosomes leave the newly synthesized proteins in the hyaloplasm (the soluble fraction; see Section 1.2B). Bound ribosomes transfer the protein into the interior of the endoplasmic reticulum. Evidence for this difference in function rests on comparative studies of different cell types.

Animal cells that secrete proteins such as digestive enzymes or hormones have a high proportion of bound ribosomes. The proteins enter the highly developed system of rough endoplasmic reticulum and, by mechanisms considered in Chapters 2.4 and 2.5, are "packaged" into membrane-delimited granules. Eventually the granules are released at the cell surface.

Similar synthesis on bound ribosomes and packaging in membrane-delimited granules occurs with lysosomal and peroxisomal enzymes. Under usual circumstances, most cells of multicellular animals do not release the contents of lysosomes and peroxisomes from the cell.

Free ribosomes are especially abundant in cells synthesizing much protein (enzymes and so forth) used internally in rapid growth but not specially packaged inside membrane-delimited structures (for example, cancer cells and most cells of embryos). In *reticulocytes*, the cells that mature into red blood cells, most of the ribosomes are of the free variety. These cells synthesize the *hemoglobin* that ultimately fills the cytoplasm. The hemoglobin is not included within special membranes and is not exported from the cell.

Hepatocytes synthesize much protein (such as serum albumin) that is secreted into the blood and other proteins for internal use (for example, in replacing proteins lost in turnover). These cells contain many free and many bound ribosomes. Presumably not only the proteins in the hyaloplasm but also some of those in nonmembrane-delimited organelles are synthesized on free ribosomes.

2.3.6   ***UNANSWERED*** Many interesting problems related to pro-
          PROBLEMS   tein synthesis are only now beginning to
                         yield to analysis. These questions include
how a ribosome "knows" where to attach to an mRNA molecule in
order to start "reading" at the beginning of the encoded information,
how the information for different proteins can be kept separate on a
single DNA or mRNA molecule or how a polysome "knows" whether
or not to attach to endoplasmic reticulum. There seem to be sequences
of bases used for "punctuation," to signal the beginning or the end of
a polypeptide chain. Some mRNA base sequences, UAA, UAG and
UGA, which do not code for amino acids, are the most likely candidates
for signaling termination, since they have been shown to terminate
growing peptide chains in experimental systems from bacteria. A current theory proposes that modified or special amino acids are used to
initiate polypeptide chains; a derivative of the amino acid *methionine*
called N-formyl methionine is a prime suspect in bacterial systems since
there is evidence that N-formyl methionine is the amino acid present
at the beginning of many growing polypeptide chains in bacteria.
Such a modified amino acid appears able to form links with only one
other amino acid rather than two, and thus provides a starting point.

The turnover rate of mRNA is variable in different cell types. A
given molecule may last a few minutes (bacteria), a few hours, or many
days (reticulocytes). In bacteria, the rapid turnover of mRNA is associated with metabolic flexibility; for example, the cell can adjust its
synthesis of some proteins to environmental conditions (Section 3.2.5).
Presumably, this is facilitated by the fact that newly synthesized
mRNA's are continually replacing old molecules in directing protein
synthesis; if the synthesis of one particular mRNA is greatly increased,
this mRNA will rapidly provide an increase in the synthesis of the
protein for which it carries the information. In reticulocytes, on the
other hand, the long-lived messenger RNA is associated with the
massive production of one protein, hemoglobin. The mechanisms controlling mRNA breakdown are only beginning to be studied; those
controlling synthesis are better understood, as will be seen in Chapters
3.2 and 4.4.

Most of molecular biology has been based on the study of viruses
and bacteria, which are much simpler than eucaryotic cells. Only
recently have nuclei from higher organisms begun to yield information
as detailed as that obtained for viruses and bacteria. It is highly likely
that the same basic mechanisms for RNA translation and protein
synthesis exist for all cells; it is well established that the genetic code
is essentially universal. (In all organisms the same set of DNA trinucleotide sequences can code for the same amino acids.) However,

bacteria have no nucleoli, and their ribosomes are smaller and of slightly different composition from the ribosomes of higher organisms. There may be other, probably many, differences in the protein-synthesizing machinery of procaryotes and eucaryotes, although as yet there is no reason to believe that these are more than differences in details.

Chapter 5.1 will present some additional information on the assembly and functioning of ribosomes.

c h a p t e r **2.4**

ENDOPLASMIC RETICULUM (ER)

A great contribution of electron microscopy was the demonstration that in many cells of eucaryotes an extensive membranous system, the endoplasmic reticulum (ER), traverses the cytoplasm. (The resolving power of the light microscope is too low for identification or analysis of this system.) The lipoprotein membranes delimit interconnecting channels that take the form of flattened sacs (known as *cisternae*) and tubules. Rough ER is studded with ribosomes most of which are probably in the form of polysomes. Rough ER and smooth (ribosome-free) ER are part of one interconnected system; the proportion of these two ER types varies in different cell types. Proteins (including enzymes), lipids, and probably other materials are transported and distributed to various parts of the cell through the ER. In some cases, these substances may accumulate and may be stored within the ER for considerable periods. In striated muscle, the ER takes a special form, the *sarcoplasmic reticulum*, probably involved in coupling nerve excitation to muscle contraction (Section 3.9.1).

The endoplasmic reticulum is much more than a passive channel for intracellular transport. It contains a variety of enzymes playing important roles in metabolic sequences, for example, in synthesis of steroids.

As mentioned earlier (Chapter 1.2.B) the ER, as such, cannot be isolated intact from cells. Fragments of ER are the major component of the *microsome* fraction. The fragments are in the form of membrane-delimited vesicles (Fig. I-22); the interior of the vesicles corresponds to the content of the cisternae and tubules of the intact cell and microsomes from rough endoplasmic reticulum retain ribosomes which stud the external surface of the vesicles. The vesicles appear to form as pinched-off portions of the ER during homogenization.

2.4.1   ***PROTEIN-***   In cells actively engaged in synthesizing
            SECRETING CELLS   and secreting proteins, *rough* ER is par-
                        ticularly highly developed.

The most intensively studied secretory cell is the type responsible
for production of digestive enzymes in the guinea pig pancreas (Figs.
II-19 and II-20); other secretory cells will be discussed in Chapter
3.6. The fate of specific proteins (precursors of digestive enzymes
made by the pancreas) has been followed by combining electron microscopy with biochemical analysis of isolated organelles and with
autoradiography. By autoradiography, labeled amino acids used in the
synthesis of these enzymes can be shown first over the rough endoplasmic reticulum, later in the region of the Golgi apparatus, still
later in secretion (*zymogen*) granules, and finally in the extracellular
space at one pole of the cell where secretions are released. (Upon
release the proteins enter ducts leading to the intestine where they
function after some modification). From the polysomes where the pro-

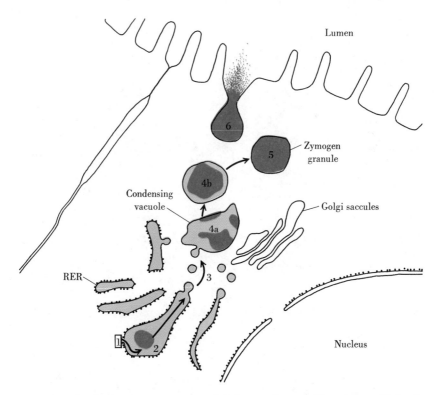

Fig. II-19 *Path of secretory material from polysome (1) to lumen (6) in the
exocrine gland cells (Section 3.6.1) of the guinea pig pancreas. After G. E.
Palade and colleagues. RER is rough endoplasmic reticulum.*

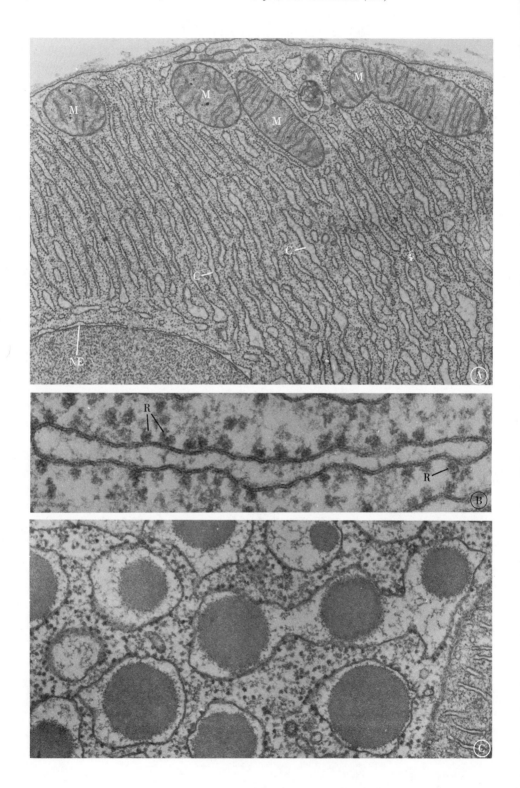

teins are manufactured (Sections 2.3.1 and 2.3.5), they enter the cisternae of the ER and move toward the Golgi apparatus. Then, apparently by transport in small vesicles, they pass from the reticulum into "condensing vacuoles"; these vacuoles lie near the Golgi apparatus and probably arise from the saccules of the apparatus. Within the vacuoles "condensation" occurs; this may involve concentration of the vacuole contents by loss of water and interactions of the proteins with other components (such as polysaccharides). As a result, dense secretion granules are formed. The vacuoles containing the secretion granules move to the cell surface where the granules eventually are discharged by a kind of reverse pinocytosis ("exocytosis") in which the vacuole membrane fuses with the plasma membrane (see Fig. II-33).

How does newly made protein get from the ribosomes to the cavities of the rough ER? One approach which may eventually answer the question is the investigation of the details of ribosome association with the ER. Some electron micrographs suggest that only one of the two ribosomal subunits is responsible for the binding of ribosomes to ER membranes (Fig. II-21). That this is probably the large subunit is indicated by microscopy and by experimental evidence. Rough-surfaced microsomes have been treated with an agent (EDTA) that complexes with ions such as Mg^{2+} and dissociates ribosomes into their two subunits. If such a dissociated preparation is centrifuged, subunits that remain attached to the membranes sediment into the pellet, while freed subunits remain suspended. Under carefully controlled conditions of dissociation, most of the small subunits can be freed from the microsomes, while most of the large subunits remain attached. This is consistent with the attachment of small subunits to microsomes by indirect linkage; that is, by virtue of their attachment to the large subunits.

Figure II-22 shows the data from an experiment designed to reveal whether newly-synthesized protein molecules go directly from the ER-bound polysomes into the cavity of the ER or whether they are first freed into the hyaloplasm and later picked up by the endoplasmic reticulum. As expected, initial incorporation of radioactive amino acids by rough-surfaced microsomes occurs in the ribosomes. Radioactivity then accumulates in the contents of the microsomal vesicles;

◀ **Fig. II-20** *Features of the rough endoplasmic reticulum. Courtesy of G. E. Palade. The three micrographs are from pancreas exocrine gland cells. **A**. Portion of a cell (rat pancreas) showing numerous flattened cisternae (C) of rough endoplasmic reticulum. Mitochondria are seen at (M) and the nuclear envelope at (NE). × 15,000. **B**. Higher magnification micrograph of a single cisterna (guinea pig pancreas). The membrane surface is studded with numerous ribosomes (R). × 120,000. **C**. Portion of a cell from a guinea pig preparation in which large granules, probably of protein, have accumulated inside the cisternae. × 50,000.*

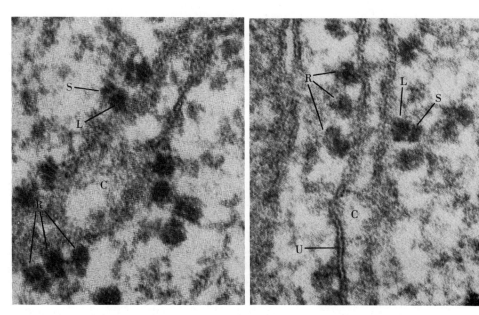

Fig. II-21 *Portions of two cisternae (C) of endoplasmic reticulum in a guinea pig hepatocyte. The delimiting membrane shows a "unit" membrane appearance (U). The ribosomes (R) consist of large (L) and small (S) subunits; the large ones apparently attach the ribosomes to the membrane. × 275,000. Courtesy of D. D. Sabatini, Y. Tashiro and G. E. Palade.*

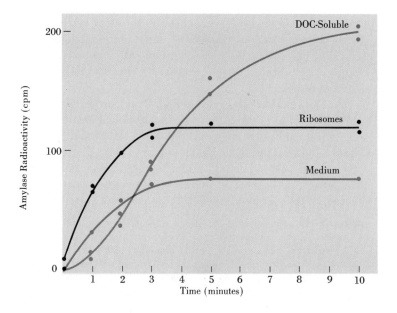

these contents are released if the microsomes are disrupted by treatment with detergents, such as deoxycholate. Note that the radioactivity in the ribosomes levels off after a few minutes, while the deoxycholate-soluble material continues to gain radioactivity; this is due to the fact that the ribosomes achieve a steady state in which the addition of new radioactive amino acids to growing polypeptides is balanced by the loss of amino acids as completed proteins are released to accumulate in the contents of the microsome vesicles. Since relatively little radioactive protein is found in the medium in which the microsomes are suspended, the most straightforward conclusion from these results is that, in the cell, protein is transferred directly from the ribosomes into the ER. The forces responsible for this transport, and for movement of protein once within the ER, are being studied. It will be of interest to determine whether movement of a given material present in the reticulum is a matter of flow within the membrane-delimited channel or whether it reflects movement of the entire cisterna — delimiting membrane plus contents — or both. From the viewpoint of future research, it is encouraging to find that ER membranes, isolated (as microsomes) from the rest of the cell, retain the organization needed to transport proteins from bound ribosomes to the ER interior (Fig. II-22). Insight into other aspects of protein transport may come from studying cases in which steps in transport can be blocked by experimental interference with cellular energy metabolism and also from instances in which specific proteins accumulate in the ER in amounts sufficient to permit microscopic visualization.

◄ *Fig. II-22 A microsome fraction derived chiefly from rough ER (pigeon pancreas) was incubated in a medium containing a radioactive amino acid (H³ leucine). At successive intervals samples were taken and treated with a detergent, deoxycholate (DOC) which frees the ribosomes and solubilizes the membranes. The deoxycholate-soluble subfraction represents the contents of the ER cisternae with solubilized portions of the ER membranes. The ribosomes can be separated from this by centrifugation. The medium is the solution in which the microsomes were suspended and presumably is the experimental equivalent of the cell's hyaloplasm. The medium, DOC-soluble material and ribosomes were studied for the extent of incorporation of radioactivity into the protein,* amylase, *an enzyme secreted by the pancreas. CPM is counts per minute.*

The fact that some radioactive amylase does appear in the medium may reflect leakage from microsomes damaged during preparation or incubation.

Additional, more detailed studies strengthen the conclusion that transport is directly from ribosomes to DOC-soluble fraction. (For example, the specific activity, *the radioactivity per unit weight, is found always to be higher for amylase in the ribosomes and DOC-soluble fractions than in the medium). From C. M. Redman, P. Siekevitz and G. E. Palade.*

2.4.2   ***PROTEIN***   The accumulation of protein within the
             VISUALIZATION   rough ER can be demonstrated in relatively
                    IN ER   few cell types. In some cells, proteins con-
                                   dense to form readily visible accumula-
tions while inside the rough ER cisternae. Granules resembling secretion granules are found in the ER of secretory cells in dog pancreas
and in the pancreas of fasting guinea pigs (Fig. II-20). This is direct
visual confirmation of the indirect evidence from autoradiography that
secretory proteins are transported in the ER. (See also Fig. III-30.)

Cytochemical methods for demonstration of enzyme activities
(Section 1.2.3) can also be used to detect proteins within the ER. For
example, cells of the thyroid gland and some cells of salivary glands
synthesize and *secrete* peroxidase enzymes. When preparations of these
tissues are incubated in the proper cytochemical medium for demonstration of peroxidase activity (Section 2.1.4), enzyme-produced reaction
product is distributed as in Fig. II-23A. As with several other micro-
scopically-demonstrable enzymes, the nuclear envelope shows reaction
product like the rest of the ER.

A method that is quite different has been used to determine the
sites of accumulation of newly synthesized antibodies. When foreign
proteins (*antigens*) are injected into a rabbit, *plasma cells* synthesize
antibody proteins (*immunoglobulins*) that can bind specifically to the
injected antigen. In the experiment with the outcome shown in Fig.
II-23B, the antigenic protein injected was *peroxidase*, purified from
horseradish. Two to four days after the last injection, the rabbit's
spleen was removed and fixed. Sections were cut and then soaked in a
solution of peroxidase (as were control spleen sections from animals
that had not been injected with the antigen). Wherever antibody to
peroxidase was present in the plasma cells, the specific antigen-anti-
body reaction caused binding of the peroxidase to these regions. When
the spleen sections were then incubated in a medium producing electron-opaque enzyme reaction product at peroxidase sites, the results
were like those in Fig. II-23; similar treatment of the control sections
resulted in no reaction product since no antibody and therefore no
peroxidase was present. Thus it may be concluded that the dilated ER
cisternae contain the antibody protein synthesized in response to the
antigen. (In interpreting this experiment, it should be noted that
peroxidase is used twice in different ways. Initially, it is an antigen
injected into the animal so that the plasma cells are induced to make
specific antibodies. Later, when the cells, which contain newly made
antibodies but no peroxidase, are soaked in a solution of peroxidase,
the enzyme is being used as a device to detect the antibodies. The
techniques of enzyme cytochemistry are employed to locate peroxidase activity in soaked cells. Where such activity is present, peroxidase

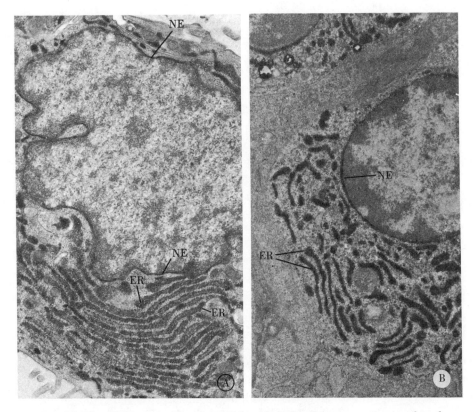

Fig. II-23 **A.** *Portion of a rat thyroid epithelial cell from a section incubated to demonstrate sites of* peroxidase enzymes *synthesized by these cells. Deposits of electron-dense reaction product are present in the nuclear envelope (NE) and in the many cisternae of rough ER. The enzyme responsible for staining is thought to be one that ultimately is released into the gland's lumen where it may participate in adding iodine to secreted protein (see pp. 207–09).* ×15,000. **B.** *Portion of a developing plasma cell treated with peroxidase as described in the text (Section 2.4.2). The ER and nuclear envelope (NE) show electron-dense reaction product formed by the enzymatic action of the peroxidase bound by the* specific antibodies *synthesized by the cell. × 7,500. Courtesy of E. Leduc.*

has been picked up from the solution and bound. The agents responsible for this binding are the specific antibody molecules, whose locations are thus demonstrated by an indirect method.)

In the study just described, it was noted that a few of the plasma cells containing detectable antibodies showed reaction product only in some ER cisternae and not in others. This raises, as an important matter for further study, the possibility that different large regions of the rough ER in a given cell may synthesize different proteins; if there are regions of rough ER specialized for synthesis of particular proteins, it will be interesting to discover how the appropriate poly-some-ER associations are controlled. A related problem is the control of movement of different components within the reticulum. For example, hepatocyte ER is thought to transport lipoproteins, lysosomal hydro-lases, peroxisomal enzymes and probably other materials such as albumin. The lipoproteins eventually are released from the cell after inclusion in vacuoles derived from ER or Golgi apparatus (Figs. I-4, II-34). The hydrolases are packaged into lysosomes, many of which form near the Golgi apparatus (Fig. II-46). Peroxisomes arise as enzyme-containing bodies that appear to form directly from the ER (Fig. II-54). How various components are kept separate, brought together or trans-ported to the appropriate cell regions are questions that will require much future study.

2.4.3   *TRIGLYCERIDE* Enzymes involved in triglyceride (Section
    SYNTHESIS IN ER 2.1.2) synthesis and some involved in phos-
pholipid synthesis are present in isolated
microsome fractions.

Autoradiography shows that radioactive lipid precursors are incorporated rapidly into material in the endoplasmic reticulum. When fats are fed to a rat, triglycerides are broken down in the intestine into smaller molecules, chiefly monoglycerides, glycerol, and fatty acids. These then cross the plasma membrane of the absorptive cells lining the intestinal cavity. Within the absorptive cells, the smaller mole-cules enter the ER, and enzymes in the ER membranes convert them into triglyceride again. In the meshwork of smooth ER and vesicles that form from it, the fat is visible in small droplets (Fig. II-24). Tri-glyceride is also seen in the cavities of the rough ER, including the nuclear envelope.

In normal rat hepatocytes relatively few small lipid droplets (probably complexed with proteins to form lipoproteins) are seen in the ER, almost exclusively in the smooth ER near the Golgi appara-tus. When rats are fed chemicals that ultimately will induce the forma-tion of large fat deposits in the hepatocytes, lipid droplets (presumably recently synthesized) very soon become more numerous in the ER. Sometimes they are fairly large. They can be seen not only in smooth ER, but also in the rough ER which is seen in the form of vesicles, each with one or more lipid droplets.

2.4.4   *SMOOTH ER* Whereas the rough ER is extensively developed in protein-secreting cells, it is the
smooth ER that is extensive in cells that secrete steroids, such as the hormone-secreting cells of the cortex of the adrenal gland and a num-

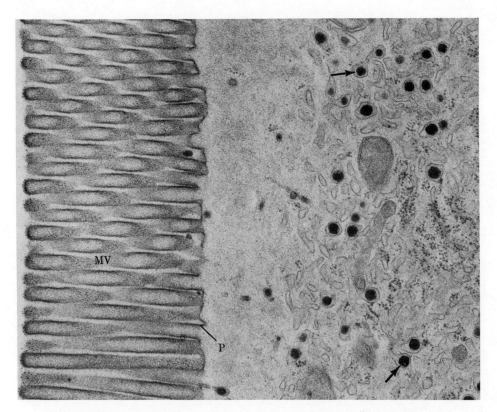

Fig. II-24 *Portion of an absorptive cell in the intestine (see Fig. II-17) of a rat fed a fat-rich meal. Many lipid droplets (arrows) are present inside the endoplasmic reticulum. At the lumen of the intestine, the plasma membrane (P) is seen to be extensively folded into closely packed microvilli (MV). × 30,000. Courtesy of S. Palay and J. P. Revel.*

ber of others (Fig. II-25). Key enzymes of steroid synthesis are found in microsome fractions which in these tissues derive mostly from fragmented smooth ER.

The intimate morphological association of smooth ER tubules and vesicles with glycogen particles in hepatocytes (Fig. I-3) has led some investigators to hypothesize that smooth ER functions in some aspect of glycogen metabolism.

In hepatocytes, both rough and smooth ER can break down ("detoxify") drugs such as *phenobarbital.* Following administration of phenobarbital to the animal, dramatic increases occur in the amounts of hepatocyte smooth ER and of the drug-degrading enzymes localized in the ER. Marked ER changes also occur in rat hepatocytes at about the time of birth; just before birth, much rough ER is formed, just after, much smooth ER appears. Such situations provide experimental

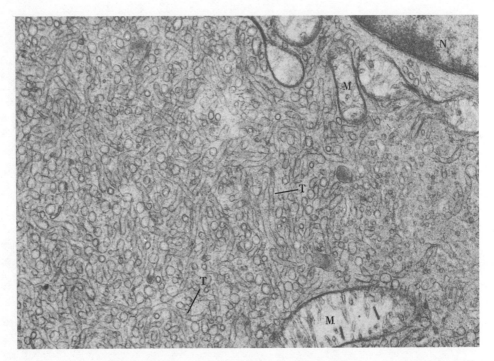

Fig. II-25 *Portion of an* interstitial cell *in opposum testis; these cells* produce *steroid hormones. Tubules of smooth ER (T) are abundant in the cytoplasm. An edge of the nucleus is seen at (N) and mitochondria are indicated by (M).* × *23,000. Courtesy of D. W. Fawcett.*

material for study of ER formation and interrelations (see also p. 67). Correlated biochemical, morphological and autoradiographic investigations suggest that rough ER "grows" by synthesizing membranes which either remain as part of the rough ER or become smooth ER by detachment of ribosomes. (For example, by administering radioactive precursors of ER membrane components and isolating rough and smooth microsomes at successive intervals thereafter, one can demonstrate that during stages of intensive smooth ER formation, label is incorporated first into rough ER and then transferred into smooth ER.) It remains to be learned if ER membranes are synthesized and broken down as units in which all components are permanently linked to one another or whether some "blank" membrane is made to which specific enzymes are subsequently attached and from which individual components (enzymes, lipids and so forth) can detach without destruction of the membrane. The distribution in the hepatocyte ER of enzymes already discussed and of other characteristic enzymes and related components (such as glucose-6-phosphatase and cytochrome b_5) is being intensively studied to determine whether ER regions differ appreciably in composition and the extent to which the ER varies under different physiological conditions.

2.4.5   **ER RELATIONS TO THE GOLGI APPARATUS**   The mechanisms by which proteins, lipids and other substances are transported from the ER to the Golgi apparatus probably vary for different substances and among different cell types or physiological states. However, the material being transported appears always to remain within membrane-delimited structures. There is no evidence to support alternatives such as the release of protein molecules from the ER to the hyaloplasm and their subsequent uptake by the Golgi apparatus.

Often, transport is attributed to vesicles that move between the ER and Golgi apparatus (Fig. III-22B). For many cell types, the vesicles are thought to bud from the ER and fuse with the stacked Golgi sacs (*saccules*; see Fig. I-4 and pp. 93–95). In guinea pig pancreas (Fig. II-19) the saccules apparently are by-passed and the vesicles transport protein from the ER to Golgi-associated "condensing vacuoles" (these may originate by budding from Golgi sacs).

Adjacent to the Golgi apparatus, there frequently are ER surfaces that are smooth (i.e. without ribosomes). Some such surfaces may be a source of the vesicles described above. However, it also has been postulated that large segments of ER cisternae can contribute directly to the Golgi apparatus, for example by losing their ribosomes and transforming into saccules that become incorporated in the stacks. Cisternae are occasionally seen as if caught in the act of transformation, with ribosomes on the surface away from the Golgi apparatus but none on the surface adjacent to the saccules (Frontispiece).

The passage of many vesicles from ER to Golgi apparatus or the transformation of ER into Golgi saccules implies much movement ("flow") of membrane from ER to Golgi apparatus. A balancing "flow" of membrane *out* of the apparatus presumably results from the budding of vesicles and vacuoles from the saccules as described in the next chapter. Present thinking about such membrane movements is based largely on tentative reconstruction of dynamic events from static electron micrographs. However, supplementary autoradiographic, cytochemical and biochemical evidence is increasingly available and should help clarify many details.

In some cell types, a special system of smooth-surfaced tubules and cisternae is present near the Golgi apparatus. Of interest is the fact that certain enzymes that are eventually packaged within membrane-delimited granules can be shown to accumulate in this system.

Figure II-26 is from tissue incubated for the demonstration of *acid phosphatase*, a lysosomal enzyme, and Fig. II-27 shows *tyrosinase*, a key enzyme in the synthesis of melanin. The system shows close association with Golgi sacs and probably is continuous with the ER. It may prove to be a special configuration involved in the concentration and packaging of some materials synthesized in the ER. (When present, it is always adjacent to the "bottom" (p. 93) of the Golgi stacks.)

c h a p t e r **2.5**
THE GOLGI APPARATUS

The importance of techniques available at a given time for the understanding of cell organelles is well illustrated by the history of what is now known about the Golgi apparatus. The structure was discovered in 1898 by Camillo Golgi, who named it the "internal reticular apparatus." His "metallic impregnation" method, involving long-term soaking in osmium tetroxide (later modified by using silver

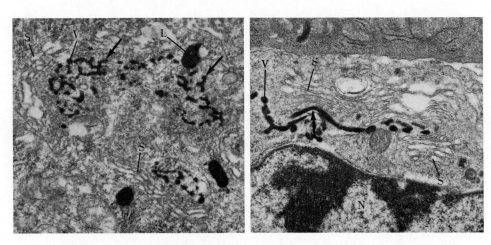

Figs. II-26 *and* **II-27** *Enzyme localizations in GERL, a special membrane system that appears to be a part of the ER that is associated closely with the Golgi apparatus. (GERL = **G**olgi-associated **ER** from which **l**ysosomes are thought to arise.) Fig. II-26 shows reaction product for* acid phosphatase *in GERL (arrows) in small vesicles (V) and in larger lysosomes (L) in a rat neuron. Golgi saccules are indicated by S. × 20,000. In Fig. II-27 (showing a similar region from a cell of mouse melanoma, a pigmented cancer) GERL shows* tyrosinase *reaction product. (N) indicates part of the nucleus. Pigment (melanin) granules appear to form from the tyrosinase-containing membrane systems by budding of tyrosinase-containing vacuoles in which melanin subsequently accumulates as a result of action of the enzyme. × 20,000.*

salt solutions) was difficult to control; despite several improvements
by Golgi and others, it failed to give consistent results. Furthermore,
the structure had not been seen in living cells. Thus, a lively controversy developed between those who thought it to be a real cell structure and those who insisted that it was simply an artifact of the fixing
and staining procedure. A flood of publications failed to clarify the
situation. With the development of electron microscopy, however,
the controversy was settled unequivocally. Wherever the classical
cytologists had located the Golgi apparatus, there the electron microscope revealed a characteristic system of membrane-bound saccules
and associated vesicles. It was soon shown that it was this system
that was darkened by metal (osmium or silver) in classical staining
procedures.

For most cytologists, these electron microscope observations
established the Golgi apparatus as an organelle rather than an artifact. A few doubters remained. They were finally convinced when a
staining procedure was developed which avoided the sources of artifact in the classical methods. This method is based on the presence
of specific phosphatases (notably thiamine pyrophosphatase, TPPase)
in the Golgi saccules (Figs. I-15, II-28, and II-29). Every detail of form
and distribution of the organelle described by early light microscopists
was duplicated by the enzyme method, and electron microscopy showed
high levels of reaction product to be localized within Golgi saccules,
and sometimes in vesicles or vacuoles derived from the saccules.

It can now be said unequivocally that all eucaryotic cells studied,
with rare exceptions such as mammalian red blood cells, possess the
Golgi apparatus. However, the physiological role of the organelle is
only now being firmly established.

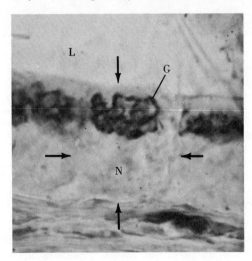

*Fig. II-28 Thiamine pyro-
phosphatase preparation (see
Fig. I-15) showing the Golgi
apparatus (G) of a cell of the
rat epididymis, a male reproductive gland. As is usually
the case for secretory and
absorptive cells, the Golgi
apparatus is situated between
the nucleus (located at N but
unstained) and the lumen (L).
Arrows indicate the approximate locations of the borders
of the cell. × 900.*

Figs. II-29 *and* **II-30** *Portions of rat hepatocytes. (M) indicates a mito-chondrion, (E) endoplasmic reticulum, (L) a lipid droplet, (S) Golgi saccules, (GV) small vesicles of the Golgi apparatus and (CV) "coated vesicles" (vesicles with a fuzzy coat on their membrane surface). The structure at (V) is a large Golgi vesicle (vacuole) filled with lipoprotein droplets; the arrow indicates one such vesicle, probably in formation as a dilatation (swollen region) of a Golgi saccule. At (R) the endoplasmic reticulum has been sectioned so that the cisterna surface is seen in face view. The ribosomes are arranged in spiral, hairpin and other patterns that probably reflect ribosome grouping in poly-somes. Fig. II-29 (lower left) was incubated for thiamine pyrophosphatase activity (see Figs. I-15 and II-28). Reaction product is seen in two Golgi saccules. II-29, × 47,000. II-30, × 30,000; courtesy of L. Biempica.*

2.5.1   **MORPHOLOGY**   There is a great diversity in the size and shape of the Golgi apparatus in different cell types. By light microscopy, it is seen to be a widespread, interconnecting network in neurons (Fig. I-15) and in cells that secrete proteins or carbohydrates. In many cells of higher plants and in some animal cells, the Golgi apparatus appears to consist of many apparently unconnected units called *dictyosomes* or better, *Golgi stacks.* (Plant cells may contain from dozens to hundreds of these units.)

In the electron microscope, the constant feature of the Golgi apparatus whatever its overall form is the presence of stacks of closely spaced membrane-delimited sacs (*saccules*). The stacks can best be seen in sections that are perpendicular to the saccules. The number of saccules varies generally from about three to seven in most higher animal and plant cell types (Figs. II-30 and III-15) to ten or twenty in some cells such as the unicellular organism *Euglena.* The saccules are separated from each other by a relatively constant distance, usually 200–300 Å. The nature of the material or force holding them together as a stack is unknown, but in a few cells, a thin layer of electron-opaque, sometimes fibrillar, material is seen between the sacs, in the middle of the 200–300 A space. The saccules themselves are of variable thickness, sometimes quite flat and sometimes dilated by materials that accumulate within. Their delimiting membranes are devoid of ribosomes. The peripheral region of the saccule may show tubules and *fenestrations* (interruptions at whose borders the sac membranes are continuous, more or less as are the membranes at pores in the nuclear envelope; see Frontispiece). Sometimes, the ER adjacent to the stack also is fenestrated but the significance of such features is presently unclear.

The Golgi apparatus often shows an obvious polarity. The saccules at the "top" of the stack may differ from those at the "bottom" in thickness, content, size and frequency of vesicles budding off, cytochemically-demonstrable enzymes or dimensions of delimiting membranes. (The apparatus frequently is curved or cuplike in overall configuration so "outer" or "convex" surface is sometimes used in place of "top" and "inner" or "concave" surface in place of "bottom.") Apparently, materials within the saccules are concentrated and otherwise altered during the passage through the apparatus. In most secretory cells, the vacuoles that contain the secretory material (often in the form of relatively dense granules) are found at the "bottom" of the stack (Fig. II-31) and may separate from one or two "bottom" saccules. How transport of material from "top" to "bottom" of the stack might occur and how the material becomes concentrated within the saccules remain to be determined. Techniques are not yet available to establish unequivocally that the sacs themselves move down the stack, with new ones being formed at the "top" of the stack and old ones being released at the "bottom." As noted in Section 2.4.5, the endoplasmic reticulum contributes material to the Golgi apparatus. Details and routes of transport between ER and Golgi apparatus may vary somewhat for different sub-

stances and different cell types. However, for many cell types it is
presumed that one route of transport is the movement of material from
the ER to the "top" of the stacks of Golgi saccules (for example, by
fusion of ER-derived vesicles with the "topmost" Golgi saccule).

Associated with the Golgi saccules is a variable number of small
vesicles, about 25–50 mμ in diameter. Apparently, the vesicles separate from the lateral edges of the saccules. At least some of these

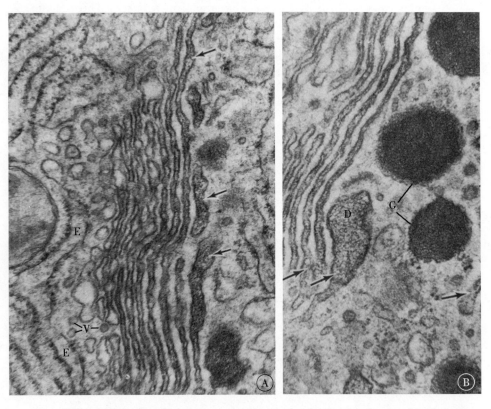

Fig. II-31 *Small portions of cells in glands (Brunner's glands) of the intestine (mouse) showing the Golgi apparatus. Dense material is present within the Golgi saccules (arrows). This material appears more concentrated in the saccules toward one surface of the stack than in the saccules near the other surface. As these figures suggest, secretory granules appear to form at least in part by accumulation of the dense material within dilated portions (D) of the Golgi saccules at one surface of the stack. Eventually the dilated regions pinch off and give rise to membrane enclosed granules (G). E indicates ER, and V, small Golgi vesicles. A. × 54,000. B. × 55,000. Courtesy of D. Friend.*

Golgi vesicles seem to carry acid phosphatase, and presumably other lysosomal enzymes, to larger vacuoles in the cell (Chapter 2.8).

A point for much future study is the fact that under some physiological and pathological conditions the Golgi apparatus shows changes in size, intracellular distribution and in such features as the number of saccules in the stack. The apparatus is a dynamic system that responds to altered levels of activity and probably to changes in related processes such as membrane "flow" from the ER (Section 2.4.5).

2.5.2 **ROLE IN SECRETION** The clearest information available on the function of the Golgi apparatus relates to secretion. The organelle is highly developed in cells secreting either proteins or complex polysaccharides. Autoradiographic evidence tentatively points to an important difference in the way in which these two types of macromolecules are handled.

As mentioned earlier (Section 2.4.1), if radioactively labeled precursors of *protein* (H^3-*amino acids*) are followed by autoradiography in many different cell types, the first organelles in which radioactive protein is seen are the ribosomes and ER (approximately 3–5 minutes after injection). Only later (20–40 minutes) is radioactive protein present in the Golgi apparatus (Fig. II-32A), and related structures. Thus, proteins are *synthesized* on the ER and can then be transported to the Golgi apparatus.

The results are markedly different with some precursors of polysaccharides (such as components of mucus, chondroitin sulfate of cartilage cells, or cell walls in plants). A number of H^3-labeled *sugars* (such as galactose) or *radioactive-sulfur labeled precursors* of sulfated polysaccharides first appear (within a few minutes) not at the ribosomes or at ER but at the Golgi stacks (Fig. II-32B). Apparently the Golgi apparatus is the site not only of concentration of polysaccharides but also of *synthesis* of the complex polysaccharides, or in the case of some carbohydrate components the site of major steps in the completion of synthesis initiated in the ER (see Section 3.6.1 and Fig. III-20).

Both with protein precursors and with polysaccharide precursors, the radioactive material eventually leaves the Golgi apparatus; in secretory cells, much ultimately is released to extracellular spaces. Appreciable release of radioactive protein from enzyme-secreting cells of the pancreas is detectable by one to two hours after administration of labeled amino acids.

In the Golgi apparatus or associated structures, materials are concentrated and packaged into vesicles and vacuoles (p. 7). This occurs in Golgi saccules of many cells [such as intestinal gland cells

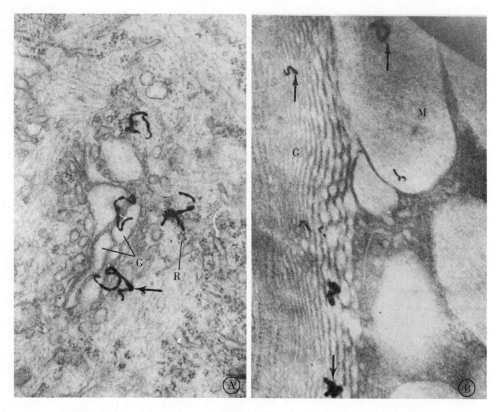

Fig. II-32 Electron microscope autoradiograms. *Each irregular dense structure (arrows) would be seen as a single grain in the* light microscope *(Fig. I-20).* **A.** *Part of a neuron in a rat spinal ganglion fixed 10 minutes after exposure to radioactive amino acid (H³-leucine). Two of the grains lie close to clusters of membrane bound-ribosomes (R) while the other two lie over Golgi saccules (G). Approx.* ×40,000. *Courtesy of B. Droz.* **B.** *Portion of a mucus-secreting cell (a goblet cell, see Fig. III-17) in rat intestine fixed 20 minutes after exposure to radioactive sugar (H³-glucose). Grains are seen over the Golgi saccules (G) and over a large Golgi vesicle (vacuole) containing mucus (M).* × 45,000. *Courtesy of M. Neutra and C. P. Leblond.*

(Fig. II-31), pituitary gland cells (Fig. II-33), hepatocytes (Figs. I-4, II-30), developing white blood cells and others], in condensing vacuoles in guinea pig pancreas (Fig. II-19) and perhaps sometimes in special configurations of interrelated ER and Golgi apparatus [melanoma cells (Fig. II-27), thyroid epithelial cells (Fig. III-20) and a few others (pp. 89–90)]. The vesicles and vacuoles produced by the Golgi apparatus often contain mixtures or complexes of components. For example,

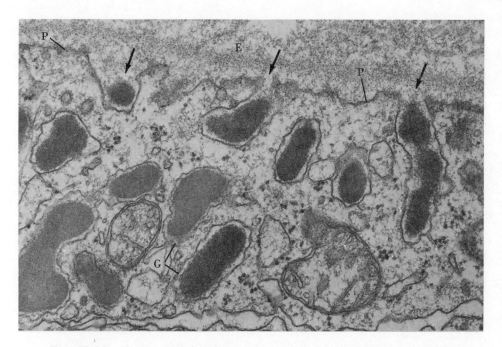

Fig. II-33 *Portion of the surface of a rat pituitary gland cell that produces the hormone,* prolactin. *Secretion granules are seen at (G). The arrows indicate granules in process of release from the cell by fusion of the membranes enclosing the granules and the plasma membrane (P). (E) indicates the extracellular space. × 23,000. Courtesy of M. Farquhar.*

several different proteins may be present along with polysaccharides and other materials; a number of important secretions are *glycoproteins*, protein molecules to which short chains of carbohydrates are linked as they pass through the ER and Golgi apparatus (Section 3.6.1). In many secretory cells, vacuole contents are in the form of electron-dense secretion granules (Figs. II-31, II-33). Section 3.6.2 will consider the *nematocysts* of *Hydra*, special Golgi-derived vacuoles that show very complex contents.

Secretory products are released at the cell surface when the vesicle or vacuole membrane fuses with the plasma membrane (Figs. II-33, II-34; see also Section 3.10.3 for discussion of the *acrosome*, a special Golgi-derived structure present in sperm cells). Other material packaged by the Golgi apparatus or associated systems functions within the cell; this is true for example of lysosomes (Chapter 2.8).

Knowledge of the functioning of the Golgi apparatus is limited, largely because purified fractions have proved very difficult to isolate.

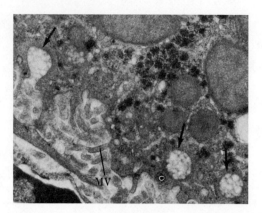

Fig. II-34 *Region of a rat hapatocyte near its sinusoidal surface (see Figs. I-1 and I-2). Arrows indicate vacuoles filled with lipoprotein droplets (see Fig. II-30). The vacuole at the upper left shows fusion of its delimiting membrane with the plasma membrane and presumably release of its contents into the extracellular space (compare with Fig. II-33). One of the surface microvilli is seen at MV. × 20,000. Courtesy of N. Quintana.*

In most cells, the volume occupied by the organelle is small and it is difficult to keep the Golgi structures intact while the cells are being homogenized. However, progress is being made in isolating "Golgi-enriched fractions." When sufficiently pure fractions are available, the chemistry of the "condensing" and "packaging" processes can be studied. Recently, fractions consisting largely of portions of the Golgi apparatus of hepatocytes have been prepared and shown to contain several enzymes, including some that participate in polysaccharide and glycoprotein synthesis.

Further discussion of secretion is presented in Chapter 3.6.

c h a p t e r **2.6**

MITOCHONDRIA

Mitochondria are present in virtually all eucaryotic cells, both plant and animal. Their number may range from a few per cell, to 1000 in a rat hepatocyte, to 10,000 in one of the giant amebae. Although mitochondrial size varies considerably in different cell types, diameters of 0.5–1.0 μ and lengths of 5–10 or more microns are common.

Mitochondria are easily isolated from homogenates in fairly pure fractions. Thus, it was learned relatively early that mitochondria play important roles in many metabolic pathways, including the breakdown and synthesis of carbohydrates, fats, and amino acids as outlined in Chapter 1.3. Mitochondria have a striking and readily recognizable fine structure. Dry weight proportions are typically 25–30 percent lipid and 60–75 percent protein. The mitochondria also contain small amounts of distinctive DNA and RNA. Like chloroplasts, mitochondria are self-reproducing; their nucleic acids and protein-synthesizing machinery are thought to be involved in elaborating some, but probably not all, mitochondrial proteins. Although both mitochondria and chloroplasts have some degree of genetic autonomy from the nucleus, the reproduction of these organelles also involves enzymes and other molecules from the rest of the cell; thus it depends in part on the hereditary information carried by the nucleus.

The extraction of mitochondrial macromolecules in separate and soluble form has been pursued for many years. Recently, progress has been made in recombining these molecules and in reconstituting considerable portions of mitochondrial structure.

In regard to terminology, several equivalent definitions of *oxidation* and *reduction* are often used. In chemical terms, oxidation-reduction reactions involve the transfer of electrons from one molecule to another. The molecule losing electrons is *oxidized*, the one gaining electrons is *reduced*. In biological systems, many such electron-transfer reactions include the exchange of hydrogen atoms; this is the case in the reactions discussed earlier (Section 1.3.1), where oxidation of a substrate takes place by the transfer of hydrogen atoms to a coenzyme which thus becomes reduced. The enzymes catalyzing such reactions are referred to as *oxidative enzymes*; included in the category is a large class known as *dehydrogenases*.

2.6.1   COMPARTMENTS WITHIN THE MITOCHONDRIA   Figure II-35 illustrates the basic structure of the mitochondrion. The organelle is delimited by an outer membrane, separated from the inner membrane by a space 60–100 Å wide. The inner membrane is thrown into many folds, or *cristae* (Fig. II-36). Enclosed by the inner membrane is the *matrix*. Techniques are available for obtaining separate outer and inner membranes and for isolating some of the matrix substances. The segregation of different enzymatic functions in the different membranes and compartments of the mitochondrion has been demonstrated with these techniques. The complexity of the structure of mitochondria is related to their function in a wide variety of metabolic pathways. Details of macromolecule arrangements have only recently become apparent.

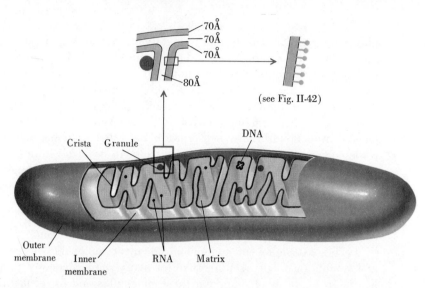

Fig. II-35 *Diagrammatic representation of a mitochondrion. After work of L. Ornstein, G. E. Palade, F. Sjostrand, H. Fernandez-Moran and others.*

As more is learned, the description presented here will undoubtedly require modification.

Most attention has been devoted to the enzymes of the Krebs cycle, components of the respiratory chain responsible for electron transport and the enzymes of oxidative phosphorylation (Chapter 1.3A). In the intact mitochondrion, electron transport and phosphorylation are *coupled:* as electron transport occurs, phosphate groups are attached to ADP to form ATP. Usually, the coupling between electron transport and oxidative phosphorylation is tight, meaning that neither ATP formation nor electron transport can occur alone; both must take place simultaneously. The enzymes for both processes are found on the inner mitochondrial membrane.

In electron micrographs of sectioned cells, the inner membrane often appears to be a "unit" membrane (Sections 2.1.1 and 2.1.2). It is thinner (50–70 Å) than most plasma membranes (70–110 Å). "Negative staining" (Fig. II-37) of isolated mitochondria (Fig. II-38)

Fig. II-36 *Mitochondria as seen by usual electron microscopic techniques:* ▶ *A, In a pancreas cell (bat). The outer (O) and inner (I) mitochondrial membranes and the flat cristae (C) are readily visible. Many intramitochondrial granules (G) are present; most cell types show fewer such granules. × 50,000. Courtesy of D. W. Fawcett. B, In the protozoan, Epistylis. As in many protozoa and algae, the cristae are tubular. × 60,000. Courtesy of P. Favard. C, In a muscle (bat). The cristae are numerous and closely packed, correlating with the high state of activity of the muscle. At the arrows it is evident that the cristae are infoldings of the inner mitochondrial membrane. × 45,000. Courtesy of S. Ito.*

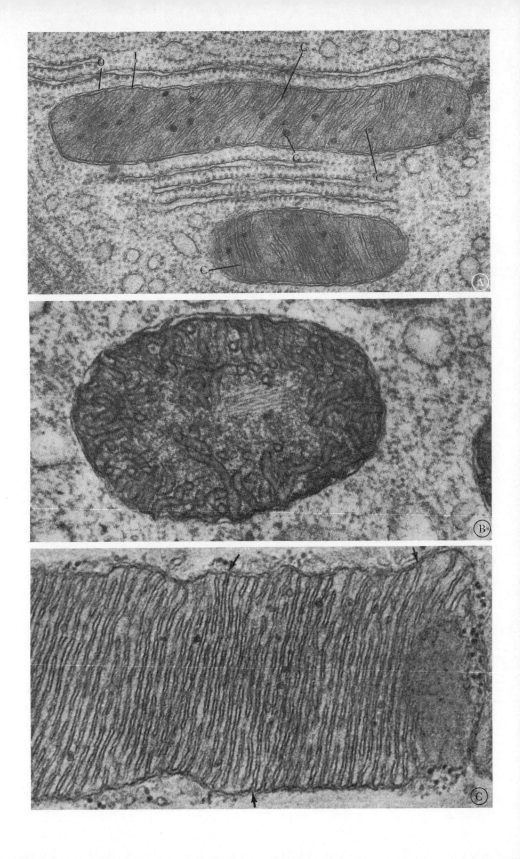

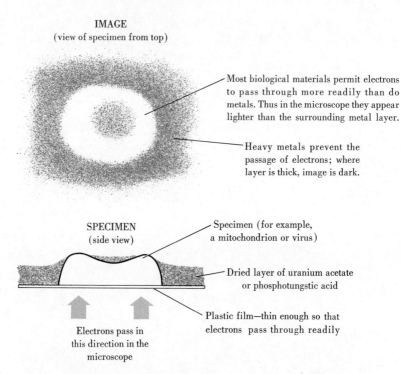

IMAGE
(view of specimen from top)

Most biological materials permit electrons to pass through more readily than do metals. Thus in the microscope they appear lighter than the surrounding metal layer.

Heavy metals prevent the passage of electrons; where layer is thick, image is dark.

SPECIMEN
(side view)

Specimen (for example, a mitochondrion or virus)

Dried layer of uranium acetate or phosphotungstic acid

Plastic film—thin enough so that electrons pass through readily

Electrons pass in this direction in the microscope

Fig. II-37 Negative staining, *one electron microscope technique for examining three-dimensional and surface aspects of cell structures. The specimen is not sectioned. It is placed on a thin plastic film and covered with a drop of solution containing heavy metal atoms (usually* uranium *or* tungsten *compounds). The solution is allowed to dry leaving the specimen in a thin layer of electron-dense material.*

shows a great many small spheres covering the surface of the inner membrane that faces the matrix; each sphere is attached by a stalk to the membrane. When these were first discovered, it was thought that each sphere might be a *respiratory assembly* containing all the enzymes responsible for electron transport and associated phosphorylation. This proved not to be the case; the spheres are too small to hold all the needed enzymes and, when the spheres are removed experimentally, the remaining membranes still possess most essential components of the respiratory chain. The spheres contain an enzyme, called F_1. When extracted from the mitochondrion in soluble form, F_1 acts as an ATPase, that is, it splits ATP into ADP and phosphate. However in the intact mitochondrion, it is considered that F_1 acts in the opposite direction and participates in coupling phosphorylation of ADP with electron transport.

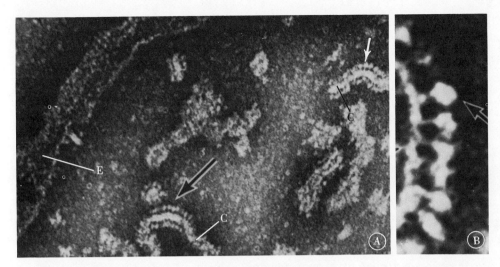

Fig. II-38 *Portion of mitochondria from heart muscle (cow). The organelles have been isolated, broken open and then negatively stained (see Fig. II-37).*

A is a view (× 80,000) showing part of the edge of a mitochondrion (E) and several cristae (C). Along the cristae surfaces facing the matrix (Fig. II-35) are arrays of spheres attached to the cristae by stalks (arrows). B is at higher magnification (× 600,000) and shows several spheres. Courtesy of H. Fernandez-Moran, T. Oda, P. B. Blair and D. E. Green.

Thus far, structures corresponding to respiratory assemblies have not been visualized by electron microscopy. However, there are at least two reasons for concluding that the enzymes of the inner membrane exist as repetitive arrays, in which enzymes of sequential metabolic steps are held in close relation with one another. First, F_1 has an intimate, functional relation to the enzymes of the respiratory chain, and it is visible in the form of a repetitive unit. Second, when mitochondrial membranes are fragmented into small pieces (by ultrasonic vibration, for example, or by agitation with glass beads), the fragments can carry out electron transport and oxidative phosphorylation; the various relevant enzyme molecules are present in the same ratios in the fragments as in the intact mitochondrion. This is as expected if the fragmentation separates the membrane into repeating assemblies; it might not occur if there were large specialized patches of membrane, each devoted to just one step in the sequence. It has been estimated that a square micron of membrane surface has 650 respiratory assemblies, each occupying a patch about 400 Å × 400 Å, and each providing a highly efficient arrangement of the enzymes responsible for electron transport and oxidative phosphorylation. The fact that *multienzyme complexes* (discrete assemblies of relatively few molecules that contain several sequentially acting enzymes; see Section 3.2.4) can be isolated from mitochondria supports such a proposal.

Much current work is directed toward interrelating electron microscope observations with biochemical studies in order to obtain a more detailed picture of molecular arrangements. For example, it remains to be determined whether or not the exact form of the F_1 units of living cells is accurately reflected in the small spheres seen with negative staining. Eventually it should be possible to specify the position and function of each molecule within an F_1 unit.

Among the major constituents of the inner membrane are lipids, particularly phospholipids, and a possibly heterogeneous group of "structural" proteins. These proteins and lipids may constitute a framework into which the enzymes of the respiratory assemblies are anchored. However, thus far "structural" proteins are defined partly by the negative fact that no enzymatic role is known for them; the details of their possible structural roles are not well understood. The lipid content of mitochondrial membranes (32 percent) is somewhat higher than that of plasma membranes (22–30 percent) or of endoplasmic reticulum membranes (25 percent).

The number of cristae per mitochondrion tends to be greater in cells with particularly intense respiratory activity (Fig. II-36); this finding is consistent with the inner membrane's structure and role in producing ATP. The mitochondria themselves are often concentrated in intracellular regions of greatest metabolic activity, for example close to the contractile fibrils in muscle cells, at the base of kidney tubule cells where molecular exchange with the blood is rapid or wrapped around the flagella responsible for movements of sperm cells.

The physiological significance of the various forms of cristae in different cell types is less clear. The cristae may appear (Fig. II-36) as simple plates (most cells of higher organisms) or as tubules (protozoa, algae, cortex of the mammalian adrenal gland); in some cases they may be organized into more complex three-dimensional structures.

Considerably less is known about the outer membrane than about the inner membrane, although several oxidative enzymes have been localized in it. The mechanisms by which mitochondria exchange molecules with their surroundings may be varied and complex. Some substances, including ATP, a few intermediates of the Krebs cycle and reduced NAD enter the mitochondrion very slowly; others, like pyruvic acid, enter rapidly. A number of components (for example calcium and phosphate) may be actively accumulated within mitochondria. The underlying mechanisms and the extent to which the inner and outer membranes each contribute to such selective permeability, and uptake characteristics are actively being studied. There are enzyme-dependent transport mechanisms for some substances. A few experiments suggest that the outer membrane is more freely permeable to inorganic ions than is the inner membrane where active "ion pump" transport systems are thought to occur.

In Section 1.3.4, the observation was made that some enzymes, such as the one that oxidizes malic acid (one of the Krebs cycle intermediates), readily become soluble when mitochondria are fragmented. It is generally assumed that this occurs because such enzymes lie in the matrix, although other alternatives have not been ruled out. For example, some enzymes may be localized between outer and inner membrane, rather then in the matrix, or they may be attached so loosely to either outer or inner membrane, that they readily detach during fragmentation of the mitochondrion.

When isolated mitochondria are permitted to take up large quantitites of calcium and phosphate ions, there is a marked increase in electron opacity and an enlargement of granules that normally measure about 500 Å in diameter. Similar effects are seen if barium (Ba^{2+}) or strontium (Sr^{2+}) is used instead of calcium (Ca^{2+}); it is inferred that the granules in the intact cell bind ions that have two positive charges (*bivalent cations*). Bivalent cations play important roles in many cell functions as will be outlined at several points (for example the Introduction to Part 4.) It will be interesting to learn whether these granules also act as biologically significant storage depots for phosphate or other ions.

Present in the matrix are fibrils that apparently contain the DNA of mitochondria, and small granules, probably ribosomes. These will be considered in a subsequent section.

2.6.2   OXIDATIVE PHOSPHO-RYLATION

Chapter 1.3 outlined the major chemical steps in the Krebs cycle and indicated those steps at which reduction of coenzymes takes place. The electrons from the coenzymes are then transported in orderly manner from one carrier to another in the respiratory chain (the electron transport system). This system is outlined in Fig. II-39; prominent among the participants are several different iron-containing proteins (*cytochromes*). One can think of the chain as starting at a position of high energy and ending at one of low energy. Each step in the chain involves loss of electrons (*oxidation*) by one molecule and gain (*reduction*) by the next molecule in the chain, and each oxidation-reduction event results in a release of energy. At three of the steps in the chain, the energy is used to form ATP from ADP and phosphate by a series of reactions that are imperfectly understood: F_1 is probably responsible for part of the sequence.

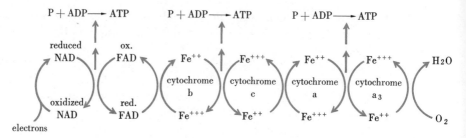

Fig. II-39 *Major steps of mitochondrial electron transport and associated oxidative phosphorylation. From A. L. Lehninger and others. Each component in the chain receives electrons from the previous component and passes them to the next. In the case of the cytochromes the receipt of electrons reduces iron ions in the molecules to Fe^{2+}. Subsequent loss of the electrons to the next component regenerates the oxidized, Fe^{3+}, form. The precise steps in oxidative phosphorylation are not yet fully established; it is widely assumed that the energy is transferred through several high-energy intermediate compounds before it is used in the phosphorylation of ADP and several investigators are studying proposals that some sort of transport across mitochondrial membranes may also be involved.*

The rate at which the respiratory chain functions is controlled by various factors, including the level of available ADP. The more ADP available for phosphorylation, the more rapid is electron transport, since the coupling of transport and phosphorylation requires that ADP be converted to ATP as electrons move down the chain. The higher the rate at which a cell utilizes ATP, the higher the ADP level, leading to higher rates of electron transport, oxygen consumption, and ATP production. Thus, an effective mechanism is provided for balancing ATP production with its utilization.

A number of chemicals are known that "uncouple" ATP production and permit the respiratory chain to function without simultaneous phosphorylation. The extent to which similar effects occur naturally in cells is unknown, although such uncoupling has been proposed as a source of cell abnormality in one or two diseases.

2.6.3   DNA, RNA AND PROTEIN SYNTHESIS; MITOCHONDRIAL DUPLICATION   The power of molecular biology and its rapid rate of progress are evident in the success of recent experiments on mitochondrial nucleic acids. Morphological observations had shown the presence of 20–30 Å wide fibrils and ribosomelike particles, and cytochemical investigations had studied the removal of the fibrils by DNAase (which hydrolyzes DNA specifically) and of the ribosome-like granules by RNAase (an RNA-hydrolyzing enzyme). The results of these efforts indicated that mitochondria might have traces of both DNA and RNA. When molecular biologists, armed with the techniques applied so successfully in the study of nucleic acids and protein synthesis in microorganisms, turned to this problem, it took only a year or two to establish the presence in purified mitochondria of DNA and RNA and of machinery for synthesizing protein. Mitochondrial nucleic acids and ribosomes were shown to be of different size and to have properties unlike comparable components of the rest of the cell. Only small amounts are present in mitochondrial fractions isolated from cells, but the distinctive properties permitted the firm conclusion that the nucleic acids or ribosomes found in isolated mitochondria are not contaminants derived from other organelles during isolation. Interestingly, mitochondrial DNA and ribosomes in some organisms resemble DNA and ribosomes of bacteria; the possible evolutionary implications of this fact will be discussed later (Section 4.5.1).

· The DNA of mitochondria is a double helix as it is in the cell nucleus. In some of the organisms studied, however, it has been shown to exist as a circular molecule (Fig. II-40) comparable to the DNA of bacteria (Section 3.2.3); several such molecules may be present in a mitochondrion. There remains some question whether circularity

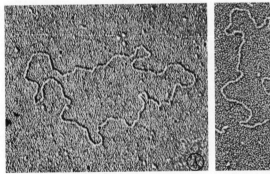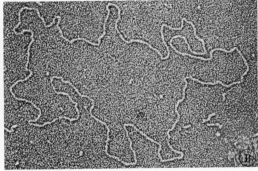

Fig. II-40 *Electron micrographs of circular DNA molecules from mitochondria of leucocytes from a leukemic human. As is true of mitochondrial DNA molecules from many animal cells, the molecule in **A** is 5 μ long. (This is the length of the thin fiber, a DNA double helix, that delineates the perimeter of the distorted circle in the micrograph). The molecule in **B** is 10 μ long and probably is a dimer made of two 5 μ molecules. Dimers and interlocked pairs of 5 circles have recently been found (in small numbers) in mitochondrial DNA preparations from several sources. It should be noted that it is in its circularity that mitochondrial DNA is said to resemble bacterial DNA. The latter are, however, much larger than mitochondrial DNA's. Eucaryotic cell nuclear DNA's appear not to be circular. × 35,000. Courtesy of D. A. Clayton and J. Vinograd.*

is universally true of mitochondrial DNA; mitochondrial DNA that is circular has been found chiefly in animal cells and it will be interesting to determine whether plant cell mitochondrial DNA differs appreciably in this regard, or whether reports that plant cell mitochondrial DNA's are not circular reflect breakage of the molecules during their isolation. Mitochondria are thought to possess the necessary machinery for replicating their DNA. Presumably, such replication is an integral part of mitochondrial duplication (see below) and transmission of hereditary information (Sect. 4.3.7).

That mitochondrial DNA codes for specific RNA's is shown by the fact that mitochondrial RNA's hybridize (Section 2.2.3) with mitochondrial DNA. It is widely assumed that some of the RNA's are translated into specific proteins. The mitochondria possess the DNA-dependent RNA polymerase enzymes that synthesize RNA. They also contain tRNA's and a series of amino acid-activating enzymes (Section 2.3.1). RNA-containing granules are also present. In a number of organisms these are about 120 Å in diameter and are probably ribosomes; such ribosomes are more similar in size and composition to the ribosomes of bacteria than to the larger non-mitochondrial ribosomes in the cytoplasm of the cell.

In a cell such as an L-cell (a type of cultured mouse fibroblast) the 250 mitochondria contain a total amount of DNA equal to 0.1–0.2% of the DNA in the nucleus. Each mitochondrion contains up to 6 (circular) DNA molecules. It is not yet known whether some may carry different information from others. In plant cells some of the mitochondrial DNA molecules are longer than animal cell mitochondrial DNA's and thus perhaps carry somewhat more information.

Although the mitochondria appear to have most, perhaps all, of the machinery needed for protein synthesis, there does not appear to be enough DNA in mitochondria to code for all mitochondrial proteins (see Section 3.1.3 for estimates of the information content of nucleic acids). Perhaps some enzymes and other proteins of the membranes are made in the mitochondria. However, a growing body of evidence suggests that other enzymes and lipids may be made elsewhere in the cell (using nucleic acids originating in the nucleus), and that these then move into the mitochondria. Genetic data (Section 4.3.7) indicates that the inheritable characteristics of mitochondria are under the control both of nuclear genes and of cytoplasmic factors presumed to be within the mitochondria. Other evidence for a dual origin of mitochondrial components is more controversial but, taken as a whole, it is fairly convincing. For example, protein synthesis by mitochondria appears to be inhibited by the drug *chloramphenicol* which also affects protein synthesis by bacteria but does not inhibit most eucaryote cell cytoplasmic protein synthesis. Growth of yeast

cells (these are eucaryotic) in chloramphenicol prevents synthesis
of many mitochondrial enzymes and produces very abnormal mitochondria. However, synthesis of cytochrome *c* (among other mitochondrial components) is not inhibited, suggesting that this protein
is made by a system that is different from the synthetic system for
other mitochondrial proteins and presumably is located outside the
mitochondria.

The question of the origin of mitochondrial proteins is part of
a more general question: How are mitochondria duplicated? Early
cytologists observed the divisions of certain cells containing only two
or three mitochondria (for example, developing sperm cells of invertebrates); they showed that new mitochondria could arise by division
of preexisting ones. Similar observations have been made recently in
electron microscope studies of algae; in these cases, mitochondrial
division is coordinated with the division of the cells. Electron microscope images of nondividing cells, such as hepatocytes, also suggest
that mitochondria can divide.

A mutant of the mold, *Neurospora*, has been used for detailed
study of mitochondrial duplication. This mutant cannot synthesize
choline, an important constituent of phospholipids in membranes;
instead the mutant obtains it from the culture medium. To label the
mitochondria, radioactive choline was placed in the medium for a
period. The cells were then allowed to continue their growth in a
medium containing *nonradioactive* choline. At intervals, mitochondrial
fractions were isolated and analyzed. Biochemically, it was learned that
the label acquired by the mitochondria during the initial period in
radioactive choline remained in the mitochondria while the cell mass
and number of mitochondria increased. By autoradiography, the labeling pattern of the individual mitochondria could be determined. The
results are diagrammed in Fig. II-41. They strongly suggest (1) that
the mitochondria grow and then undergo division or fragmentation and
(2) that such growth probably accounts for all of the increase in mitochondrial number. The key finding was that the choline label per mitochondrion *decreased* as the number of mitochondria *increased*; the
mitochondrial population as a whole always showed a random distribution of the label among all the members of the population. This was
the result expected if, on the average, the mitochondria passed half
of their phospholipid components to each daughter mitochondrion.
Thus, the total label remained almost constant. (Degradation of mitochondria was not an important factor over the time period studied
and under the conditions used.) However, the label was increasingly
dispersed and diluted among more and more individual mitochondria.

The results of this experiment would have been different if mitochondria did not divide and if new mitochondria arose as newly formed

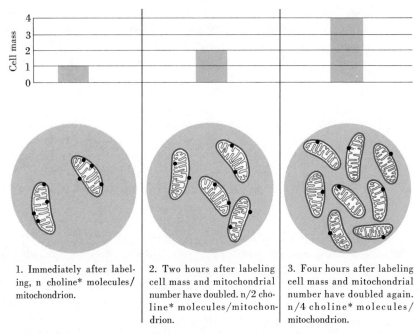

1. Immediately after labeling, n choline* molecules/mitochondrion.

2. Two hours after labeling cell mass and mitochondrial number have doubled. n/2 choline* molecules/mitochondrion.

3. Four hours after labeling cell mass and mitochondrial number have doubled again. n/4 choline* molecules/mitochondrion.

Fig. II-41 D. Luck's experiment on Neurospora. *This fungus consists of elongate multinucleate filaments called* hyphae. *Thus, as the mass of cellular material doubles the total volume and number of organelles also doubles but there need not be division into twice as many cells. In the experiment on which this figure is based the hyphae were grown in labeled choline (choline*) for 14 hours (to insure extensive initial labeling of mitochondria) then transferred to unlabeled medium and studied during subsequent growth. The black dots represent autoradiographic grains.*

structures assembled from recently synthesized precursors without the inclusion of portions of old organelles. The experimental outcome might then have shown (1) the appearance of a large and increasing percentage of unlabeled mitochondria as the mitochondrial number increased after labeling; and (2) the retention by each labeled organelle of the entire amount of radioactivity initially obtained. Only the mitochondria assembled during the exposure to radioactive choline would be labeled; those made later (or earlier) would not.

Biochemical and autoradiographic analysis with labeled DNA precursors also show a distribution of labeled "parental" DNA among daughter mitochondria consistent with the conclusion that mitochondria duplicate by growth and division.

Although it is likely that in most cells new mitochondria arise as they do in *Neurospora;* the possibility remains that in some instances they may arise differently. Yeast cells grown in the presence of oxygen have mitochondria with typical structure and biochemistry. When they are grown in the absence of oxygen, they lack almost all cytochromes and the electron microscope shows few, if any, typical mitochondria. They possess, instead, a few small membrane-bounded bodies with little internal structure. Upon addition of oxygen to the medium, there is both a marked increase in number of such bodies and an extensive synthesis of cytochromes; eventually the bodies become typical mitochondria. The precise manner of this transformation is variable, depending upon concentrations of glucose and lipid precursors in the growth medium. The source of the new membrane-bounded bodies is not clear, and it is possible that they arise from the division of preexisting bodies. However, it is reported occasionally that mitochondria arise in yeast upon transfer to aerobic conditions of cells that have no readily recognizable premitochondrial bodies. It may be that mitochondria can arise from primordia with very simple structure. Most likely, even if such is the case, genetic continuity with previously existing mitochondria is maintained; one would expect, for example, that even the simplest primordium will be found to contain mitochondrial DNA. (There are those who argue that the few reports of mitochondrial origin in the apparent absence of membrane-delimited premitochondrial bodies reflect technical difficulties in fixation and other preparative procedures for microscopy and that appropriate techniques permit demonstration of such bodies. There is not yet a fully documented and universally accepted case of mitochondrial origin from non-membrane-delimited structures.)

2.6.4   PLASTICITY IN THE LIVING CELL   The study of living cells by phase contrast microscopy and by microcinematography (Section 1.2.2) reveals that in many cells mitochondria are continually in motion. They show dramatic changes in shape and volume, and they can probably fuse with one another or separate into several parts. The static images of the electron microscope cannot reveal such movement. In some muscles of the rat, for example, highly branched mitochondria are seen in electron micrographs; these may actually be snapshots of mitochondria in the process of fusing or fragmenting.

The thyroid hormone, *thyroxine*, calcium ions, and other physiological substances can induce mitochondrial swelling in the test tube. Form alterations in isolated mitochondria are also observed upon addition of *dinitrophenol* and other nonphysiological agents that uncouple phosphorylation from oxidation. Changes in respiratory rate and other metabolic features of isolated mitochondria have been correlated with changes in the volume of the mitochondria and with other structural changes. For example, when isolated hepatocytic mito- chondria are stimulated to full respiratory activity, the structure of the organelles is reversibly altered from one similar to that shown in Fig. II-36. The matrix becomes more electron dense, and the inner membrane is somewhat unfolded so that the cristae are of less regu- lar appearance. Eventually, it should be possible to explain these alterations of structure and function in terms of mitochondrial macromolecules, to determine the mechanisms of the various types of form change (there has been some study, for example, of possible "contractile" systems) and to extend the few observations suggesting that changes like those seen in isolated mitochondria also occur in the cell.

2.6.5   **RECONSTI-**   Initially, reconstitution experiments in-
            TUTION   volved only a few enzymes, but as successes
       EXPERIMENTS   increased, the attempts became more ambi-
                            tious. The reconstitution of mitochondria
begins with fragmentation; from the fragments separate extracts are made containing a few enzymes linked in complexes, single enzymes, and proteins or lipids. Attempts are being made to reconstitute entire sequences, such as the respiratory assembly or oxidative phospho- rylation, by mixing extracted components together under carefully controlled conditions. Results of one impressive series of experiments are diagrammed in Fig. II-42. It may not be surprising that certain properties of the enzyme F_1 (such as its sensitivity to inhibition by oligomycin) differ in isolated enzymes from the intact organelle. How- ever, it is particularly interesting that the original pre-isolation prop- erties of F_1 return when other mitochondria components are added to it; portions of the initial structure are then restored. This is one of many examples of the intimate relations of mitochondrial structure and function.

Reconstitution experiments may provide information about the way in which organelles form or grow. In revealing the manner in which complex structures can be disassembled and reassembled in the test tube, they may furnish clues to the types of bonds that hold structures together and the the modes of assembly in the cell.

Fig. II-42 *Disruption of isolated beef heart mitochondria and separation* ▶ *of the resulting fragments into membranous and "soluble" components followed by reconstitution of some mitochondrial structure and function. The drug* oligomycin *is known to inhibit* F_1-*related functions in intact mitochondria. Thus, sensitivity to this drug is one measure of the extent to which reconstitution succeeds in reproducing conditions similar to those in mitochondria. Based on experiments of E. Racker and colleagues.*

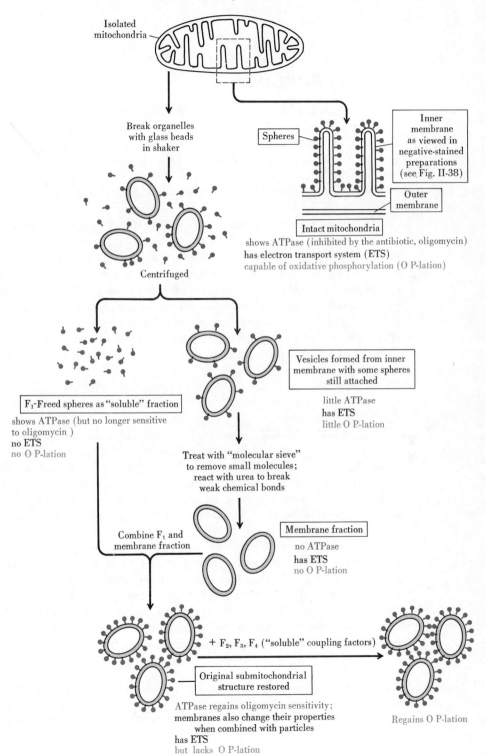

Isolated mitochondria

Break organelles with glass beads in shaker

Spheres

Inner membrane as viewed in negative-stained preparations (see Fig. II-38)

Outer membrane

Centrifuged

Intact mitochondria

shows ATPase (inhibited by the antibiotic, oligomycin)
has electron transport system (ETS)
capable of oxidative phosphorylation (O P-lation)

F_1-Freed spheres as "soluble" fraction

shows ATPase (but no longer sensitive to oligomycin)
no ETS
no O P-lation

Vesicles formed from inner membrane with some spheres still attached

little ATPase
has ETS
little O P-lation

Treat with "molecular sieve" to remove small molecules; react with urea to break weak chemical bonds

Combine F_1 and membrane fraction

Membrane fraction

no ATPase
has ETS
no O P-lation

+ F_2, F_3, F_4 ("soluble" coupling factors)

Original submitochondrial structure restored

ATPase regains oligomycin sensitivity;
membranes also change their properties when combined with particles
has ETS
but lacks O P-lation

Regains O P-lation

c h a p t e r **2.7**

CHLOROPLASTS

The plastids of plant cells are of various types and they contain different proportions of several pigments and of components such as starch. The most familiar and abundant plastids are *chloroplasts*, which contain large quantities of the green pigment *chlorophyll*. They are responsible for the photosynthetic use of the energy of sunlight to effect the transformation of carbon dioxide and water into carbohydrates with the simultaneous release of oxygen.

Choloroplasts are relatively large organelles that vary in size and shape from species to species. In some algae, the one or two chloroplasts are in the form of cups or elongate spirals that fill much of the cytoplasm. In higher plants, a great many chloroplasts may be present in each cell in the form of ovoid or disclike bodies. Leaf cells contain several dozen chloroplasts, each measuring $2-4 \mu \times 5-10 \mu$; typical dry weight figures for chloroplast composition are: 40–60 percent protein, 25–35 percent lipid, 5–10 percent chlorophyll, 1 percent pigments other than chlorophyll, and small amounts of DNA and RNA.

2.7.1   **STRUCTURE** Choloroplasts are bounded by two membranes. Within, they contain additional membranes surrounded by a matrix, the *stroma*. Granules containing starch may be scattered in the stroma. In many algae, however, starch may accumulate near a special region known as the *pyrenoid* (see Fig. II-43 and Section 3.4.1), in which the starch is apparently synthesized from glucose.

The internal membrane systems of the chloroplast are usually in the form of flattened sacs called *lamellae* or *thylakoids*. In many algae, the sacs are arranged in parallel arrays and run the length of the plastid (Figs. II-43 and III-11). In higher plants, the structure varies somewhat but usually resembles the arrangement shown in Fig. II-44. The *grana* consist of stacked sacs resembling a pile of coins, and they are connected to each other by membranes running in the stroma.

Fig. II-43 *Schematic representation of some structural and developmental* ▶ *features of* chloroplasts. *The diagram (A-1) of particle distribution in the grana membranes shows one of several plausible proposals (after the work of C. R. Arntzen, R. A. Dilley and F. L. Crane).*

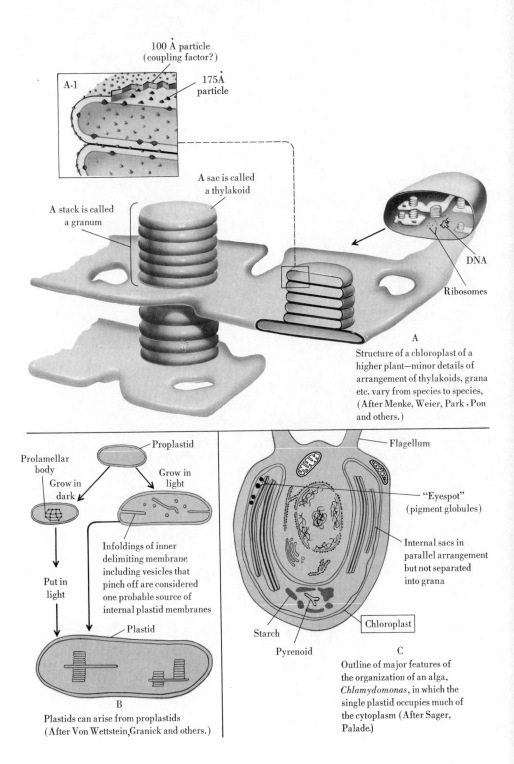

A-1

100 Å particle
(coupling factor?)

175Å
particle

A sac is called
a thylakoid

A stack is called
a granum

DNA

Ribosomes

A

Structure of a chloroplast of a
higher plant—minor details of
arrangement of thylakoids, grana
etc. vary from species to species,
(After Menke, Weier, Park , Pon
and others.)

Proplastid

Prolamellar
body

Grow in
dark

Grow in
light

Infoldings of inner
delimiting membrane
including vesicles that
pinch off are considered
one probable source of
internal plastid membranes

Put in
light

Plastid

B

Plastids can arise from proplastids
(After Von Wettstein, Granick and others.)

Flagellum

"Eyespot"
(pigment globules)

Internal sacs in
parallel arrangement
but not separated
into grana

Chloroplast

Starch

Pyrenoid

C

Outline of major features of
the organization of an alga,
Chlamydomonas, in which the
single plastid occupies much of
the cytoplasm (After Sager,
Palade.)

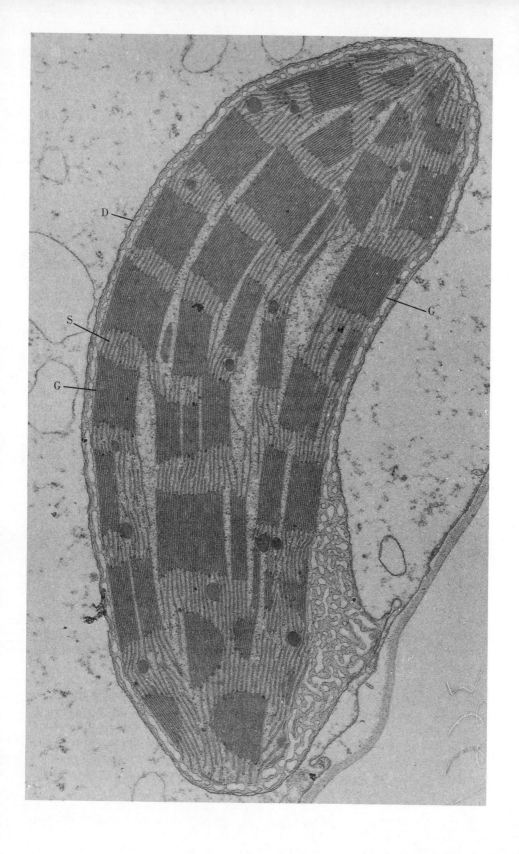

2.7.2   ***FUNCTION*** Photosynthesis involves two major sets of reactions, each set consisting of many steps. One set depends on light and cannot occur in the dark. The other set, referred to as the *dark reactions*, is not directly dependent on light. The light-dependent reactions result in the production of ATP from ADP (photophosphorylation), the formation of reduced NADP from NADP (a coenzyme related to NAD), and the release of oxygen derived from water. In the dark reactions, carbon dioxide is builtup into carbohydrates using the ATP and reduced NADP from the light-dependent reactions.

2.7.3   ***RELATION OF*** As is true of pigments in general, the color
         STRUCTURE TO of chlorophyll results from its selective
           FUNCTION absorption of light with certain wavelengths
                          (Section 1.2.1). The transformation of light energy into chemical energy in chloroplasts depends upon the ability of chlorophyll to enter an "excited" state when it absorbs light energy. The energy in light exists in discrete packets or *photons*. The absorption of photons by chlorophyll molecules can raise the "energy levels" of electrons in the molecules; under some circumstances it results in ejection of an electron from a chlorophyll (photoionization). The generation of ATP and reduced NADP depends on electron transport, beginning with light-excited chlorophyll and proceeding through a chain that involves iron-containing proteins (cytochromes, ferredoxin), a copper-containing protein (plastocyanin), and compounds related to lipids (plastiquinone).

This sequence of events implies a high degree of molecular organization. Like the mitochondria, plastids may be isolated and fragmented and portions of the structure separated (by centrifugation) into distinct fractions. When chloroplasts are broken open, many of the enzymes of the dark reactions are released, but most components involved in the light-dependent reactions are not. Thus it appears that, while much of carbohydrate synthesis occurs in the stroma, the formation of ATP, reduced NADP, and O_2 takes place in association with the sacs. Further subfractionation indicates that the sac membranes contain chlorophyll and the components involved in electron transport. Chlorophyll molecules are similar in size and solubilities to lipid

◄ ***Fig. II-44*** *A chloroplast from a leaf cell of corn. The two delimiting membranes are seen at D. The grana (G) consist of stacks of sac-like thylakoids. Stroma membranes (S) run between the grana. × 75,000. Courtesy of D. L. K. Shumway. See Fig. II-43A for a diagrammatic interpretation.*

molecules and could participate in formation of membranes of the various types discussed in Chapter 2.1.

Beyond this lies a realm of uncertainty. How are the pigments and enzymes put together on the membranes providing a large surface for light absorption and the molecular arrangements that result in efficient function? In dry-weight, the membranes are about one-half protein, one-eighth chlorophyll and other pigments (including the orange *carotenoids*), and three-eighths lipids similar to those discussed earlier (Section 2.1.2). Extremely small fragments of chloroplast membranes can engage in multistep reactions requiring the sequential operation of several enzymes and other components. This suggests that at least some of the plastid macromolecules are in repeating multienzyme arrays. Analysis of the membranes indicates a great excess of chlorophyll; the number of chlorophyll molecules is 10–100 times greater than that of other key participants (notably, electron transport enzymes) in the light-dependent reaction. This raises the question of how the membranes are arranged so that energy absorbed by large numbers of chlorophyll molecules is efficiently transferred to a relatively small number of electron transport systems.

One theory proposes that the membranes contain units each containing up to several hundred closely packed chlorophyll molecules and other pigments associated with one or a few reaction centers. A reaction center (sometimes called an *active center*) contains special chlorophyll molecules linked to an electron transport system. Light may be absorbed by any one of the pigment molecules. The energy from light absorbed by a given pigment molecule is carried to the reaction center by passage from one molecule to another. The physical processes underlying this transfer are based on the fact that one chlorophyll (or other pigment) molecule in an excited state can transfer energy to another nearby chlorophyll molecule; it raises the second molecule to the excited state and returns to its normal, unexcited state. It is possible that energy is passed essentially at random from molecule to molecule in the unit until, by chance, it reaches the reaction center. It is also possible that some organization exists that guides the passage. In either event, when the energy reaches the reaction center, electron transport commences with the ejection of an electron from a special chlorophyll molecule. In the unit, then, most of the pigments provide a large "light-harvesting" system or "antenna" which funnels energy to chlorophylls at the reaction center; here electron transport releases the energy for conversion into usable form, that is, into ATP and reduced NADP.

There is evidence that at least some of this analysis is reasonably accurate. Some of the chlorophyll molecules (5–10 percent) in a plastid have properties expected of reaction center molecules that would accept the energy from other "light-harvesting" pigment molecules and pass it to the electron transport system. Due perhaps to unique orientation in the membrane, these chlorophylls are found to differ slightly in light absorption characteristics. (Their maximal absorption is of light with a wavelength about 700 mμ, in contrast to the other chlorophyll molecules that absorb maximally at 675 mμ.) From physical theory this can be taken to mean that the "excited" state of the special chlorophyll molecules is at a relatively slightly lower energy level. This would imply that once energy is transferred to the special chlorophylls, it is "trapped" there and cannot be transferred to other chlorophyll; it can, however, be transferred further "downhill" via the electron transport enzymes.

However, perhaps not all of the pigments and enzymes of the membrane are bundled into units. Future research might show, for example, that the reaction centers do occur as discrete assemblies on the membranes, but that the bulk of the chlorophyll is not separated into well-defined units. Chlorophyll might be spread more or less uniformly in the membrane, and a given molecule might transfer the energy of absorbed light to any one of a number of nearby reaction centers.

Electron microscopy often shows regular repeating structures 100–200 Å in diameter in the chloroplast sac membranes; these have sometimes been called *quantasomes*. The relations of such morphological units to the hypothetical functional units are being intensively studied. Probably the electron microscope units are not arrays each having molecules of all the light reaction enzymes. When isolated, some of the units can be shown to have properties analogous to the mitochondrial F_1 units (Section 2.6.1); thus they are thought to act as a coupling factor between photosynthetic electron transport and phosphorylation. There are reports that the coupling factor can be reversibly removed from the plastid membranes leaving behind the electron transport enzymes. Also being actively explored is the possibility that several different kinds of subunits are present on the chloroplast membranes, because careful measurements indicate the probable presence of morphological units of at least two different sizes. Much current interest is focused on the suggestion that one type of subunit is present on the outer surface of the thylakoid and another, deeper within the sac membrane (Fig. II-43).

2.7.4 *REPRODUCTION* Like mitochondria, plastids contain their own DNA and RNA distinct from both nuclear and other cytoplasmic nucleic acids. Plastids appear to be capable of protein and RNA synthesis, but their ribosomes, like mitochondrial ribosomes, are smaller than other cytoplasmic ribosomes and show distinctive sensitivity to inhibitors of protein synthesis such as chloramphenicol (Section 2.6.3). It is probable that plastids (like mitochondria) contain some components made within the plastid, and others, made in the cytoplasm outside.

In algae, plastids are seen to divide regularly as part of the cell division cycle and thus to maintain a constant number. In higher plants also, plastids can divide. In addition, plastids can develop from much simpler organelles, the *proplastids*. These appear to be self-duplicating. They are membrane-bounded, but do not contain the elaborate patterns of internal membranes found in mature plastids (see Fig. II-43). Plants grown in the dark provide an interesting system for studying plastid formation from proplastids. Their tissues are usually colorless, containing no fully developed plastids and little chlorophyll. However, modified proplastids are present; they contain arrays of membranous tubules, often in highly regular arrangements called *prolamellar bodies*. The latter appear to contribute to the formation of normal internal membranes when the plants are illuminated. On exposure to light, rapid synthesis of chlorophyll takes place, and this is paralleled by development of the usual internal chloroplast structure. In this and a number of other situations, it appears that chlorophyll incorporation is normally an integral and perhaps necessary or controlling step in the assembly of chloroplast internal membranes.

Rare mutant algae can form some normal-appearing internal plastid membranes that lack one or a few of the components normally present; a few of the mutants may contain little chlorophyll, but most lack membrane proteins or lipids. Eventually, study of such mutants will aid in establishing, for example, whether chloroplast membranes are made by more or less simultaneous assembly of all components into the final membrane structure or whether some components assemble to form a "framework" membrane into which others are inserted later. The limited evidence that normal appearing (though functionally deficient) membranes can form even if some of the usual components are lacking is being actively followed up for clues to the kinds of assembly steps that can produce the complex and precisely ordered array of chloroplast membrane components. One interesting proposal applied to both chloroplasts and mitochondria is that some of the enzymes assemble together as multimolecular complexes before associating with other components in the membrane (see Chapter 4.1 and Section 3.2.4).

In higher plants grown under normal conditions, division of formed chloroplasts and differentiation of proplastids probably both contribute to the increase in plastids during growth.

chapter **2.8**

LYSOSOMES

Historically, study of organelles begins usually with the accumulation of morphological observations and then passes to the isolation of the organelle in relatively pure fraction and biochemical study. For lysosomes, however, this pattern was reversed. Their discovery began with an investigation of *hydrolytic enzymes,* enzymes catalyzing reactions of the type $A_1—A_2 + H_2O \rightarrow A_1—H + A_2—OH$. Biochemical analyses of cell fractions separated from rat liver homogenates revealed that five such enzymes, all acting optimally in *acid* media, sedimented together in centrifugation. Further, it appeared that these enzymes were inactive toward their potential substrates if the fraction was carefully prepared to avoid disruption of the fragile organelles. This led to the hypothesis that the *acid hydrolases* were packaged together in an organelle previously undescribed; because the enzymes it contained were hydro*lytic,* the organelle was called a *lysosome.* From their sedimentation characteristics, it was predicted that hepatocyte lysosomes had a certain size, about 0.4μ in diameter. From the observation of *latency,* i.e, the fact that substrates added to isolated lysosomes were not split unless disruptive procedures were used, it was predicted that the organelle was surrounded by a membrane. When the acid hydrolase-containing fractions were examined by electron microscopy, morphologically distinctive cytoplasmic particles were indeed found, and the predictions concerning both size and outer delimiting membranes proved correct.

In subsequent years, many acid hydrolases were shown to be present in lysosomes (Fig. II-45). Because the activities of one of the lysosomal enzymes, *acid phosphatase,* could be localized cytochemically in microscopic preparations, evidence accumulated for the existence of a wide variety of forms of lysosomes. It soon became evident that lysosomes are versatile organelles involved in a variety of cell functions and showing different morphologies.

2.8.1   ***FORMS AND***   With the exception of a few cells, such as
          FUNCTIONS   mammalian red blood cells, all animal cells
                    so far examined possess lysosomes. Evidence for the presence of lysosome-like structures in plant cells is beginning to accumulate. Both biochemical and cytochemical studies

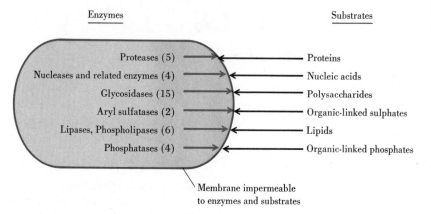

Fig. II-45 *The biochemical concept of lysosomes according to the work of C. deDuve, his collaborators and others. There are 36 known hydrolases in lysosomes. The numbers in parentheses indicate how many different enzymes hydrolyze each class of substrates shown to the right; the several enzymes within each group differ in specificity and other properties.*

have already demonstrated lysosomes in protozoa, insects, amphibia, mammals, and a variety of other vertebrates; it is probable that in animal cells at least lysosomes are ubiquitous.

A tentative determination that a cytoplasmic particle is a lysosome is possible (1) if electron microscopy shows it to be membrane delimited, and (2) if cytochemistry shows it to have one or more of the hydrolase activities found in lysosomes as studied biochemically. Identification from cytochemistry is tentative since reliable methods are available for only a few of the lysosomal hydrolases, so the presence of other hydrolases can only be *assumed*. Acid phosphatase activity is the most widely used "marker" enzyme demonstrable by staining procedures, although fairly reliable methods are also available for two or three other lysosomal enzymes. By such cytochemical methods, and in a growing number of instances by parallel biochemical studies, a variety of morphologically different structures may be identified as lysosomes.

It would be of obvious advantage to be able to study all suspected lysosomes by biochemical methods and thus confirm their status. However, for one reason or another, many tissues are poorly suited for organelle isolation procedures. For example, some organs have large numbers of intermingled cell types, so that one cannot easily identify the source of isolated organelles as cells of a given type. In other cases, it is difficult to obtain amounts of tissue large enough or to achieve adequate purity of fractions. Cytochemistry with its direct access to single cells can avoid many such problems. Ideally, cytochemistry and biochemical studies on isolated organelles are used together.

Many lysosomes are in the same size range as the first liver lysosomes studied, about 0.5 μ in diameter. The largest however, such as the "protein droplets" in cells of mammalian kidney (Section 3.5.4), are several micra in diameter. The smallest are the *Golgi vesicles;* these are only 25–50 mμ in diameter. The identification of some of the vesicles of the Golgi apparatus as probable lysosomes rests upon the presence of a delimiting membrane and upon cytochemical evidence that they possess acid phosphatase activity; in addition, some autoradiographic observations suggest that the vesicles carry proteins, presumably acid hydrolases, from the Golgi apparatus to vacuoles responsible for enzymatic digestion of material phagocytosed by some types of white blood cells.

The size heterogeneity of lysosomes is paralleled by heterogeneity in form, origin, and function. All lysosomes are related, directly or indirectly, to *intracellular digestion.* The material to be digested may be of exogenous (extracellular) or endogenous (intracellular) origin. Collectively, the lysosomal enzymes are capable of hydrolyzing all the classes of macromolecules in cells (Fig. II-45) and presumably would do so if they were not confined in structures delimited by membranes. There is considerable speculation about the possibility that in abnormal situations, lysosomal enzymes may escape into the cytoplasm and either kill the cell or bring about dramatic metabolic changes. However, in most of the known functions of lysosomes, the material on which the acid hydrolases act must gain access to the interior of the lysosome, and the enzymes remain confined within the organelle.

There is much variation in the mechanisms by which material to be digested becomes enclosed inside a lysosome with the acid hydrolases. Figure II-46 is a diagrammatic representation of the major types of lysosomes and their possible modes of origin. The diagram illustrates the proposal that lysosomal enzymes, manufactured at the ribosomes, are transported by the endoplasmic reticulum to lysosomes, either directly or via packaging in the Golgi apparatus. In many cells, *primary lysosomes* appear to be produced by the Golgi apparatus. These lysosomes are thought of as packages that transport hydrolases to other membrane-delimited bodies with which the primary lysosomes fuse; the Golgi vesicles referred to above probably fall in this category. When both the enzymes and the material to be digested, being digested, or already digested are present within a lysosome, the lysosome is referred to as a *secondary lysosome.* Secondary lysosomes may accumulate large quantities of undigested or indigestible molecules; the resulting structures are known as *residual bodies.* The classification of lysosomes in functional categories such

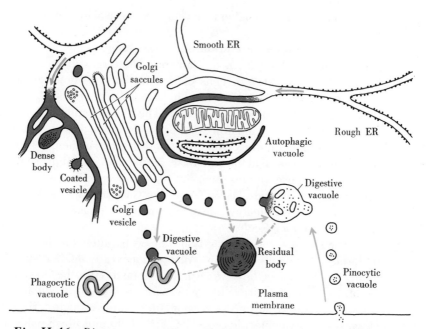

Fig. II-46 *Diagram suggesting probable interrelations of some organelles in the transport of lysosomal hydrolases (red) and the formation of lysosomes. Illustrated are: (1) Fusion of primary lysosomes (Golgi vesicles) with phagocytic and pinocytic vesicles resulting in the formation of digestive vacuoles. Some of the vesicles are "coated" (see Section 2.1.4). (2) Movement of hydrolases within the RER and SER to the vicinity of the Golgi apparatus and to lysosomes. Dense bodies (bodies containing electron dense grains and other material) and autophagic vacuoles are among the structures that may form from the ER or from closely interrelated components of the ER and Golgi apparatus; (3) Formation of residual bodies by the accumulation of indigestible residues within lysosomes.*

as "primary," "secondary," "residual body," and so forth is useful in establishing relationships among structures that may differ considerably in morphology. This morphological variation reflects the variety of materials digested and the fact that lysosomes may form in several ways.

2.8.2   HYDROLYSIS OF EXOGENOUS MACRO-MOLECULES   Although many materials, especially small molecules, enter cells individually, some enter *in bulk*, by either *phagocytosis* or *pinocytosis*. Both of the latter processes involve uptake into vacuoles formed by infolding of the plasma membrane (Section 2.1.4) and both can contribute material to lysosomes. In protozoa (see Chapter 3.3), hydrolysis of material taken up by phagocytosis and pinocytosis is an important mechanism in feeding. Pinocytosis can be observed to occur in many cell types of multicellular animals, and phagocytosis is seen in a few specialized cell types.

Phagocytic white blood cells of mammals have been intensively studied by light and electron microscopy, phase contrast microcinematography, and by biochemistry. These cells are filled with cytoplasmic granules, some of which are primary lysosomes packaged by the endoplasmic reticulum–Golgi apparatus systems in the same way as secretion granules. The cells engulf bacteria by including them in phagocytic vacuoles (Fig. II-47). The cytoplasmic granules quickly approach the vacuoles, the membranes of granules and vacuoles fuse, and the granule contents are emptied into the vacuoles. Since the phagocytic vacuoles have now both lysosomal enzymes and the macromolecules (bacteria) to be digested, they are, by definition, secondary lysosomes. Under the influence of lysosomal hydrolases, and of other enzymes and nonenzymatic antibacterial materials emptied into the vacuoles by other granules, most bacteria are destroyed (see Section 3.2.3). The soluble products of hydrolysis enter the cytoplasm. A very few bacteria (for example, those responsible for tuberculosis) are coated with materials that resist distruction by lysosomal enzymes; thus, the organisms can live and multiply inside phagocytic vacuoles.

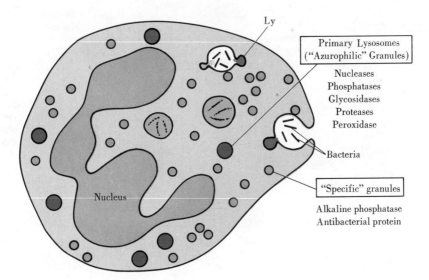

Fig. II-47 *Diagram of a white blood cell (neutrophil) ingesting bacteria. Two classes of granules fuse with the phagocytosis vacuoles and contribute digestive enzymes and other components; the granules colored red are lysosomes. After the work of M. Baggiolini, D. F. Bainton, Z. A. Cohn, C. de Duve, M. G. Farquhar and J. G. Hirsch.*

The lysosomes of phagocytic white blood cells are of decisive importance in the defense mechanisms of multicellular animals. Other phagocytes are found in many organs such as lung, liver, and spleen; all have large lysosomes whose enzymes break down foreign materials that are picked up by phagocytosis or pinocytosis. An interesting system has been described recently in which the material engulfed (by pinocytosis) is partially degraded inside lysosomes but evades complete destruction. As shown diagrammatically in Fig. II-48, when the virus known as *Reovirus* is taken up by cultured cells into pinocytic vacuoles, its protein "coat" is hydrolyzed, but its RNA "core" is not. When the protein coat is removed, the RNA leaves the vacuole and enters the cytoplasm where it gains control of host cell metabolism. In the case of other viruses studied, lysosomal enzymes degrade the nucleic acids as well as the protein, and thus destroy the infectivity of the virus; presumably, this contributes to defense against viral infection. Probably the evolution of Reovirus resistance to lysosomal degradation was based on the formation of double-stranded RNA rather than the single-stranded form found in other viral RNA. The RNA-digesting enzyme *ribonuclease* of lysosomes can hydrolyze the RNA of Reovirus when the RNA is experimentally altered to single-

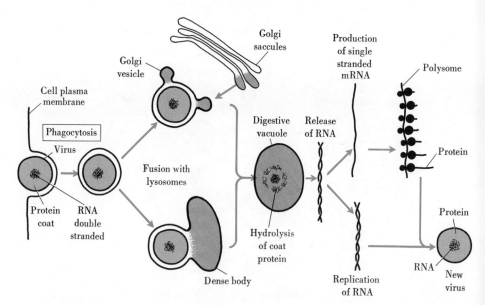

Fig. II-48 *Diagram showing* Reovirus *uptake by cells (mouse fibroblasts grown in culture) and the fate of the virus within the cell. The viral RNA is replicated and also acts as a messenger RNA carrying information for the synthesis of virus proteins (see Chap. 3.1). After the work of S. Silverstein and S. Dales.*

stranded form, but not when it is present in the double-stranded form found naturally in the virus. It should be noted that while both incorporation of Reovirus into lysosomes and multiplication of Reovirus within the cell are readily observable, there are those who hesitate to conclude that it is the RNA from lysosomally-digested viruses that is responsible for subsequent multiplication (see Fig. II-48). It is very difficult to rule out the possibility that a few Reoviruses enter the cell and are uncoated by an unknown nonlysosomal route and that it is these viruses that go on to multiply.

Another example thought to involve hydrolysis in lysosomes is of considerable physiological importance. The protein *thyroglobulin* is secreted by thyroid gland cells into the lumen of the gland. This initial phase of thyroid function is outlined in Figure II-49 and will be

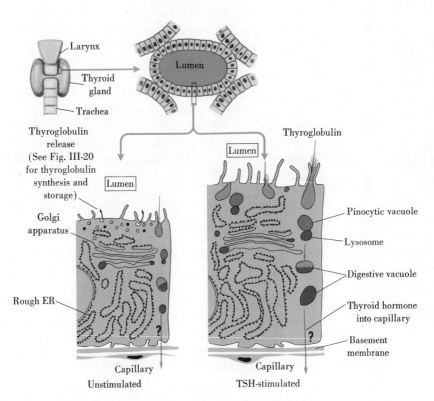

Fig. II-49 *The uptake of thyroglobulin from its storage sites (colloid in the lumen) has been studied under normal conditions but is more readily observed when the thyroid gland cells are stimulated by hormonal treatment (TSH) to increase their rate of release of thyroid hormone. These schematic diagrams illustrate the suggestion that lysosomal hydrolases split thyroglobulin to produce hormones that are then released into capillaries.*

The synthesis and storage of thyroglobulin are diagrammed in Fig. III-20.

considered in greater detail in Section 3.6.1. It is generally believed that when thyroid-hormone release is stimulated (for example, by hormones from other glands) the thyroglobulin stored in the lumen is engulfed by the secretory cells in large pinocytosis vacuoles. Electron microscopic and cytochemical observations show that lysosomes fuse with these pinocytic vacuoles. Apparently the thyroglobulin is partially hydrolyzed within the vacuoles and the thyroid hormone thyroxine is one of the digestion products. It is as thyroxine and related componds that the hormone passes out of the cell and into the blood capillaries at the base of the cell, the opposite pole of the cell from the lumen where the initial release of thyroglobulin takes place. Lysosomes isolated from thyroid secretory cells can hydrolyze thyroglobulin; further study of this hydrolysis should provide a direct biochemical determination of whether it can account for hormone release.

A final example of the uptake of exogenous materials is the uptake of protein (from calf serum included in the growth medium) by a type of phagocytic cell (*macrophages*) obtained from the abdominal cavity of the mouse and grown in tissue culture. This is an excellent system for eventual elucidation of the control mechanisms that regulate acid hydrolase synthesis and lysosome production. When exposed to high concentrations of serum, the cells pinocytose very actively and accumulate large numbers of both primary and secondary lysosomes. Accompanying this change in lysosome number, dramatic rises can be measured in the levels of lysosomal enzyme activities (Fig. II-50). If the cells are then returned to a low serum medium most lysosomes disappear and levels of lysosomal enzymes return to normal.

Several interesting problems relating to hydrolysis of exogenous molecules are actively being studied. Which mechanisms control lysosomal movement within the cell and fusion with other bodies? What features of their membranes permit lysosomes to fuse with a phagocytic or pinocytic vacuole, but apparently prevent fusion with the mitochondria, ER, or nucleus? Are primary lysosomes the only source of the acid hydrolases that enter vacuoles containing exogenous material? In response to the last question, it is known that fusion of secondary lysosomes with newly formed pinocytosis or phagocytosis vacuoles commonly occurs in many cells. A striking experiment demonstrates such fusions. Tissue culture cells (mouse fibroblasts) were permitted to engulf an electron-opaque material, finely dispersed iron particles. Like other engulfed material, the iron particles accumulate in secondary lysosomes. Having "marked" these lysosomes with iron, the cells were then permitted to engulf a second "marker," finely dispersed gold particles in a mixture of DNA and protein. Since the gold particles are readily distinguishable from the iron particles used in the

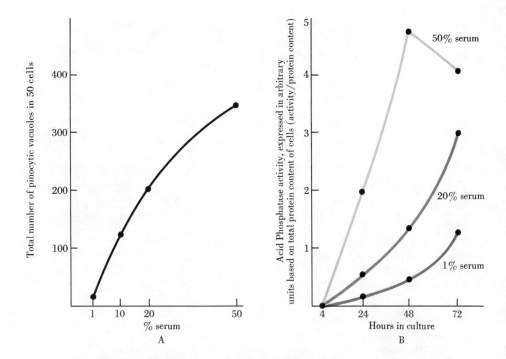

Fig. II-50   *Results of studies by Z. A. Cohn and colleagues on macrophages from the body cavity of a mouse. The cells were cultured in media containing varying proportions of newborn calf blood serum (a complex mixture of proteins and other components). Similar, though less dramatic effects are seen with sera from three other mammals. The higher the concentration of serum, the greater the number of pinocytosis vesicles (counted by phase microscopy, **A**), the greater the number and size of lysosomes (many of the lysosomes accumulate near the Golgi apparatus), and the higher the levels of activities of acid phosphatase **B** and of the two other acid hydrolases studied (β-glucuronidase and cathepsin). The specific mechanisms and factors responsible for the induction of pinocytosis and for the effects on enzyme synthesis and transport are currently being sought. Also of interest is the fact that when the cultured cells are transferred from a medium with 50% serum to one with 1% serum, the number of lysosomes and the concentrations of acid hydrolases decrease markedly, although neither lysosomes nor acid hydrolases have been found to be released from the cells to the culture medium.*

first "feeding," the results were unequivocal (Fig. II-51). The presence of both iron and gold in the same lysosomes demonstrates that fusions occur between old secondary lysosomes and newly formed digestive vacuoles. The same lysosomal enzymes apparently may be used for more than one round of digestion.

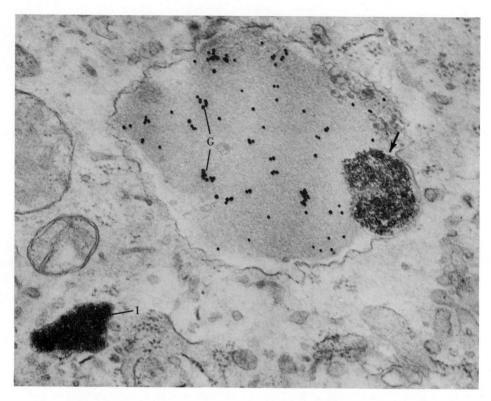

Fig. II-51 *Small portion of a cell (mouse fibroblast grown in culture) that was exposed first to colloidal iron and several hours later to colloidal gold (see text). A residual body (I) has accumulated a mass of electron-dense iron particles. The larger gold particles are seen within a digestive vacuole (G). The presence of iron particles (arrow) in the vacuole containing gold strongly suggests that a lysosome formed as a digestive vacuole during the exposure to iron has fused with the more recently formed gold-containing vacuole. × 50,000. Courtesy of G. B. Gordon, L. R. Miller and K. Bensch.*

2.8.3 HYDROLYSIS OF ENDOGENOUS MACRO-MOLECULES

Most or all eucaryotic cells have a mechanism by which bits of their own cytoplasm are surrounded by a membrane and subsequently degraded. This process has been named *autophagy* to suggest self- (*auto-*) phagocytosis. In protozoa, in many multicellular vertebrates and invertebrates, and in some plant cells, "autophagic vacuoles," a type of secondary lysosome, have been described. Autophagic vacuoles are identified by their content of one or more recognizable cytoplasmic structures — mitochondria, plastids, bits of ER, ribosomes, glycogen, or peroxisomes (Fig. II-52).

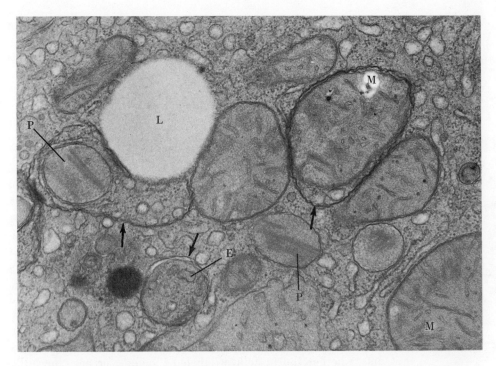

Fig. II-52   *Region of cytoplasm in a rat hepatocyte. Three autophagic vacuoles are present; their delimiting membranes are indicated by the arrows. Within one vacuole a mitochondrion is present (M), a second contains a peroxisome (P) and the third, fragments of ER (E). Other organelles, not included in autophagic vacuoles lie nearby (M, P). (L) indicates a lipid droplet in the cytoplasm near one of the vacuoles. × 30,000. Courtesy of L. Biempica.*

Does autophagy account significantly for the destructive aspect of the normal turnover of cell constituents as discussed in the introduction to Part 2? The incorporation of an organelle into an autophagic vacuole presumably results in the simultaneous degradation of all the components of the organelle. If such incorporation plays a major role in the turnover of ribosomes, then one might expect the RNA of ribosomes to turn over at the same rate as the protein of ribosomes; initial scanty data indicates this may be the case. Similarly, the various proteins of mitochondria seem, from the limited information available, to be destroyed at the same rate; this is also true of the several proteins studied in hepatocyte peroxisomes. The half-life (time required for half of the molecules of a population to be replaced) of hepatocyte mitochondria is about 10 days; for peroxisomes, the figure is about 2 to 3 days; hepatocyte ribosomal RNA has a half-life of about 5 days. The formation of one autophagic vacuole, containing one mitochondrion, per 15 minutes per liver cell could account for the observed rate of mitochondrial turnover.

On the other hand, the lipid components of the ER membranes (and perhaps of other membranes as well) may turn over at a different rate from the protein components, and the various "soluble" enzymes of the hyaloplasm may differ from one another in turnover rates.

Do the organelles degraded by autophagy have defects or other features perhaps related to the "age" of the organelle that initiate the autophagic process? Or are autophagic vacuoles random "gulps" of cytoplasm? There are cells in insects that appear to form vacuoles containing predominantly one type of organelle at one stage in development and another type at a later stage; this argues for selectivity at least in these instances. In apparent contrast are findings suggesting random degradation of hepatocyte peroxisomes. The proteins of peroxisomes in hepatocytes can be radioactively labeled, and the loss of label can be followed by isolating the peroxisomes at successive intervals during post-labeling growth, in the absence of radioactive precursors. Under conditions of probable constancy in peroxisome number and mass, the replacement of radioactive protein by nonradioactive protein follows an "exponential" pattern: at any given time, the probability of the degradation of a given peroxisome protein molecule is the same whether the protein is labeled ("older") or not ("newer"). Presumably this results from a similarly random breakdown of whole peroxisomes. The rate of loss of radioactivity from the peroxisomes does not increase with time, as might be expected if the older organelles were broken down preferentially. Hepatocyte ribosomal RNA shows similar behavior.

In sum, while it is likely that autophagy does play an important role in turnover, the details of the role remain to be determined and it cannot be asserted yet that the role is a predominant one. Much additional information should be forthcoming in the next few years, since experiments and observations of the types just discussed provide important clues to lines of further investigation that are apt to prove fruitful.

Although autophagy is a normal physiological phenomenon, proceeding probably at low rates in normal cells, many more autophagic vacuoles are found when cells are under metabolic stress (when they are either undergoing extensive remodeling or responding to hormonal stimulation or to injury). Autophagy may be a means by which cells digest bits of their own cytoplasm, surviving periods of stress or metabolic urgency without self-destruction. Several situations resulting in an abundance of autophagic vacuoles are known in embryonic development, when cells die as part of normal "modeling" of organs and tissues ("programmed cell death"). This should not necessarily be construed to mean that autophagic vacuoles *lead* to the death of the cells. Indeed, it may be that the cells die, under these circumstances, *despite* a "defense" mechanism involving autophagy.

How do portions of the cytoplasm come to be enclosed within the membrane of autophagic vacuoles, and how do acid hydrolases enter the vacuoles? Figures II-46 and II-53 outline one hypothesis which suggests that the endoplasmic reticulum, particularly in the Golgi zone, is involved in forming the delimiting membrane of the vacuole and perhaps in directly providing lysosomal hydrolases. It has also been proposed that the delimiting membranes derive from the Golgi apparatus and that the hydrolases come from lysosomes that fuse with the vacuoles. Fusion of other lysosomes with autophagic vacuoles has been demonstrated in hepatocytes.

2.8.4   *CELL INJURY* It was suggested soon after the discovery
**   *AND CELL DEATH*** of lysosomes that abnormal conditions such
as the uptake of injurious materials might
lead to the leakage of lysosomal enzymes into the cell cytoplasm and to consequent cell injury or death. Although unequivocal evidence for this suggestion is very difficult to obtain, it remains a potentially important mechanism that may aid in understanding cell pathology. There is little reason to doubt that after a cell has died, its lysosomal hydrolases are released to participate in destroying cellular and extracellular materials.

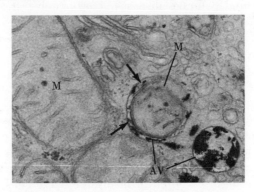

Fig. II-53 *Two autophagic vacuoles (AV) in a hepatocyte from a section of rat liver that was incubated to show sites of* acid phosphatase *activity. The reaction product appears black. The vacuole at the left contains a mitochondrion (M). Arrows indicate appearances suggesting that the membranes surrounding this mitochondrion may be forming by the flattening and transformation of an acid phosphatase-containing sac as proposed in Fig. II-46. A second mitochondrion (M), at the left of the micrograph is not within an autophagic vacuole.* ×30,000.

Certain compounds, including drugs and hormones, are known to affect the membranes of lysosomes isolated from the cell. Some, such as vitamin A, appear to *labilize* the lysosomes (that is, make them less resistant to disruption) and others, such as cortisone, seem to *stabilize* the organelles. Although definitive evidence is still unavailable, it may be that the physiological effects of such compounds on living cells are mediated in part through lysosomes. One hypothesis argues that the effects are results of changes in the leakage of hydrolases into the cytoplasm (labilizers might promote leakage) whereas another proposes that labilizers and stabilizers affect the fusion of lysosomes with other structures.

2.8.5 STORAGE OF INDIGESTIBLE RESIDUES IN LYSOSOMES

Many indigestible substances, such as iron, finely dispersed gold, and some large carbohydrate polymers experimentally administered to cells, accumulate within the lysosomes. Among the most interesting natural accumulations are those that occur with aging in several tissues of man, and those that apparently result from genetically caused absence of lysosomal hydrolases. The frequency of granules called "lipofuscin" pigment granules increase with age in human nerve, heart, and liver cells. The granules are secondary lysosomes (residual bodies) in which lipids and other materials accumulate. Many congenital "storage diseases," fortunately rare but often fatal early in childhood, show abnormally large lysosomes. Most of those thus far studied appear to result from the inherited absence of a gene required for synthesis of a specific hydrolase normally present in lysosomes.

The first studied of these "lysosomal diseases" is a generalized *glycogen* storage disease called Pompe's disease, in which the missing enzyme involved is normally found in liver lysosomes and is involved in degradation of glycogen. In the absence of this enzyme the liver lysosomes become engorged with glycogen (see Chapter 5.2, for unsuccessful attempts at cure of this disease). A variety of other inherited storage diseases, often involving the nerve cells, show accumulations of *mucopolysaccharide* or *lipid* inside lysosomes due to a deficiency of a specific hydrolase.

Not all abnormally stuffed lysosomes, however, need arise from enzyme deficiency. One disease is already known in which the genetically abnormal gene leads first to deposition of abnormal hemoglobin molecules in the hyaloplasm. Then, probably by autophagy, these end up in lysosomes which thus become enlarged. The complement of lysosomal hydrolases is apparently normal but because the *material* is abnormal it cannot be degraded and it accumulates within the lysosomes.

2.8.6   **"MICRO-**   It is quite possible that normal turnover of
           AUTOPHAGY"   organelles and macromolecules involves nonlysosomal enzymes that are present in the hyaloplasm. However, as outlined in Section 2.8.4, there is direct evidence that lysosomes do participate to some extent in turnover; organelle breakdown within autophagic vacuoles is seen. Can the same be said for breakdown of individual molecules when they are not part of visible structures? Some hyaloplasm presumably is included along with the organelles in autophagic vacuoles. Are there also mechanisms for the formation of minute vacuoles containing one or a few protein, RNA, polysaccharide, or lipid molecules? Is there also a mechanism for the digestion of the contents of such vacuoles by lysosomes? The answers are not known, since such *"microautophagy"* would probably not be detectable with current techniques.

c h a p t e r 2.9

PEROXISOMES (MICROBODIES, GLYOXYSOMES)

Peroxisomes were seen in rodent kidney and liver in the early 1950s, when electron microscopy of sectioned biological materials was in its infancy; they were then called microbodies. One of the peroxisome enzymes, urate oxidase, was among the first enzymes to be studied in isolated fractions of rat liver cells. The distribution of this enzyme in cell fractions was markedly similar to, but not identical with, that of the lysosomal enzyme, acid phosphatase. It proved to be situated in peroxisomes rather than lysosomes.

A number of procedures have been developed to separate peroxisomes from other organelles, and cytochemical methods are now available for light microscope and electron microscope visualization of peroxisomes (see Figs. I-18 and II-54). It is now established that peroxisomes are distinct organelles of widespread occurrence. They are present not only in the liver (hepatocytes) and kidney [cells of the "proximal convoluted tubules" (Section 3.5.4)] of a wide variety of vertebrates, but also in protozoa, yeast, and many cell types of higher plants. Details of peroxisome enzyme content and fine structure have

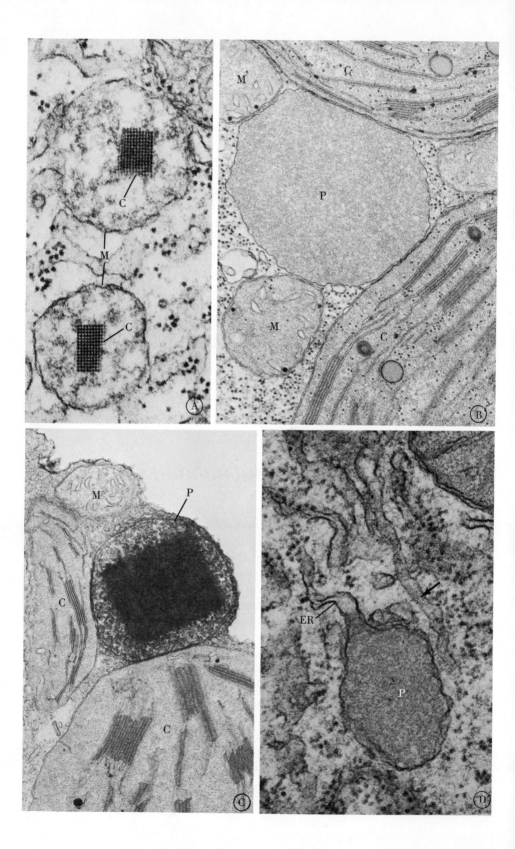

been studied in rat and mouse liver, rat kidney, green leaves of tobacco and other plants, and fat-laden "endosperm" containing stored nutrients of castor bean seedlings. Although the peroxisomes of these different cell types differ considerably, they are sufficiently similar to be considered variants of one type of organelle. The variations among peroxisomes of different tissues has interesting evolutionary implications which will be briefly explored later (Section 4.5.1).

2.9.1   OCCURRENCE   AND MORPHOLOGY   Among the unicellular organisms in which peroxisomes have been described are various amebae, the ciliate *Tetrahymena*, and yeast. In multicellular animals, their presence is clearly established at this time only in liver and kidney. In rat hepatocytes, there is roughly one peroxisome for every two to four mitochondria. They show no apparent concentration in special regions of hepatocyte cytoplasm, unlike lysosomes which are concentrated near the bile canaliculus (see Chapter 1.1; Figs. 1-3 through 1-5). Among higher plants they are found in some tissues of embryos and in green tissues capable of a metabolic process known as *photorespiration*.

Most peroxisomes are roughly spherical particles ranging in diameter from about 0.3–1.5 μ. Judging by their sedimentation in sucrose gradients, peroxisomes are denser than either mitochondria or lysosomes. Each is delimited by a single membrane about 65–80 Å thick and contains a finely granular matrix. In many tissues, peroxisomes show *cores* or *nucleoids*; these vary in morphology in different cell types; they may show crystal-like arrangement of tubular subunits (Fig. II-54). In comparing livers of different mammals, there is a general correlation between the presence of cores or nucleoids and of the enzyme urate oxidase. Isolated cores show urate oxidase activity, but the precise nature of the association between the core, or its tubules, and the enzyme is not established.

Peroxisomes show intimate relations with the endoplasmic reticulum. In both plant and animal cells, they seem to form as dilatations

◀ **Fig. II-54**   *Peroxisomes. **A**. Two peroxisomes from a cell of the grass plant,* Avena. *(M) indicates the delimiting membranes and (C), the crystal-like cores.* × 75,000. ***B**. Portion of a cell of tobacco leaf showing a peroxisome (P) closely associated with chloroplasts (C) and mitochondria (M).* × 40,000. ***C**. Portion of a cell similar to B in a section incubated in the cytochemical medium used for Fig. I-18. The density in the peroxisome (P) is due to deposition of reaction product. Mitochondria (M) and chloroplasts (C) are free of reaction product. **D**. Portion of a hepatocyte (rat) showing a peroxisome (P) attached to ER. Other cisternae of ER lie nearby (arrow).* × 60,000. *A–C courtesy of S. E. Frederick, E. H. Neucomb, E. Virgil and W. Wergin. D courtesy of L. Biempica.*

of the endoplasmic reticulum; the ER swells and fills with electron-dense material. Continuities of the membranes of peroxisomes with ER are frequently observed. In addition, endoplasmic reticulum often lies close to the peroxisome periphery, the membranes of the two being separated by 150–200 Å. There may prove to be some continuing functional relation between the two organelles, ER and peroxisomes.

In green cells of plants, peroxisomes are often found in close contact with chloroplasts and mitochondria. This, too, may reflect a close metabolic relation among the different organelles, as outlined below.

2.9.2   **ENZYME ACTIVITIES AND FUNCTIONAL SIGNIFICANCE**   A familiar pattern of cell organization – the location of several metabolically related enzymes within one organelle – is well illustrated by peroxisomes.

The peroxisomes of rat liver and kidney, the first to be isolated by centrifugation, contain four enzymes related to hydrogen peroxide (H_2O_2) metabolism; thus the name *peroxisomes*. Three of the enzymes produce H_2O_2 (urate oxidase, D-amino acid oxidase, and α-hydroxy acid oxidase) and one destroys H_2O_2 (catalase). Catalase makes up about 40 percent of the total peroxisome protein in rat liver and it appears to be localized in the matrix. All peroxisomes thus far studied, in both plants and animals, contain α-hydroxy acid oxidase and catalase. Since H_2O_2 is toxic to cells, catalase may have an important protective function, but it appears probable that the enzyme has other metabolic roles as well.

The peroxisomes of liver and kidney may be important in the degradation of purines (a class of nitrogen containing organic compounds of which the adenine and guanine bases of nucleic acids are familiar examples). At least in frogs and chickens, not only is urate oxidase present in peroxisomes, but also two other enzymes of purine catabolism.

In mammals, in *Tetrahymena*, and in castor bean seedlings, peroxisomes play important roles in converting fat to carbohydrate (*gluconeogenesis*). In the early germination of castor bean seedlings, the fat stored in the endosperm of the seed is rapidly converted to carbohydrate (including the sugar *sucrose*) which is utilized in the growth of the plant. The conversion proceeds by a metabolic sequence known as the *glyoxylate cycle*, somewhat resembling the Krebs cycle (Fig. I-23). During germination, when fat conversion begins, the key enzymes of the cycle are synthesized. Peroxisomes isolated from the endosperm at this time contain glyoxylate cycle enzymes and are referred to as *glyoxysomes.* The enzymes are not found in most peroxisomes from other plant tissues or from animal tissues.

The green leaf cells in which peroxisomes are present are capable of photorespiration, a light-dependent metabolic sequence in which oxygen is consumed. This sequence is stimulated by high light intensity, low CO_2 concentrations, and high O_2 concentrations. Under these conditions, excess molecules of the 2-carbon compound *glycolate*, a major metabolic product of the chloroplasts, apparently pass out of the chloroplasts. The glycolate gains access to the peroxisomes and is oxidized. Light-dependent glycolate production and oxygen-dependent glycolate oxidation comprise *photorespiration*, which thus results from joint action of chloroplasts and peroxisomes. Perhaps the close spatial associations observed between the two types of organelles is related to this metabolic interaction. In addition, metabolic interactions between peroxisomes and mitochondria are known and may also be facilitated by close associations. Clearly, it must be remembered that organelles move around within the cell; the close association of one type of organelle with another could conceivably be a matter of chance. However this is unlikely to be the sole explanation for all such associations as those between peroxisomes and other organelles.

c h a p t e r **2.10**

CENTRIOLES, CILIA, AND FLAGELLA

Structures with the characteristic morphology of centrioles occur in most animal cells and in some plant cells. They participate in cell division as will be discussed in Chapter 4.2. Of their other roles, the best established is their relationship to *cilia* and *flagella*, the long motile structures that extend from the surface of many unicellular organisms, from the sperm of a wide variety of plants and animals, and from other cell types of multicellular organisms. Cilia and flagella invariably form in association with *basal bodies*, organelles widely regarded as alternate forms of centrioles.

2.10.1   **STRUCTURE OF CENTRIOLES**   Centrioles usually occur in pairs. In typical nondividing cells, there is one pair per cell (Fig. II-55). Often this pair is located close to the Golgi apparatus. Characteristically, the centrioles are cylindrical structures approximately 0.15 μ in diameter and 0.3–0.5 μ

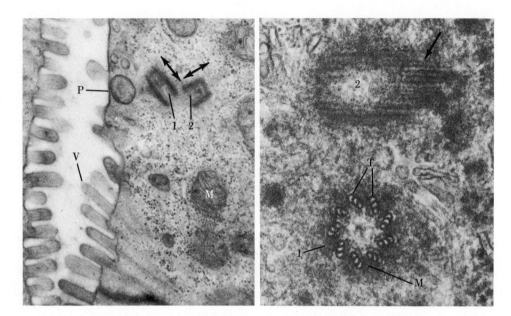

Fig. II-55 *Portion of the absorptive surface of a cell of the intestine (chick embryo) showing* centrioles. *Both centrioles of the pair (1, 2) have been cut longitudinally. The long axes of the centrioles (arrows) are almost perpendicular to one another. (M), indicates a mitochondrion, (P), the plasma membrane and (V), microvilli. × 35,000. Courtesy of S. Sorokin and D. W. Fawcett.*

Fig. II-56 *Portion of a cultured cell from the Chinese hamster showing a* centriole pair *sectioned at a different angle from Fig. II-55. Centriole 1 is cut transversely; it shows the characteristic arrangement of tubules (T) as 9 triplets embedded in a cylinder of dense matrix (M) surrounding a less dense central region. Centriole 2 is cut obliquely thus obscuring the pattern of tubules. Part of one tubule cut in longitudinal section is seen at the arrow. Approx. × 70,000. Courtesy of B. R. Brinkley, and E. Stubblefield.*

long. This is just at the limit of resolution of the light microscope, so that little was learned of their detailed structure until electron microscopy developed.

Centrioles have a characteristic appearance in the electron microscope. The two members of the pair often lie with their long axes perpendicular to each other (Fig. II-55). Each is made of nine sets of tubulelike structures, arranged as the walls of a cylinder; each set is a *triplet* composed of three closely associated tubular elements (Fig. II-56). These triplets are embedded in an amorphous matrix. At one end of the centriole, a "cartwheel" structure is often found (Fig. II-59). The presence of this structure at only one end defines a polarity of the centriole which will be outlined later (Section 3.3.3).

2.10.2   ***STRUCTURE OF***   In typical ciliated cells, each cell contains
            CILIA AND   large numbers of cilia about $2-10\ \mu$ long.
            FLAGELLA   In flagellated cells, there are usually only
one or two flagella per cell; they are often
as long as $100-200\ \mu$. The diameters of both cilia and flagella usually
are less than $0.5\ \mu$.

Cilia and flagella are of similar structure. They resemble centrioles in having nine sets of tubules arranged in a cylinder $0.15-0.2\ \mu$
in diameter (Fig. II-57). Unlike centrioles, an additional pair of tubules
is found in the center of the cylinder. Also, the peripheral sets are
"doublets" of two tubular elements each, rather than the triplets of
centrioles; "arms" often extend from the doublets as shown in Fig.
II-57. The pattern of cilia and flagella is thus "9 + 2" instead of "9 + 0"
like that of the centrioles. In addition, whereas centrioles lie within
the cytoplasm and have no surrounding membrane, mature cilia and
flagella are bounded by a membrane that is an extension of the plasma
membrane.

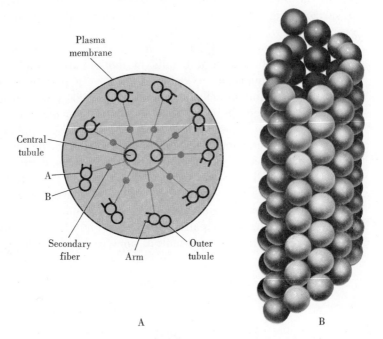

Fig. II-57   *A. Diagrammatic representation of a cross section of a cilium or
flagellum. After I. R. Gibbons and others. The A and B tubules of each doublet
are distinguishable by the presence of arms, on the A tubule. In addition to
the major pattern of tubules, a system of fine secondary fibers is often noted.
B. Interpretation of the structure of a microtubule as an assembly of globular
subunits. After J. Randall and others.*

2.10.3   *CILIA AND FLAGELLA; FUNCTIONAL ASPECTS*

Cilia beat (Fig. II-58) by producing an effective "downstroke" followed by recovery, that is, by a return to the original position; the rate of motion is rapid enough that the tip may attain velocities of several thousand microns per minute. Flagella move with an undulating snakelike motion. Sometimes the beat has helical or circular patterns. In protozoa and sperm cells, cilia or flagella endow the cells with motility. In cells lining the gills of molluscs, their movement induces circulation of water past the gills. In ciliated epithelia such as those lining the trachea of mammals, collective movement of several hundred cilia on each cell provides a strong upward current that helps carry mucus and trapped dust out of the lungs. (Cigarette smoking inhibits normal ciliary action in the respiratory tract and thus affects the normal mechanisms for removal of foreign matter.)

Cilia typically beat in coordinated waves, so that at any given moment some cilia of a cell are in the downstroke of the cycle and others are in the recovery position. This assures a steady flow of fluid past a surface (Fig. II-58). The limited available information on ciliary coordination will be presented in Section 3.3.3.

Although cilia and flagella are relatively simple in structure, the relations of structural details to motion are complex, and the significance of many observations is elusive. In part because certain specialized, nonmotile cilia (Section 3.8.1) lack the central pair of tubules, it has been proposed that the central pair plays a special role in motility. Initial observations were interpreted as indicating that almost all cilia and flagella bend at right angles to the plane defined by the central pair. But careful reevaluation casts doubt on this; at least in some cells bending may actually be at an angle to this plane. Mutants of a few unicellular organisms, and sperm or other cells of some multicellular forms may have a central tubule arrangement quite different from the usual pattern shown in Fig. II-57. In some cases, a single modified tubule is present; in some mutants the central tubules

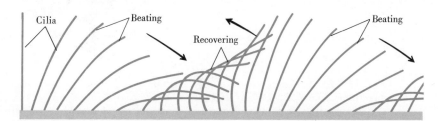

Fig. II-58 *The cilia on the surface of a cell beat in coordinated waves such that at any given moment some are in the effective ("downstroke") portion of the cycle and some are recovering.*

are absent. The effects of these differences in structure on the pattern of motion of the organelle are being intensively studied for clues to the roles of the central tubules: many of the deviant forms are nonmotile, but it is already clear that the "normal" pattern of two central tubules is not an absolute requirement for motion.

In sperm tails, the pattern of motion may be influenced by additional fibers that run down the tail alongside the doublets (see Fig. III-32).

Does contraction of the tubules of cilia and flagella provide the force for motion? If the tubules could shorten by 10 percent of their length, much of the known pattern of motion could be accounted for. But there is no strong evidence for contraction of the tubules. Careful observations on cilia during bending indicate that there may be a cycle of tubule sliding along the length of the cilium; for example, at some stages during the cycle of bending one can find a short zone near the very tip of a cilium where only eight peripheral tubule sets are present, as if the ninth set had temporarily slid down a short distance. Such longitudinal movement of some tubules with respect to others could conceivably give rise to bending or it could be a result of bending. (In the next chapter, it will be shown that sliding of one set of components with respect to another is involved in the contraction of muscle.)

Since cilia or flagella that have been detached and isolated from the cell can beat if ATP is added to the suspending medium, it is likely that ATP provides the energy used in motility. This observation also indicates that cilia and flagella are not passive organelles moved primarily by the action of the rest of the cell. Beyond this, there is little known about the chemical events underlying ciliary or flagellar motion. As outlined in the next chapter and also in Section 4.1.4, the proteins of the organelles are now being studied, and this should soon provide much of the needed information. It is well to remember that, while the tubule systems are the most obvious structural features of cilia and flagella, they are surrounded by a considerable amount of material that lacks obvious structure. This material may contain molecules important to motion, perhaps through their interaction with the tubules.

2.10.4   **BASAL BODIES;** A basal body is present at the base of each
         THE FORMATION cilium or flagellum (Fig. II-59). This is true
         OF CILIA AND from the onset of formation of a given cilium
         FLAGELLA or flagellum. Often fibers (*rootlets*) extend
         from the basal body deep into the cytoplasm, perhaps as part of a system that anchors the basal bodies (Figs. II-59 and III-22A).

Each basal body has the same $9 + 0$ structure as a centriole. The nine peripheral doublets of the associated cilium or flagellum appear almost as extensions of two of the three tubules of basal bodies triplets. That basal bodies and centrioles are closely related, probably as alternate forms of the same organelle, is indicated by several lines of evidence. (The distinction between the two is functional and derives from the preelectron microscope era when the identity in structure could not be known.) For example, in some developing sperm cells, a flagellum grows in association with a centriole during division, while the centriole is still in its customary position for dividing cells, at the poles of the mitotic spindle. (Section 4.2.6 will show that centrioles are often involved in formation of the spindle.) Thus the same body can function simultaneously, as both a centriole (in cell division) and a basal body (in flagellum formation).

It has been proposed that at least some of the growth of cilia and flagella depends upon the assembly of structural subunits which are "preformed," in the sense that subunit synthesis is a separable process from subsequent assembly into structures. If the two flagella of the unicellular organism *Chlamydomonas* are broken off by mechanical agitation or other means, new ones regenerate. Preventing protein synthesis by the addition of inhibitors (such as *cycloheximide*) after amputation does not prevent regeneration, although the regenerates do not reach full length. If only one flagellum is removed, the other shortens for a while as regeneration occurs, then both return to normal length. When the numerous cilia at the surface of sea urchin embryos are removed, regeneration occurs even in the presence of inhibitors of protein synthesis.

Presumably, free subunits are usually present in the cells studied in these experiments and are available for assembly. Thus, although the subunits are presumed to be chiefly protein, newly made proteins are not needed for regeneration of considerable lengths of cilia and flagella, and the inhibition of protein synthesis does not prevent regeneration. The experiment in which one flagellum is removed is interpreted as indicating that subunits can be freed from one flagellum to support the growth of another.

Apparently, the basal bodies in some way control assembly of ciliary and flagellar subunits. Of interest is the fact that the $9 + 0$ structure of the basal body can control the formation of the somewhat different $9 + 2$ structure. Precisely what the basal body does is not known. In one investigation cells were fed radioactive amino acids at different points during regeneration of flagella and were then studied by autoradiography; results of the study indicated that as a flagellum grows it adds much of its new protein (identifiable by radioactivity) at the tip rather than at the base. A disconcerting possible implication is that the assembly of new flagellar structure occurs at that end of the organelle opposite to the region attached to the basal body. However, techniques are not precise enough at present to support the conclusion that *all* the protein is added from the tip. It is still possible, for example, that the tubules grow from the base while the plasma membrane or the apparently unstructured material is added to at the tip.

2.10.5   **CENTRIOLE** The study of centriole duplication faces **DUPLICATION** two main difficulties. First, centriole chemistry is poorly understood; the organelles are too small for adequate cytochemical study, and they have not yet been isolated in a form pure enough to permit biochemical study. Second, the mode of formation may vary from cell to cell. When ciliated protozoa, such as *Paramecium*, divide by binary fission, the precise pattern of basal bodies near the surface (Fig. III-10) is transmitted to both daughter cells. Probably this involves the duplication of each basal body in coordination with cell division. In other cases, however, basal bodies appear at one point in the cell life cycle despite the previous absence of microscopically visible centrioles or basal bodies. This is true of the unicellular organism *Nagelaria* (Fig. II-59). This organism may exist in a nonflagellated form in which centrioles, basal bodies, or flagella have not been found, or it may become a flagellated cell with two or more flagella and basal bodies. This sudden appearance of basal bodies in cells that previously lacked visible centrioles or basal bodies also occurs in the sperm of some plants. It remains to be determined whether some self-duplicating structure is present that gives rise to basal bodies; it is possible that such structures have not been recognized in the microscope because of their small size or lack of a distinctive appearance.

Despite these difficulties, progress has been made. In the ciliate *Tetrahymena*, DNA has been shown cytochemically to be present in the cortex where the basal bodies are aligned. (The arrangement is similar to that shown in Fig. III-10.) Some believe that RNA is also present. Technical problems have hindered detailed biochemical analysis of these presumed basal body nucleic acids, and there still are questions as to the validity of the procedures used in the cytochemical studies. However, it appears quite possible that basal bodies do have at least some hereditary machinery partly independent of the DNA in the nucleus (see also Section 3.3.3).

In many cells, centrioles appear to duplicate by the growth of a new *procentriole* near the "old" centriole. The procentriole contains the $9 + 0$ pattern but is much shorter than the "old" centriole.

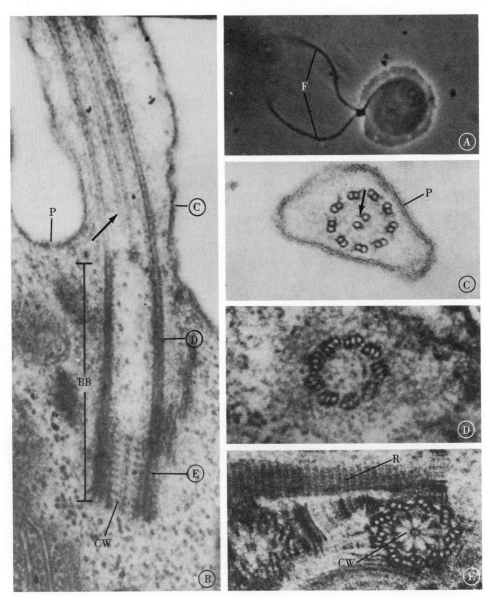

Fig. II-59 *Flagella and basal bodies of the unicellular organism* Nagelaria
*during its flagellate phase. **A** is a phase contrast micrograph; the two flagella
are seen at (F). **B** to **E** are electron micrographs. **B** is a longitudinal section
and **C**, **D** and **E** are transverse sections at the levels indicated by the corresponding letters in **B**. (P) indicates the plasma membrane, (CW) the "cartwheel" structure at one end of the basal body (BB) and the arrows, the central
pair of tubules of the flagellum. In **E** a cross-striated rootlet (R) is seen. Such
structures are often found extending from basal bodies into the cytoplasm;
perhaps they help to anchor the basal body. A ×2,000. B–E approx. ×90,000.
Courtesy of A. D. Dingle and C. Fulton.*

It is characteristically oriented at right angles to the "old" centriole (Fig. II-60) and eventually grows into a complete centriole. This occurs in cells where centriole duplication is part of cell division, as well as in some cells where many basal bodies form. In some of the cells forming basal bodies, several procentrioles appear more or less simultaneously, radially arranged around the "old" centriole. Procentriole formation appears to involve neither fission nor direct budding from preexisting centrioles; how it does occur remains to be explained.

In cells of the respiratory and reproductive tracts of mammals, only two centrioles are present early in development; eventually large numbers of basal bodies are formed and produce the cilia that line the tracts. In the respiratory system, some of the basal bodies have been seen to form near the centrioles but many arise from small aggregations of amorphous or finely fibrillar material with no obvious resemblances to centrioles. Although these aggregates are considered by some to be "generative" structures, originally formed by centrioles, it is possible that they arise from some unrecognizable precursor material that originates elsewhere. In the cells that develop into the sperm of certain ferns and cycads, large compound bodies are formed from an unknown source and then fragment to produce numerous procentrioles which ultimately give rise to the basal bodies of the many flagella found on each cell.

"Self-duplication," in the sense that an old structure gives rise to a new one, more or less directly, may well be involved in the instances where procentrioles form near "old" centrioles. But how formation of centrioles occurs in cells initially without centrioles is unclear. Is some special self-duplicating structure present that may,

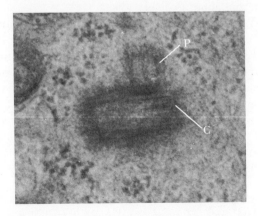

Fig. II-60 *Portion of a human cultured cell* (HeLa, *see Chapter 3.11*) *showing a centriole (C) and an associated "procentriole" (P). It is thought that the procentriole grows at right angles to the centrioles eventually producing a configuration resembling Fig. II-55. × 64,000. Courtesy of E. Robbins.*

in some circumstances, produce centrioles but, under other circumstances, does not do so?

c h a p t e r   **2.11**

MICROTUBULES AND MICROFILAMENTS

2.11.1   MICROTUBULES As preparative methods for electron microscopy improved, microtubules became evident as a regular component of the cytoplasm of most cells. Microtubules are best preserved by fixatives like glutaraldehyde or formaldehyde and only in the past few years have electron microscopists begun to use these agents to treat tissue before it is exposed to osmium tetroxide (Chapter 1.2) which, used alone, may disrupt the tubules. Microtubules are extremely thin cylinders, approximately 200–300 Å in diameter (Figs. II-61, II-62, IV-13); it is not known whether there are any limits to possible lengths since beyond several microns the measurements are very difficult to make in the thin section used for electron microscopy. They are not membrane delimited. From images such as Fig. II-61, it is concluded that the microtubule wall is made of some form of repeated basic structural building block. In cross section, a microtubule has the appearance of a circular array of about a dozen subunits (13 are commonly reported). In longitudinal view, the walls sometimes appear to be made of parallel filaments, again numbering about twelve.

Varied interpretations of these observations on microtubule substructure have been made. One interesting and widely accepted model suggests that the wall is actually composed of globular subunits approximately 50 Å in diameter, roughly the dimensions of a protein molecule of moderate size. Supposedly, the subunits are helically arranged around the wall as shown in Fig. II-57; at the same time, they fall into rows running parallel to the tubule long axis. The longitudinal rows of subunits are probably the structures that appear as filaments in the wall.

Microtubules are prominent in the elongate processes of many cells such as the axons or dendrites of nerve cells (see Fig. III-23) or the rigid, modified pseudopods of some protozoa (Fig. II-62). The red blood cells of some fish have a flattened disc shape; microtubules are arranged in a circular band just under the edge of the disc. A similar distribution is found in *blood platelets*, cell fragments that function in blood clotting. In many cells, microtubules are numerous

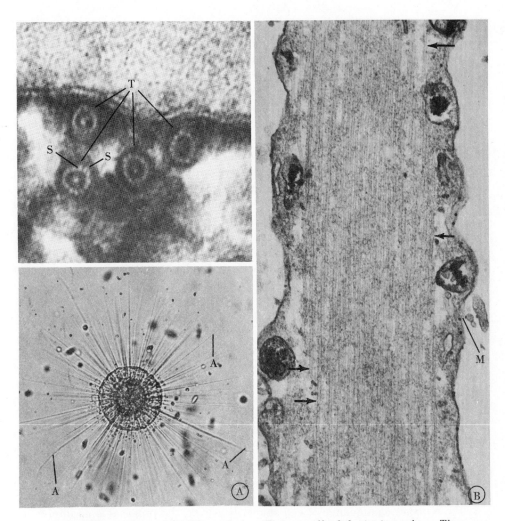

Fig. II-61 *(Upper left)* Microtubules *(T)* in a cell of the juniper plant. The tubules are embedded in an electron-dense matrix to produce a natural negative "staining" in which they appear light. This unusual situation permits direct observation of the subunits *(S)* within the tubule wall. *(The tubules are seen in cross-section.)* ×315,000. Courtesy of M. Ledbetter.

Fig. II-62 *A.* The protozoan, Actinospherium, *shows many modified pseudopods (axopods, A) that extend radially from the body of the cell.* × 100. *B.* A longitudinal section of an axopod. Its membrane (M) is part of the plasma membrane. Numerous microtubules (arrows indicate four) are found longitudinally arranged inside the axopod. Transverse sections would show many circular outlines (see Figs. I-9, II-61, III-29).* × 45,000. *Courtesy of L. G. Tilney and K. R. Porter.*

in structures that have a firmer, more gel-like consistency than the rest of the cytoplasm; perhaps the most universal of these structures is the mitotic spindle (see Fig. IV-13). With the exception of features such as the arrangement in doublets, the tubules of cilia and flagella seem quite similar to other microtubules (see Chapter 2.10).

Such information tends to support the proposal of two major interrelated roles for microtubules: maintenance or control of cell shape (a "cytoskeletal" function) and participation in cellular motion (for example, motion of cilia and flagella and movement of chromosomes by the spindle). A third possibility, that microtubules act in some sense as pipes or channels along which material may be transported, remains an interesting speculation with little evidence to support it at the level of transport along individual tubules. (However, there is some reason to believe that *groups* of tubules may define channels of oriented flow within the cell. For example, rapid changes in color of fish can result from extensive intracellular migration of pigment granules along pathways delineated by microtubules within *melanophores* and other pigment cells at the surface of the fish. The pigment can spread throughout the cell or reversibly be concentrated in a small area in response to environmental factors).

High pressure, low temperature, and several chemicals (such as the drug *colchicine*) can be used to dissolve microtubules and to disrupt structures in which microtubules are numerous. For example, the special protozoon pseudopods shown in Fig. II-62 retract into the cell, and their microtubules disappear as pressure is elevated or temperature lowered. Both pseudopods and microtubules reappear when the cells are returned to normal conditions. This experiment and other similar ones are the bases for the suggestion that microtubules can play a controlling role in the formation of asymmetric cell shapes. Microtubules are hardly the only relevant factors; for example, attachments and other interactions among neighboring cells and the cell walls of plants and bacteria (as Part 3 will outline) are important in the maintenance of cell shape. However, the proposal is an interesting one and currently under active study. If, as is thought likely, microtubules arise by relatively simple processes that assemble many identical protein subunits (Section 4.1.4), then the reversible assembly and disassembly of the tubules could provide a flexible control over cell shape. Still to be determined are the nature and mechanisms of the intracellular factors that normally influence assembly and distribution of microtubules.

Much remains unclear about possible microtubule roles in motion. For example, a layer of microtubules is present near the cell surface of many plant cells, and the constant flow (*streaming*) of cytoplasm characteristic of many plant cells (Fig. II-65 and Section 3.4.1) is especially prominent close to this region. The streaming may occur in directions that parallel the long axes of the microtubules. However, this does not establish a causal relationship between microtubules and streaming. It is possible that the microtubules do not cause streaming, but rather that the tubules align in response to streaming or, conversely, that they orient and direct the flow of cytoplasm that is moving in response to other influences. Another possibility is that the microtubules provide a mechanical framework which transmits or responds to forces generated elsewhere. There are those who believe that the tubules in cilia and flagella are relatively rigid but that they interact with each other and with other surrounding material. Section 4.2.7 will outline an interesting proposal that the spindle moves the chromosomes attached to it by processes dependent on the rapid assembly and disassembly of structures in which microtubules are involved. The contraction of tubular structures consisting of repeating subunits can occur when the subunits are rearranged to form a shorter, wider tubule. This has been demonstrated for the shortening of the "tails" of bacterial viruses (Section 3.1.1) and is conceivable in systems of cytoplasmic tubules although such shortening has yet to be seen.

Although it is likely that some microtubules in different cells have different functions, it is less clear whether they differ substantially in chemistry. Only a few systems are suitable for the isolation of microtubule components; the mitotic spindle and some cilia and sperm flagella have been studied most successfully. There is growing belief that the tubules in each of these systems are composed largely of one type of protein subunit; the subunit proteins isolated thus far have been as small as 50,000–60,000 in molecular weight. This size agrees well with the dimensions of the globular subunits seen by microscopy. Further analysis is needed of the proteins from spindles, cilia, flagella, and other systems, to determine how the protein subunits are linked together in the tubules. (A successful test-tube reconstitution experiment is outlined in Section 4.1.4.) The extent to which the proteins differ from one microtubule to another remains also a matter for further study.

Another matter for further study is the existence of bridges between adjacent microtubules; thin bridges have recently been noted in some situations where microtubule interactions are thought to be important for control of structure or for motion. Also, there are intriguing hints of slight differences among microtubule subunits; for example, tubules appear to vary somewhat in the ease with which they may be experimentally dissolved, and some investigators report slight differences in composition of the proteins when the two tubules of ciliary doublets are compared.

Microtubules and structures composed of microtubules are often associated with centrioles or basal bodies. Sometimes the tubules appear to radiate from electron-dense "satellite" bodies that lie close to the centrioles but have no marked substructure. Association of tubules and centrioles is clear in several of the specialized structures already mentioned (cilia, flagella, and spindle); it is found also in nondividing, nonciliated cells. In such cells, the centrioles often lie close to the Golgi apparatus, and many microtubules may be seen nearby; the area is somewhat more gel-like than the rest of the cell. Such observations have led to the suggestion that centrioles might be the chief source of cellular microtubules. If this notion is correct, however, the abundance of microtubules in plant cells having no visible centrioles remains unexplained.

As mentioned above, the drug *colchicine* can be used to disrupt structures in which microtubules are prominent; it also prevents regeneration of amputated flagella probably by inhibiting the assembly of subunits (Section 2.10.4). The ability to bind colchicine has been shown to be a relatively unique characteristic of microtubule proteins isolated from several different kinds of structures. This binding ability is proving useful in facilitating the identification of microtubule subunits and in studying the synthesis and assembly of such subunits.

2.11.2   ***MICRO-***   Cells often contain fine cytoplasmic fila-
          FILAMENTS   ments that are widely assumed to be pro-
                             teins. It is likely that these filaments differ in composition and function from cell to cell, as differences in thickness are readily observable. Like microtubules, filaments are considered to function in the maintenance of structure and possibly in cell motion.

Microvilli of absorptive cells in the intestine of vertebrates often have a core of filaments that are 40–50 Å in diameter and arranged parallel to their long axis. These filaments are probably involved in stabilizing the structure of the microvillus (see Fig. III-18). They are extensions of a zone of packed filaments at the base of microvilli, the *terminal web*, which may provide rigidity to this region of the cell. In skin cells and other *epithelial* (Section 3.5.1) cells, bundles of fine filaments, known as *tonofilaments*, are present; often the bundles terminate at or near special regions of plasma membrane (desmosomes) that anchor one cell to another (Fig. III-18).

2.11.3   ***FILAMENTS***   The association of filaments with motion
        AND MOTION: is most evident in striated muscle (Figures
           STRIATED   II-63, II-64). The striated muscle fiber con-
            MUSCLE   tains within it numerous elongate myo-
                             fibrils, which are made of regularly arranged fine filaments and which form the basis of muscle contraction. Each

myofibril may be regarded as being composed of repeating contractile units or *sarcomeres* (Fig. II-63). Within the sarcomere, there is a regular pattern of thick (100 Å) and thin (60 Å) filaments as shown in Fig. II-64. This alignment of filaments gives these muscle fibers their cross-banded (striated) appearance; the zones of thin filaments (I-bands) can readily be distinguished from those of thick filaments (A-bands) by light microscopy. (The individual filaments, however, are below the limit of resolution of the light microscope.)

The polarizing microscope (Section 1.2.2) shows that a regular organization of longitudinal elements is present in living muscle and in preparations fixed for electron microscopy. The interference and phase microscopes can be used for measuring the amount of material present in different regions of the muscle cell. The comparison of muscle fibers in their normal state with fibers that have been treated with solutions that extract the muscle protein *myosin* shows that only the A-bands lose mass. This study and similar ones demonstrate that myosin is found in thick filaments and that another muscle protein, *actin*, is present in thin filaments.

Figure II-63 diagrams a widely accepted theory about the arrangement of protein molecules in the filaments. According to this theory, muscle contraction takes place when the two sets of thin filaments of each sarcomere slide toward the center, past the central set of thick filaments. The sliding of filaments shortens the sarcomere. If the muscle is stretched and held at fixed length, the extent of the overlap of thick and thin filaments is reduced, and the force the muscle can generate is correspondingly reduced. This indicates that the interaction of the two sets of filaments is responsible for the force of contraction.

Although many details of muscle contraction have been explained some still remain obscure. How the sliding itself is accomplished, for example, has been attributed to the action of bridges between thick and thin filaments (Fig. II-64) which might serve to pull filaments past one another. As Fig. II-63 indicates, the myosin molecules in one-half of the thick filament appear to be oriented with opposite polarity to the molecules in the other half. Presumably this reflects the fact that the two halves of the thick filament "pull" thin filaments "in" toward the center; the thin filaments associated with one end of a thick filament move in the opposite direction to those associated with the other end. It is known that myosin can split ATP and that actin has ATP or derivatives bound to it. However, the mechanism by which the chemical energy of ATP is converted into the mechanical energy of muscle contraction remains unclear.

Models have been proposed that translate filament interaction into molecular terms. It is proposed that the bridges are parts of thick filaments that protrude from the filaments in a helical arrange-

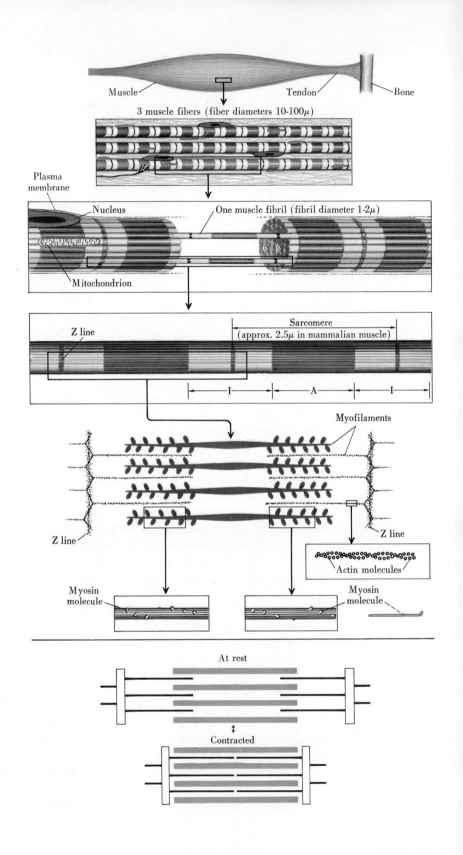

Muscle — Tendon — Bone

3 muscle fibers (fiber diameters 10-100μ)

Plasma membrane

Nucleus — One muscle fibril (fibril diameter 1-2μ)

Mitochondrion

Z line — Sarcomere (approx. 2.5μ in mammalian muscle)

I — A — I

Myofilaments

Z line — Z line

Actin molecules

Myosin molecule — Myosin molecule

At rest

⇕

Contracted

◄ *Fig. II-63* *The structure of vertebrate striated muscle (see also Fig. III-28). The unit comparable to a more conventional cell is a fiber; it is surrounded by a single plasma membrane but contains several to many nuclei. Within the fiber, myofibrils are present; these consist of filaments which are ordered in such fashion as to produce the appearance of alternating A and I bands along the fibrils. Each filament is of several hundred protein molecules. The diagram illustrates the probable arrangement of* myosin *molecules in the thick filaments and of* actin *molecules in the thin filaments; there is some evidence that a second protein,* tropomyosin *also is present in the thin filaments. Many mitochondria are present throughout the fiber; they are found between adjacent myofibrils.*

The region between 2 successive Z-lines along a myofibril is referred to as a sarcomere; *The sarcomere can be thought of as a repeating unit of function. The sliding of filaments within a sarcomere during contraction is illustrated at the bottom of the diagram. Modified from Loewy and Siekevitz; after the work of H. Huxley and others.*

ment (Fig. II-63). Studies of isolated myosin molecules indicate that each molecule has a portion referred to as the "heavy meromyosin portion," which contains sites that can bind to isolated actin molecules. The heavy meromyosin portion is able to split ATP and presumably it participates in the bridges between actin and myosin filaments.

2.11.4 OTHER MOTILE SYSTEMS At present research is beginning to describe in some detail how a long molecule in solution can reversibly contract into a shorter configuration. If the molecule has many positively or negatively charged groups along its length, the similarly charged groups repel

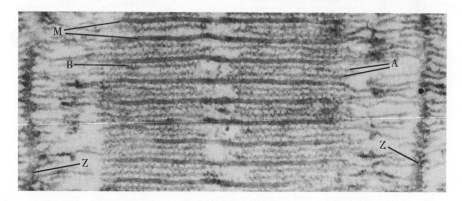

Fig. II-64 *A* sarcomere *from a rabbit muscle. The thin filaments (A) extend inward from the Z line; in the center they overlap with the thick filaments (M). Indications of fine bridges between thin and thick filaments are seen in the zone of overlap (B). × 100,000. Courtesy of H. Huxley.*

each other and tend to keep the molecule stretched out, preventing it from coiling into a more random configuration. If the charges are neutralized, the molecule folds up spontaneously by the random molecular motion that characterizes all molecules in solution. It is a long step from this simple system to the contraction of a muscle and the beating of a flagellum, which result from the joint activity of many macromolecules. Many of the relevant structures and some important processes of motile cell systems have been described, and relatively simple changes in macromolecules are central aspects of these processes. However, motion of both cells and organelles remains only partially understood in molecular terms; the most fully comprehended of the many kinds of motion is the contraction of striated muscle.

In most cells, many organelles are probably in constant motion, both random and nonrandom. Among other functions, this speeds the transport of molecules from one region to another, and probably facilitates exchanges among organelles. In some cases, the motion is highly oriented; for example, vesicles formed by pinocytosis at the surface of cells tend to move toward the region of Golgi apparatus. In Section 3.4.1, consideration will be given to the highly oriented streaming of plant cell cytoplasm (Fig. II-65).

Ameboid motion, involving the extension of pseudopods, characterizes amebae, white blood cells, and many tissue culture cells. (Amebae can move at rates greater than 100 μ per minute.) In pseu-

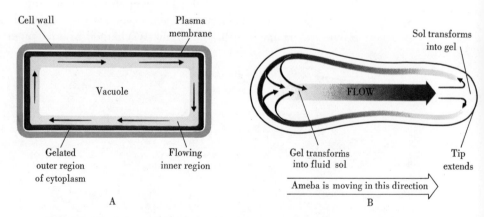

Fig. II-65 *Two important types of cytoplasmic motion.* **A.** *Cyclosis in a plant cell—the rapid streaming of cytoplasm around a central vacuole.* **B.** *One type of* ameboid motion. *The extension of a pseudopod by an ameba is accompanied by constant flowing of cytoplasm in the direction of extension and probably by continual transformation from gelated ectoplasm or* cortex *(outer region of cytoplasm) to fluid* endoplasm *at the posterior end with the reverse transformation at the anterior end.*

dopod formation by amebae, a gel-like region (cortex) is formed at the periphery of the cell, and more fluid cytoplasm flows forward into the pseudopod as it extends (Fig. II-65). One theory holds that contraction of the cortex is responsible for motion of the inner cytoplasm. Contraction at the rear might push cytoplasm forward; contraction toward the front might exert a pull.

Microtubules and microfilaments have been thought to participate in these forms of motion and in several other forms. However, while these structures are often visible in the cells being studied, explanations of their functions are largely hypothetical. In a number of cell types, filaments 50–60 Å in diameter are present. This is the diameter of muscle actin filaments, and like actin filaments, these filaments seem to bind the purified "heavy meromyosin" fragments of myosin molecules. These observations are too recent for their interpretations to be clear. Microtubule proteins are similar to actin in size, amino acid composition, and other features. [For example, both have a so-called *bound nucleotide* attached; in actin, this is ATP or its derivatives, while in microtubule proteins, it is GTP (guanosine triphosphate) or its derivatives.] However, while such similarities are often of interest from the viewpoint of evolution (Section 4.5.1), they need not point to identity or close similarity in function.

Proteins similar to myosin are also found sometimes in motile cells other than muscle. The search for proteins resembling those of striated muscle is based on the hope that the other types of motion may be similar in basic mechanism to the contraction of muscle. This is, however, only a hope; it may well be that protein interactions and other basic mechanisms will vary in different forms of motility.

Thus, microtubules and microfilaments are a potential cast of characters for oriented motion, but the play has yet to be translated into a language we understand.

chapter **2.12**

THE HEPATOCYTE: A CENSUS

This is a convenient point to return to the hepatocyte for a summary of its content of major cytoplasmic organelles.

It is difficult to obtain even moderately precise quantitative information about the numbers and volumes of organelles in a cell as large and complex as the hepatocyte. Estimates of the relative numbers of different organelles can be made from light microscope preparations such as those shown in Figs. I-10 through I-18, but they are limited by several factors, such as the resolving power of the light microscope (Section 1.2.1). Sections for the electron microscope are thin and a given section often includes only a small portion of a given structure. Organelles rarely have simple shapes; thus treatment by the usual geometric formulae is imprecise. A section may be regarded as a sample of the cell, the contents of which depend on the size, number, shape and distribution of organelles and on the section thickness and angle. Statistical formulae are often used to relate the information from sections to the cell as a whole.

For organelles such as the lysosomes, the variety of morphological types complicates estimates; some types may be difficult to identify in the microscope. Isolated cell fractions are of limited use, unless extreme care is taken to insure that the fraction is reasonably pure and unless it is possible to determine what proportion of the total cell content of the organelle is present in the fraction studied. Sometimes this proportion can be approximated by comparing the fraction with the unfractionated cell in terms of the amounts of characteristic components, such as the mitochondrial respiratory enzymes or nuclear DNA.

Even the cells of a single type may vary considerably within an organ. Within a hepatic lobule (see Chap. 1.1), for example, the hepatocytes near the periphery are exposed to blood that is richer in oxygen and nutrients than the cells near the central vein; by the time the blood reaches the center of the lobule it has undergone considerable exchange of material with the peripheral cells. Probably related to this is the fact that mitochondria, lysosomes, ER, and glycogen differ in both size and amount when peripheral and central cells are compared. For such reasons, the estimates that have been made are very rough and can convey only an overall impression of an "average hepatocyte."

A rat hepatocyte averaging 20 μ in diameter has a cytoplasmic volume of roughly 5000 cubic microns (μ^3) and a nuclear volume that is 5–10 percent of this. Of the cytoplasmic volume, mitochondria normally occupy 15–20 percent, peroxisomes 1–2 percent, and lysosomes 0.2–0.5 percent. (This excludes the small Golgi vesicles that are probably primary lysosomes, but whose number and volume is difficult to estimate.) Glycogen varies greatly in amount but may occupy 5–10 percent or more of the cytoplasmic volume. The volume of the ER is virtually impossible to estimate because of its very irregular shape, but the areas of its membranes may approach 25,000 μ^2 for rough ER and 15,000–20,000 μ^2 for smooth ER. (These last figures probably include some of the Golgi apparatus membranes; the Golgi apparatus is sometimes said to occupy 5–15 percent as much of the cell as the ER, but the diversity of components of the apparatus makes such estimates uncertain.)

The number of mitochondria per cell is usually thought to be about 1000–2000. A few hundred peroxisomes (perhaps 500) are present per cell and, excluding Golgi vesicles, there are possibly one-quarter to one-half as many lysosomes as there are peroxisomes. There are several million ribosomes, one pair of centrioles, and a large but unknown number of microtubules and microfilaments.

FURTHER READING

Fawcett, F., *The Cell, An Atlas of Fine Structure*. Philadelphia: Saunders, 1966. 448 pp. An extensive collection of electron micrographs of organelles with explanatory material.

Gibor, A., "Acetabularia: A Useful Giant Cell," *Scientific American*, Nov. 1966, 215:5, p. 118.

Goldstein, ed., *Cell Biology*, Dubuque: Wm. Brown and Co., 1966. 212 pp. A collection of accounts of important experiments on macromolecule synthesis, cell division, and other problems.

Haggis, G. H., *The Electron Microscope in Molecular Biology*. New York: Wiley, 1967. 84 pp. A short collection of electron micrographs; some stress on microscopic analyses in terms of molecular structure.

Haynes, R. H. and Hanawalt, P. C., ed., *The Molecular Basis of Life*. San Francisco: W. H. Freeman and Co., 1968. 368 pp. A collection of *Scientific American* articles on molecular biology.

Jensen, W. A. and Park, R. B., *Cell Ultrastructure*. Belmont: Wadsworth, 1967. 60 pp. A short collection of electron micrographs with some stress on plant material; there is a useful appendix on methods.

Levine, R. P., "The Mechanism of Photosynthesis," *Scientific American*, Dec. 1969, 221:6, p. 58.

Neutra, M. and Leblond, C. P., "The Golgi Apparatus," *Scientific American*, Feb. 1969, 220:2, p. 100.

Nomura, M., "Ribosomes," *Scientific American*, Oct. 1969, 221:4, p. 28.

Racker, E., "The Membrane of the Mitochondrion," *Scientific American*, Feb. 1968, 218:2, p. 32.

Watson, J. D., *The Molecular Biology of the Gene*. New York: W. A. Benjamin, 1965. 494 pp. A comprehensive introduction to modern molecular biology that also covers much basic biochemistry.

Additional *Scientific American* articles relating to organelles are in the Living Cell collection (see references at the end of Part I).

ADVANCED READING

Allen, J. M., ed., *Molecular Organization and Biological Function*. New York: Harper and Row, 1967. 243 pp. A collection of articles reviewing several topics including, among others, protein synthesis, visual reception and the structure and function of cilia and flagella and chloroplasts.

Annals of the New York Academy of Sciences, 168: 209–381. *The Nature and Function of Peroxisomes.* A collection of articles covering many areas of current research on peroxisomes. (December 1969).

Branton, D. and Park, R. B., eds., *Papers on Biological Membrane Structure.* Boston: Little Brown and Co., 1968. 311 pp. A collection of articles describing research on cellular membranes including some of the key early work.

Dingle, J. T. and Fell, H., eds., *Lysosomes in Biology and Pathology*, Netherlands: North Holland Co., 1969. A two volume collection of articles describing recent research on lysosomes.

DuPraw, E. J., *Cell and Molecular Biology.* New York: Academic Press, Inc., 1968. 739 pp. Useful information on the nucleus and other organelles.

Hruban, Z. and Recheigl, M., Jr., *Microbodies and Related Particles, Morphology, Biochemistry and Physiology.* New York: Academic Press, Inc., 1969. 296 pp. Particularly useful for descriptions of structure of peroxisomes in a wide variety of organisms.

Kirk, J. T. O. and Tilney-Bassett, R. A. E., *The Plastids, Their Chemistry, Growth, Structure and Inheritance.* San Francisco: W. H. Freeman and Co., 1967. 608 pp. A review of many aspects of plastid structure, function and development.

Lehninger, A. L. *The Mitochondrion.* New York: W. A. Benjamin and Co. 1964. 263 pp. In places slightly out of date but overall a good review of basic information. See also Lehninger's *Bioenergetics* referred to at the end of Part 1.

Lima-de-Faria, A., ed., *Handbook of Molecular Cytology.* Amsterdam, London: North Holland Publishing Co., 1969. 1508 pp. An excellent reference work containing 53 chapters covering all major aspects of cell structure; each chapter was contributed by a specialist in the field discussed.

Proceedings of the Royal Society of London, Series B, 173: 1030 (April 15, 1969. 111 pp. is a special discussion issue containing summaries of research on cytoplasmic organelles. Particularly useful are the contributions on peroxisomes (C. DeDuve), growth of flagella (J. Randall), and interactions of nucleus and cytoplasm (J. B. Gurdon and H. R. Woodard).

Rabinowitch, E. and Govindjee, *Photosynthesis.* New York: Wiley and Sons Inc., 1969. 273 pp. An introductory text.

Roodyn, D. B. and Wilkie, D., *The Biogenesis of Mitochondria.* London: Methuen and Co., 1968. 123 pp. A brief review of recent research on the origin, division, etc. of mitochondria.

Vincent, W. S. and Miller, O. L., *The Nucleolus, Its Structure and Function.* National Cancer Institute Monograph #23 (Dec. 1966). Bethesda: National Institutes of Health. 610 pp. A collection of articles reviewing recent research on the nucleolus.

Wolstenholme, G. E. and O'Conner, M., eds., *Principles of Biomolecular Organization.* Boston: Little Brown & Co., 1966, 491 pp. A CIBA Foundation Symposium. Many useful articles summarizing recent research at the molecular level on the structure of viruses, microtubules, mitochondria, muscle filaments and other systems.

p a r t **3**

CELL TYPES: CONSTANCY AND DIVERSITY

The diameters of known cells range from 0.1 μ for the simplest procaryotes to 1–100 μ for most protozoa, algae and cells of multi-cellular organisms; occasional specialized cells, such as some egg cells, have diameters in the millimeter or centimeter range. This diversity in size is matched by diversity in morphology and metabolism. The simplest cells are not much more complicated in structure than a mito-chondrion or a plastid. The most complex show intricate patterns of thousands of organelles.

Each cell type is characterized by a particular organization of its organelles and macromolecules. Structural differences among cell types are related to differences in function. Sometimes this relation is obvious; for example, chloroplasts are present in plant cells, which are capable of photosynthesis, and absent in animal cells, which are incapable of photosynthesis. Often, however, the relations between structure and function are more subtle. Thus, in some insect flight muscle fibers, the thick filaments extend throughout the length of the entire sarcomere, almost from Z line to Z line. The sliding of thick and thin filaments during contraction involves only short, probably oscillatory, movements producing length changes of less than 5 percent. (These muscles control the extraordinarily rapid wing motion.) By contrast, in the skeletal muscle of vertebrates where movements are slower, much greater length changes occur, often exceeding 20 percent. The thick filaments occupy only part of the sarcomere length, near the center. Presumably this is associated with the extensive filament sliding that takes place (See Section 2.11.3.)

While different organization characterizes cells doing different things, cells doing similar things generally show similar organizations. For example, most animal cells engaging in the extensive synthesis of *proteins* that are secreted at the cell surface have large nucleoli and extensive rough endoplasmic reticulum. The Golgi apparatus in these cells is large and usually located between the nucleus and the region of the cell surface where the secretion is released. In contrast, a variety of cells producing steroids have abundant smooth ER.

These similarities permit one to speak of cell types (protozoa, algae, absorptive cells, gland cells, and so forth). Findings made in the study of one cell can often be extrapolated to others of similar structure, but this must be done with caution. Although similarities often result from common evolutionary origin, evolutionary changes do not stop once a particular organization develops; consequently, cells with general features in common often differ in detail. Also, comparable organization can be achieved from very different evolutionary starting points (*convergent evolution*).

The great diversity of cell types, with different organization and functions, has provided favorable experimental materials. The experiments on nuclear function described earlier (Chapter 2.2) were possible because the alga *Acetabularia* and the protozoon *Ameba* are large unicellular organisms, relatively simple in structure, easy to maintain in the laboratory, and capable of surviving drastic microsurgery. The mitochondrial replication experiments in the fungus *Neurospora* (Section 2.6.3) were possible because mutants were available that depended entirely upon external sources of choline, an essential component of phospholipids in mitochondria. This greatly facilitated labeling. Studies of the extraordinarily large neurons of lobsters and squids have supplied much information about nerve impulse conduction. And bacteria have been used extensively in studies of heredity and biochemistry, because of their relative simplicity, their rapid growth in media of simple and readily controlled composition, and the availability of many mutants.

Partial catalogues of cell types exist. Thus, histology textbooks (*histology*: the study of tissues) may describe virtually all cell types of man or of other organisms; a few such books are listed at the end of this section. This text will not attempt any such catalogue. Rather, a number of cell types have been selected to illustrate major differences in the organization and features of procaryotes and eucaryotes, animal cells and plant cells, and specialized cells of multicellular animals. In the last group, mammalian cells have been chosen more often than others because they have been most extensively studied. Some discussion of the specialized cells of invertebrates, plants, and nonmammalian vertebrates has been included, as well as a brief consideration of the viruses. The latter are not cells and their evolutionary status in the biological world is still uncertain. There has been much controversy about whether viruses are truly "living," for they depend on cells for their reproduction, synthesis of macromolecules, and most other activities. However, because of their simplicity, viruses have been of incalculable value in the study of many important biological problems. Like the bacteria, they have been used to provide much of the present insight into the molecular mechanisms of heredity.

c h a p t e r **3.1**

VIRUSES

Viruses have none of the organelles discussed in Part 2, and, strictly speaking, they have no metabolism. From one point of view, they represent the ultimate in specialization. For their duplication, they depend entirely on their ability to enter cells of living organisms and to redirect the metabolism of the cells (by substituting their nucleic acid for the cell's DNA). Different viruses enter different cells, from bacteria to man. In extreme cases, the host cell very rapidly becomes diverted from its normal functions to the production of new viruses; in these cases, the host cell ultimately dies, releasing the viral progeny to infect other cells. Their dependence on cells for reproduction argues against the consideration of viruses as *primitive* in the evolutionary sense, that is, as the ancestors of cellular life. Perhaps the simplest notion is that they arose as a sort of parasite at the same time as, or subsequent to, the origin of cellular life; possibly some are degenerate forms which have evolved from structures that were once capable of independent life.

3.1.1 **STRUCTURE** A great variety of viruses are known (Fig. III-1). Many are extremely small, about the size of a ribosome; the largest have maximal dimensions of 0.1–0.3 μ.

Basically, viruses consist of a nucleic acid core (DNA or RNA) wrapped in a coat made of fewer than a hundred to several thousand protein molecules (see Figs. III-2 and IV-4); in the more complex viruses, some lipid and polysaccharide components are also present. The RNA viruses (for example, polio virus and type A influenza virus) are the only biological systems known in which RNA and not DNA is the hereditary material. Often the RNA is present in its usual single-stranded form, but as mentioned earlier about Reovirus (Section 2.8.2), some viruses have a core of double-stranded RNA similar in properties to DNA. The viral protein coat protects the nucleic acid during the extracellular phase of the virus life cycle. Also, some of the proteins in the coat probably bind the virus to the cell surface, prior to the entry of viruses into cells. Some viral surface proteins include

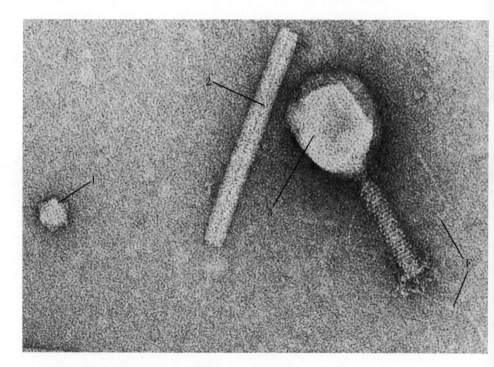

Fig. III-1 *Negatively stained preparation (Fig. II-37) showing three different viruses. 1 is ϕX-174, a single-stranded DNA virus infecting bacteria. 2 is* tobacco mosaic virus *that infects tobacco plants. 3 is the bacteriophage* T_4*; the head and tail are clearly visible but the tail fibers are difficult to preserve and only some of them are seen (F; see Fig. IV-4). $\times$ 250,000. Courtesy of F. Eiserling, W. Wood and R. S. Edgar.*

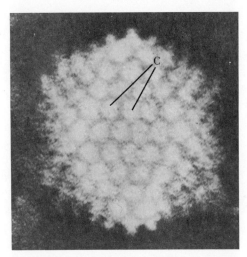

Fig. III-2   *Negatively-stained adenovirus, a DNA virus that infects human and other animal cells. The micrograph shows the capsomeres (C), protein subunits composing the "coat"; these have a polyhedral arrangement. (Careful study indicates that the coat consists of 252 capsomeres arranged as an icosahedron). × 650,000. Courtesy of R. A. Valentine and H. G. Pereira.*

enzymes that aid in penetrating surface layers of cells. For some bacterial viruses, only the nucleic acid enters the host cell. For the animal viruses, mechanisms of entry and of nucleic acid uncoating are less clearly understood and may vary for different viruses.

Figure III-1 illustrates the diversity of viral form. In the simpler viruses, the coat consists of a number of protein units, of one or a few types, geometrically arranged around the nucleic acid. In many small viruses, the protein subunits surround a DNA (or RNA) core to form a polyhedron (Fig. III-2). In the case of tobacco mosaic virus (TMV), the core of RNA is arranged as an elongate spiral and is surrounded by protein units in a helical pattern (see Fig. IV-4); the overall structure is an elongate rod. Some viruses have a much more complex structure involving several different kinds of protein. For example, T_4, one of the *bacteriophages* (viruses that attack bacteria), has a head composed of DNA surrounded by protein subunits; in addition, it has a "tail" made up of several sets of other types of subunits including a set of fibers at the end (see Fig. IV-4). The tail attaches the virus to the cell surface; enzymes in the tail help digest the bacterial surface. The DNA passes through the tail and then through the hole in the bacterial wall produced by the tail enzymes; a contraction of the tail apparently aids the entry of the DNA into the cell.

Animal viruses may enter cells by *attaching* to the surface; some are then *engulfed* by the cell through pinocytosis or phagocytosis. In such cases, uncoating of nucleic acid might occur within the cell. (The uncoating of Reovirus inside lysosomes has been discussed earlier in Section 2.8.2.) The attachment of viruses to the cell surface depends, at least in part, on the presence of specific cell-surface components to which particular strains of viruses can attach. The mechanisms of subsequent entry and uncoating are being sought. Uptake by pinocytosis is readily observed, but investigators disagree as to whether viruses so engulfed go on to multiply. There may be other, less obvious entry mechanisms that are yet to be elucidated.

Virus *release* from cells may occur with the death and rupture of host cells (*lysis*). Sometimes, host cell destruction is less rapid, and the virus is released by inclusion in a bud pinched off at the cell surface. The bud remains around the virus. It is generally believed that in this way membrane-delimited viruses derive their membrane (or at least major lipid, polysaccharide, and protein components) from the host cell surface membrane and associated structures. However, in a few cases some of the proteins, and perhaps other constituents in the membrane surrounding the virus, have been shown to be specific molecules that are made only in infected cells, under the influence of the virus genome.

3.1.2   ***EFFECTS ON***   Within the cell, different viruses concen-
            THE HOST   trate in different regions. Some types are
                           found in the nucleus, some in the endo-
plasmic reticulum (in Fig. III-3), and others elsewhere in the cytoplasm.

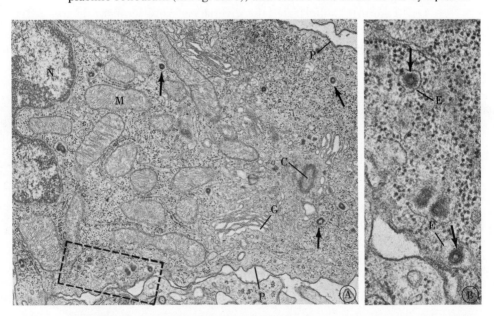

Fig. III-3 *Portion of a cell in the spleen of a leukemic mouse. In **A** (× 18,000) (N) indicates the nucleus, (P) the plasma membrane, (C) a centriole and (G) a part of the Golgi apparatus. The electron-dense bodies within the endoplasmic reticulum (arrows) are enlarged in **B** (× 56,000). The delimiting membrane of the ER is seen at (E). The bodies within the reticulum are viruses. Courtesy of E. de Harven.*

The viruses somehow evade or overwhelm mechanisms that tend to protect the cell. The resistance of Reovirus RNA to lysosomal hydrolysis is mentioned in Section 2.8.2. Some of the bacteriophages have DNA modified in such manner that they resist degradation by nucleases in the host bacteria; unusual DNA bases are sometimes present including some with glucose molecules attached. Effects on cell metabolism vary greatly. Some viruses, such as some bacteriophages and the viruses causing polio and influenza, drastically alter cell metabolism and may lead to cell death. Host cell functions may be entirely diverted to the production of new viral components, for which the nucleic acid of the infecting virus acts as a template. In the case of some viruses with single-stranded RNA as their genetic material, a special "replicating form" of RNA, apparently double-stranded, is synthesized. The virus first brings about formation of a complementary copy of its RNA, and then uses this in turn to produce new RNA's which are complementary to this copy. This results in the production of RNA's that are identical to the original RNA. The same process occurs with the few viruses that have single-stranded DNA. Viral-infected cells also often synthesize a variety of enzymes not normally present in cells but involved in the production of specific viral constituents. For example, normal cells contain no enzymes capable of using RNA templates rather than DNA templates for synthesis of RNA, but cells infected with RNA viruses do have such enzymes. The viral nucleic acid specifies production of these enzymes as well as production of the viral coat proteins. Host cell ribosomes and other components are used in the synthesis.

In other cases, viruses have less obvious effects. They may enter into a coexistence with the host cell and remain reproducing in long-term residency more or less in synchrony with the cell. The extreme form of this is found in some bacteria where certain (*lysogenic*) viruses become closely associated with the bacterial DNA. The viruses behave very much like part of the host chromosome; viral and bacterial DNA replicate at the same time. Such long-term relationships may break down under some circumstances. The virus becomes virulent, duplicates much more rapidly than the host cell, and soon kills the cell. It is speculated that some diseases of higher organisms are due to such activation of latent viruses.

As discussed later (Section 3.12.3) viruses are known that cause tumors when injected into animals or that bring about rapid and abnormal growth when added to cell cultures.

Nonviral infective agents may also influence cell behavior. Thus, a strain of the fly, *Drosophila*, is known in which the viable offspring are almost exclusively female; males die as embryos. This trait is due to the presence of an infectious microorganism, a spirochete, transmitted through the ova. Infective particles are found in the cytoplasm of some strains of *Paramecium*. Bacterial infection of higher organisms influences cell activities in diverse ways—witness the varieties of bacterial diseases.

3.1.3 **VIRUSES AS** Since genetic information is stored as DNA
EXPERIMENTAL base sequences, the amount of DNA in a
TOOLS cell is a rough measure of the genetic information available to the cell. The "code words" for amino acids are each three bases long. A *very* rough estimate of the number of different proteins (see p. 71 for sizes) for which a cell might carry information is based on dividing the number of base pairs in its DNA by three. The estimate is rough since not all DNA carries information for proteins and because some DNA sequences occur in multiple copies. Base *pairs* are used since the two DNA strands of the double helix are complementary and only one is probably used as a template for RNA transcription; for single-stranded RNA molecules, the number of bases can be used as a direct estimate of "information content." Cells of vertebrates and higher plants contain several billion (10^9) DNA base pairs, bacteria on the order of 10^6, and viruses between 10^3 and 10^5. In TMV, the RNA is a single strand with slightly more than 6000 bases. T_4 bacteriophage has roughly 200,000 DNA base pairs specifying about 25–50 different proteins. Several viruses are known that contain only a few thousand base pairs, enough to specify only three or four proteins, including the coat proteins and those required for function in a host cell. The DNA of *polyoma* virus (Sect. 3.12.3), whose coat is of 72 similar protein subunits, is a circular molecule with 5000 base pairs; the molecule is 1.6 μ long. By contrast, the circular DNA of the bacterium *Escherichia coli* is about 1000 μ long, and a single chromosome of a eucaryote cell may contain enough DNA to make a double helix that is millimeters or centimeters in length. A single mitochondrion may contain circular molecules of DNA only 5 μ in length (Fig. II-40).

These comparisons dramatize the relative simplicity of viruses. One can now hope to identify the function of every gene in the viral "chromosome" (nucleic acid molecule) and to identify every protein for which the nucleic acid codes—a perspective that is presently unrealistic for chromosomes of cells. As simple, self-reproducing structures, viruses have been of great value as experimental tools. The demonstration that only the DNA of some bacteriophages enters bacteria was an important piece of evidence in establishing the primacy of nucleic acids in heredity. Reconstitution of certain viruses from their isolated components (Section 4.1.5) is relatively easy; this is being exploited to unravel the mode of formation of more complex biological structures. The involvement of some viruses in the transformation of normal cells into cancer cells is providing insight into the role of nucleic acids in cancer (Section 3.12.3). As it is better understood how in viruses, nucleic acid molecules many thousands of Å long are packed into spaces a few thousand Å or less in length, valuable clues will be gained to the rules governing nucleic acid arrangement in plant and animal cells.

c h a p t e r **3.2**

PROCARYOTES

The small size of procaryotes and, until recently, difficulties in preserving their structure for electron microscopy have hampered extensive analysis by microscopy. Today the structure of some bacteria and blue-green algae is known in fair detail, and a beginning has been made in the study of procaryotes such as the pleuropneumonialike organisms (PPLO). The situation is quite different with respect to biochemistry and genetics; some bacteria, *Escherichia coli* and related forms, are probably the most intensively studied and best understood organisms. Because of the wealth of such biochemical information, many questions about the relations of structure and function may be studied to advantage in bacteria.

The overall cellular dimensions of procaryotes are usually on the order of a fraction of a micron to a micron or two. Evidence is appearing for a higher degree of intracellular organization than was once thought to be present in procaryotes, but this organization is simpler than that of eucaryotic cells. In some cases, the cells contain moderately elaborate arrangements of membranes, but these usually are not organized as discrete organelles with a special surrounding membrane separating them from the rest of the cell.

3.2A

PPLO (PLEUROPNEUMONIA-LIKE ORGANISMS)

The pleuropneumonialike organisms are the simplest known cells. The diameters of the smallest are 0.2–0.3 μ, and of the largest on the order of a micron. Thus, in size, the PPLO's overlap with the largest viruses and with the smallest bacteria. Some PPLO's cause respiratory and other diseases in animals and man. The genus, *Mycoplasma*, is the most widely known.

3.2.1   **STRUCTURE AND FUNCTION**   Figure III-4 is a diagram of a PPLO cell. The DNA is contained in a nuclear region. Surrounding it are ribosomes and the other components involved in protein synthesis. A plasma membrane 75 Å thick surrounds the cell. The enzymes present in the cell include, among others, those required for DNA replication, the transcription and translation involved in protein synthesis, and the generation of ATP by anaerobic breakdown of sugar along pathways similar to those described earlier (Chapter 1.3). Although PPLO biochemistry has hardly been studied, it is evident that these cells can live and grow in media containing only lifeless material. Unlike viruses, they are free living; they do not require host cells for duplication.

3.2.2   **MACROMOLE-CULES AND DUPLICATION**   The smallest of the PPLO's contains DNA with about 500,000 base pairs, and a few hundred ribosomes. (By comparison, bacteria contain roughly 10 times as many DNA base pairs and about 50–100 times as many ribosomes). Such cells probably can make no more than 500–1000 different kinds of proteins. Because of the small size, division is difficult to study and little is well established. In one type of PPLO (*Mycoplasma gallisepticum*) it is thought that a type of binary fission takes place. A second bleb forms at the end of the cell opposite to the original bleb. The cell then pinches into daughter cells, each with a nuclear region, ribosomes and a bleb. For other PPLO's, various possible alternative reproductive mechanisms are being explored including fission, budding, formation of small sporelike bodies and growth of large branched filaments that ultimately fragment.

PPLO is probably far more complex than the primitive self-duplicating systems from which life arose. Yet it may be close to the minimum size and complexity required for independent cellular life and reproduction under present-day conditions. Thus, knowledge of PPLO metabolism and reproduction may shed light on control mechanisms and of other aspects of both macromolecular synthesis and structural duplication in more complex cells.

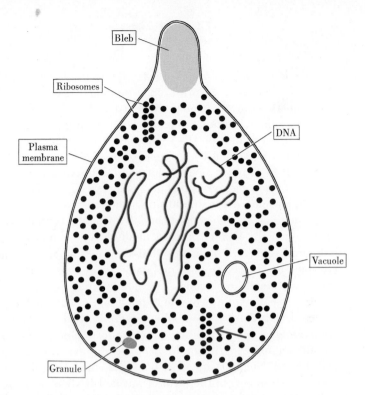

Fig. III-4 *Diagram of PPLO (after the work of D. R. Anderson and J. Mani-loff and H. J. Morowitz). Of the great variety of PPLO's only a few have been studied by electron microscopy. These differ in cell shape, arrangement of DNA fibrils, patterns of ribosome distribution (in one study, many of the ribosomes were found associated in long helical arrays; arrow in diagram), and other features. Some of these differences may reflect intrinsic differences among different PPLO types. Others probably result from differences in growth conditions and in methods of preparation of cells for microscopy.*

The PPLO's that have been studied all show the presence of a delimiting plasma membrane (usually, as diagrammed, with "unit" membrane structure), and of ribosomes and DNA fibrils (in at least some cases, a single circular double helix of DNA per cell is thought to be present). They have no cell wall and no nuclear envelope or other elaborate intracellular membrane systems. The granule and vacuole represent structures seen in some types and are of uncertain significance. For example, some investigators claim that the vacuole is actually a deep infolding of the plasma membrane that appears free from the plasma membrane in some preparations only because connections were not included in the thin section being studied. Blebs also have been noted only in some types of PPLO's.

3.2B

BACTERIA

There are many different types and species of bacteria, but only a few have been thoroughly studied with modern techniques. One of these is *E. coli* (*Escherichia coli*), found in the mammalian digestive tract. *E. coli* is a favorite experimental organism for geneticists, biochemists, and molecular biologists. An *E. coli* cell measuring $1\mu \times 2\mu$ may contain roughly 5000 distinguishable components, ranging from water to DNA and other macromolecules. Its genetic information is carried by a single DNA molecule; enough information is present to code for several thousand different proteins; on the order of 10,000 ribosomes are present. A single cell, by doubling at rates of more than once every 15–30 min, can give rise to millions of essentially identical cells in a short time. Growth and reproduction can occur in a chemically defined growth medium, containing only glucose and inorganic salts; this indicates that the cell contains all the necessary enzymes to synthesize both metabolic precursors and macromolecules from very simple molecules. The combination of relative genetic simplicity, ease of experimental manipulation (for example, production and isolation of mutants), metabolic versatility, and the ability to grow on simple media and to produce large uniform populations makes *E. coli* an outstanding experimental tool. In recent experiments, it has proved possible to isolate a specific bacterial gene (of the β-galactosidase system; see below) in the form of purified DNA molecules. This exemplifies the usefulness of bacteria in analysis of the organization and functioning of genetic systems.

Bacterial metabolism is extraordinarily diverse. Many bacteria have pathways that differ only in some details from those of higher organisms. Aerobic and anaerobic pathways for sugar breakdown and mechanisms of electron transport and ATP formation show many features in common with comparable processes in eucaryotic plant and animal cells. But some metabolic pathways are essentially unique to certain bacteria; an example is the utilization in photosynthesis of H_2S and other sulfur compounds rather than H_2O. The varied biochemistry of different bacteria is correlated with survival of different bacteria in virtually every type of environment. This variety has provided opportunity for insight into metabolic controls of many kinds.

3.2.3   **STRUCTURE**   Electron microscopy shows that bacterial cells possess a plasma membrane under a cell wall, ribosomes and polysomes, and one or several nuclear regions (Fig. III-5). Under most growth conditions, *E. coli* contains two or more nuclear regions, each containing apparently identical copies of the one chromosome — a circular structure made of a thin fiber about 1000–1500 μ long. As far as is known, the fiber is a single continuous double-stranded DNA molecule. In contrast to eucaryotic cells, no

Fig. III-5 *A Gram-positive bacterium,* Bacillus subtilis, *showing the prominent wall (W) surrounding the cell, mesosomes (M) and nuclear area (N). The mesosomes are infoldings of the plasma membrane (C) and also are closely associated (arrows) with the nuclear region. The small dense granules (R) in the "cytoplasm" are ribosomes.*

As is true of procaryotes in general, the arrangement of DNA in the nuclear region cannot readily be analyzed by microscopy of sectioned cells; many short fibrils are seen in thin sections and one cannot determine visually whether they are part of one folded molecule or represent some other, unknown arrangement. Study of the structure of procaryote chromosomes (or "genophores" as they sometimes are called to distinguish them from the chromosomes of eucaryotes) has thus depended heavily on investigation of material isolated from the cell. Approx. × 75,000. Courtesy of A. Ryter.

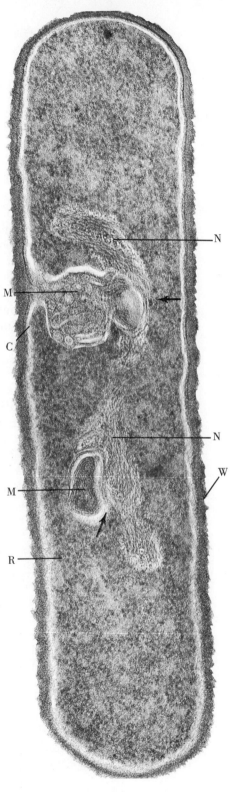

histones are associated with the DNA, although *polyamines* (small organic molecules that contain amino groups) may be bound to some of the phosphate groups of the bacterial DNA. As has been stressed before, the procaryote "nucleus" is not separated from the "cytoplasm" by a special membrane system.

Though the differences have no known major functional significance, bacterial ribosomes are slightly smaller than the ribosomes of the cytoplasm of eucaryotes (they sediment at 70 S as opposed to 80 S) and contain slightly more RNA (60 percent) than protein (eucaryotic cell ribosomes contain 40–50 percent RNA). No endoplasmic reticulum or Golgi apparatus is present.

The extracellular wall that surrounds the cell is usually a moderately rigid structure responsible for maintenance of cell shape. Its removal leads to alteration of the cell into a fragile "protoplast," which may burst from water influx unless osmotic conditions in the surrounding medium are carefully adjusted (see Section 2.1.1). Bacteriologists have long used *Gram stains*, which enabled them to distinguish two different major bacterial classes by light microscopy. Many bacteria, such as the pneumonia-producing organism *Pneumococcus*, are colored by the Gram stains and are referred to as Gram-positive. Other bacteria, such as *E. coli*, are Gram-negative. The cell walls of Gram-positive bacteria contain polysaccharides and *mucopeptides*, polymers of sugar derivatives (amino sugars) plus amino acids. The macromolecules are extensively linked to one another, producing a complex multimolecular network that provides considerable strength to the wall. The walls of Gram-negative bacteria contain a similar network but have in addition an outer layer of lipids, complexed with protein and polysaccharides and often appearing as a unit membrane in the electron microscope. Some of the outer-layer components in Gram-negative cell walls are toxic to animals and may account for certain effects of bacterial infection.

The differences in cell wall chemistry between Gram-positive and Gram-negative bacteria are associated with differences in sensitivity to important antibacterial agents. For example, Gram-positive bacteria are usually more sensitive to penicillin. Penicillin interferes with the interlinking of cell wall macromolecules during cell growth and thus reduces the rigidity of the wall and the resistance of the cell to osmotic rupture. The polysaccharide-splitting enzyme *lysozyme* also affects Gram-positive species more readily than Gram-negative ones; in the latter, the lipid-containing layer probably acts as a barrier to penetration of the enzyme into the wall. When white blood cells phagocytose bacteria, cytoplasmic granules rich in lysozyme fuse with the phagocytic vacuoles, as do the lysosomes (Fig. II-47).

Some Gram-negative bacteria and some Gram-positive ones that make special coats survive uptake by white blood cells and other phago-cytes probably because their cell walls remain intact and protect the cell itself from osmotic and enzymatic disruption.

The structure of some bacterial cell walls involves repeating sub-units packed like tiles in a floor. The subunits resemble structures that may be formed in the test tube from mixtures of lipids and other molecules, suggesting that the walls may form their structure spontane-ously when the appropriate components are released by the cell (see Section 4.1.1).

Some bacteria can transform into dormant spores that are meta-bolically inactive and extraordinarily resistant to extremes of tempera-ture, dehydration, and other drastic environmental conditions. The spores contain a single nuclear region, and relatively little cytoplasm. Each is enclosed in a very thick cell wall. The control of spore formation and germination are being actively studied; these processes are analogous to the differentiation of specialized cells of the multicellular eucaryotes. During spore formation, new cell products are made. During germination, synthetic machinery that previously was inactive is turned on; a similar process occurs upon fertilization of egg cells (Chapter 4.4). Problems being investigated include the identification of the control mechanisms that operate to produce orderly synthesis of the right amounts of the right components at the right time, insuring bac-terial transformation into spores under some conditions and conversion back into active cells under other conditions.

3.2.4   ***SEGREGATION OF FUNCTIONS; CELL DIVISION***   Figure III-6 shows a multienzyme complex that has been isolated from *E. coli*. Several dozen protein subunits are associated in a structure that carries out a number of sequential steps in pyruvate metabolism (Chapter 1.3). The complex can be dissociated into three distinct enzymes and can then be recon-stituted in the test tube. The ability of the complex to form spontane-ously from mixtures of its components is of great interest for under-standing the mode of formation of biological structures (Chapter 4.1). Similar complexes are present in eucaryotic cells.

The existence in bacteria of multienzyme complexes indicates the high degree of organization that underlies the apparent morpho-logical simplicity. Many enzymes, some probably in the form of multi-enzyme complexes, are attached to the plasma membrane or to special internal membranes thought to derive from the plasma membrane. For example, there are a variety of photosynthetic bacteria. *Bacterio-chlorophyll* and other photosynthetic pigments differ somewhat from the

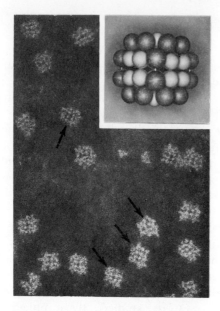

Fig. III-6 *Pyruvate dehydrogenase* multienzyme complexes *(arrows) isolated from E. coli and negatively stained. The diameter of the complex is 300 Å, roughly twice that of a ribosome (see Fig. III-5). Each complex contains several molecules of three types of enzymes which in turn are made of subunits (probably a total of 72) arranged as indicated in the model at the upper right. × 200,000. Courtesy of L. J. Reed, R. M. Oliver and D. J. Cox.*

comparable pigments of eucaryotes. (Some bacteria utilize hydrogen sulfide in photosynthesis and do not evolve oxygen, the H_2S is oxidized to produce sulfur as higher plants oxidize H_2O to produce oxygen.) The photosynthetic pigments and enzymes are associated with internal membranes that are arranged as lamellae, tubules, or vesicles in different species; these are not separated from the rest of the cell as a discrete chloroplast by a surrounding membrane. (See Fig. III-7 for a comparable organization.)

Interest has centered on another membranous structure, the *mesosome*, formed by infolding of the plasma membrane (see Fig. III-5). These are found chiefly in Gram-positive forms, although some Gram-negative forms show similar but simpler infoldings of the plasma membrane. Respiratory enzymes in bacteria are believed to be bound to the plasma membrane and to be concentrated at the mesosomes.

Mesosomes, or similar special regions of plasma membrane, may play roles in bacterial duplication and division. Such division involves DNA replication, growth, and separation into two cells by formation of a *septum* across the cell. The septum grows in from the surface; it consists of a plasma membrane and cell wall. The central question is: How is DNA behavior controlled so that daughter cells receive the proper share of nuclear regions? By analogy with eucaryotic cells (Chapter 4.2), it might be expected that the chromosome is anchored to another structure that controls the chromosome's position during division. In electron micrographs the DNA of each nuclear region appears to be attached at one point to a mesosome, and the mesosomes are often seen near the forming septum and attached to daughter nuclei, as if controlling their separation. Further, if a population of dividing bacteria is exposed for very brief intervals to radioactively labeled precursors of DNA and if the cells are immediately disrupted and cell fractions prepared, the small portion of DNA made during the brief interval (identifiable by its radioactivity) is isolated in a fraction containing many fragments of membranes. These findings have been interpreted as indicating that mesosomes are sites of the DNA-replication enzymes and that they function in chromosome duplication and in distributing daughter chromosomes (DNA molecules) to each daughter cell.

Phagocytosis, pinocytosis, and lysosomes have not been observed in bacteria. However, some bacteria produce *exoenzymes* that act outside the plasma membrane and break down macromolecules into smaller units that can pass into the cell. Such enzymes can facilitate passage of bacteria into or within tissues of organisms they infect, as well as providing nutrients for the bacteria. The release of extracellular lytic enzymes has been likened to the emptying of lysosomal enzymes into a phagocytic vacuole of an animal cell; a phagocytic vacuole is, after all, a bit of the external milieu that has been enclosed within a membrane and taken into the cell. However, the release of enzymes in bacteria does not, as far as is known, involve fusion of a membrane-delimited, enzyme-containing vacuole with the plasma membrane. Enzymes can enter the extracellular medium when some of the individuals in the population die and liberate their contents. But this seems an unlikely general mechanism for release. One proposal is that bacterial exoenzymes pass directly through the plasma membrane. Possibly the polysomes that synthesize the exoenzymes are attached to the plasma membrane, and newly synthesized enzymes leave the cell by mechanisms like the (unknown) mechanisms responsible for passage of proteins from bound polysomes into the ER cavities of eucaryotic cells (see Section 2.4.1). It should be noted that controlled breakdown and resynthesis of regions of the extracellular wall is a necessary step in bacterial growth and division. Part of the wall must be dismantled to permit an increase in cell volume and new wall must be made to separate daughter cells. Presumably, enzymes released locally from the cell participate in such remodeling. However, the extent to which localized breakdown of wall structure occurs and the controls and mechanisms of remodeling are still incompletely understood though several alternative proposals are being investigated.

There is no evident attachment of ribosomes and polysomes to other structures in the bacterium, perhaps with the exception of the plasma membrane. However, the possibility must be left open that some organization of ribosomes exists, which is either broken down when cells are fixed for microscopy or is not readily recognizable in the microscope.

Many bacteria, including *E. coli*, have flagella but these are quite unlike the flagella of eucaryotic cells. There may be many flagella per cell. Each consists of a single fiber 100–200 Å thick and several microns long, protruding from an intracellular "basal granule" through the cell membrane and cell wall. The fiber is made largely of the protein *flagellin*. Wavelike or rotatory movement of its flagella move the bacterium.

3.2.5   GENETICS AND Study of *enzyme induction* and *repression*
METABOLISM in bacteria has elucidated mechanisms by
which genes can be turned on or off. Many
E. coli enzymes are *constitutive*, that is, they are synthesized irrespective of the growth medium. Others are *inducible*; they are produced in some growth media but not in others. The most intensive work on induction has been done on an *E. coli* metabolic pathway responsible for steps in the metabolism of certain sugars and referred to as the *β-galactosidase system*. In the absence of the sugar *lactose* (or other "inducing" compounds of similar molecular structure), the bacteria make very few molecules of the enzymes that metabolize lactose. When inducing compounds are added to the growth medium, extensive and coordinated production of three related enzymes starts: *β-galactosidase* which splits lactose into simpler sugars—galactose and glucose; a *permease* involved in entry of lactose into the cell; and an *acetylase* involved in certain reactions of sugars. The proteins which show these enzymatic activities are coded for by genes (DNA nucleotide sequences) that are adjacent to one another along the chromosome; a single mRNA is produced that contains the information for all three proteins. Several thousand β-galactosidase molecules rapidly accumulate in the cell following exposure to lactose.

A theoretical explanation for these findings suggests that there are *structural* genes responsible for specifying the amino acid sequences of the enzymatic proteins, and that two other genes are involved in control of the rate of mRNA production from the structural genes. One, the *operator*, turns the structural genes "on" or "off" as a group. The other, the *regulator*, is responsible for control of the operator. The operator plus the structural genes it controls are collectively referred to as an *operon*. When turned off, the genes of the operon are inoperative and are said to be *repressed*.

This theoretical framework has recently been given a molecular explanation. Several proteins isolated from bacteria have exhibited properties expected of specific genetic repressor substances. One, from *E. coli*, apparently controls the β-galactosidase system. This protein can bind specifically to DNA known to contain the β-galactosidase operon and also can bind to β-galactosidase-inducing compounds. Probably production of the repressor protein is controlled by the regulator gene. In the absence of inducer substances, the protein is thought to bind to the DNA at the operator site on the chromosome, which is adjacent to the structural genes. The affinity of the protein for inducers is much greater than for DNA, so that when inducers are present they bind to the repressor protein and free the operator. This leads in some way to *derepression* of the structural genes which are then transcribed to form mRNA.

Several other metabolic pathways have also been shown to be under their own operator-regulator controls. For the β-galactosidase system, which is involved in *breaking down* components of the growth medium, enzyme synthesis depends on the presence of inducers. For other pathways, such as one responsible for *synthesizing* the amino acid histidine, production of the enzymes is shut off if the end product molecule resulting from operation of the pathway is provided in the growth medium. The evolution of such mechanisms has provided flexibility in metabolism; the synthesis of certain enzymes is adjusted to the particular environmental conditions. Since β-galactosidase may contribute as much as 3 percent of the total protein in a bacterium, it is clearly advantageous that its synthesis is halted when appropriate substrates are not available.

Another important control mechanism is known as *feedback inhibition*. This controls enzyme *activity*, rather than enzyme *amount*. If enzyme E catalyzes the production of b from a, and b is further metabolized by other enzymes to produce d, then in many cases d can be shown to inhibit the activity of E. Thus, as excess d builds up, further synthesis of b and d is slowed down. The simplest explanation for this is that d binds to E and alters its enzymatic properties. It is possible that enzymes subject to feedback control have special binding sites separate from the "active sites" at which they bind and act upon their own substrates. Binding of an inhibiting compound to the protein at the special site might result in change of the three-dimensional shape of the protein molecule, altering the three-dimensional arrangement of amino acids at the "active site," and thus inhibiting enzymatic activity. Such effects of binding at one site upon the properties of another site on the same molecule are known as *allosteric* effects.

Feedback inhibition is a well-established phenomenon. The involvement of shape changes in proteins ("allosteric" effects) has recently been proposed. Suggestive findings of shape changes associated with inhibition of enzyme activity have been made for only a few enzymes, but the proposal is a plausible one from what is known of protein structure and function.

Also of interest from the viewpoint of metabolic control is the fact that the degradative phase of cell macromolecule turnover appears to be shut off in rapidly growing bacteria; there is little detectable breakdown of protein molecules. When the cells stop growing, degradative enzymes apparently are made or are activated and a balance between protein synthesis and breakdown is re-established.

A final example of important genetic phenomena encountered in bacteria is the existence of *episomes*. These are DNA structures with the unusual ability to exist either free in the cytoplasm or in close structural and functional association with the bacterial chromosome. One example was discussed in Section 3.1.2; lysogenic bacteriophages can replicate, essentially as part of the bacterial chromosome or they may become independent of the chromosome, undergo rapid replication, and kill the host cell. Another episome is the bacterial sex factor known as *F*. This is a circular DNA molecule which can replicate independently in the bacterial cytoplasm or as part of the bacterial chromosome. In the latter case, the addition of *F* to the bacterial chromosome results in the ability of the bacterium to transfer parts or all of its chromosome to another cell as part of a mechanism of sexual reproduction. This takes place during conjugation when cell-to-cell continuity is established by membrane fusion, as will be briefly outlined in Section 4.3.3.

3.2C

BLUE-GREEN ALGAE

About 2000 species of blue-green algae are known. They occupy a wide variety of habitats and some are of considerable ecological or economic significance. In the context of this book, their most interesting characteristic is that the blue-green algae are *procaryotes*, resembling bacteria in many regards. A few grow as single separate cells and many form filamentous multicellular colonies.

The metabolism of blue-green algae is based on photosynthesis. They are the most primitive plants to possess chlorophyll and in which photosynthesis produces oxygen. In addition to chlorophyll, these algae contain unique pigments, collectively called *phycobilin;* one of these pigments (*phycocyanin*) is blue and another (*phycoerythrin*) is red. The variability in color of different species of blue-green algae usually results from differing amounts of green (chlorophyll), blue, and red pigments. The phycobilin pigments along with other special metabolic features and the ability to form spores that are quite resistant to the environment, facilitate growth of the algae under conditions of temperature, water salinity, dryness and light intensity that would preclude survival of most higher plants.

As in all procaryotes the DNA (as usual, a double-stranded helix) is not segregated into a nucleus separated from the cytoplasm by a nuclear envelope. No endoplasmic reticulum, Golgi apparatus, or mito- chondria have been observed. The ribosomes and polysomes are "free" and are similar to those of bacteria. The photosynthetic apparatus (Figure III-7) is not segregated into a membrane-delimited chloroplast. The cell walls resemble the walls of bacteria in containing lipoproteins, lipopolysaccharides and mucopeptides, and related mucoprotein (car- bohydrate-protein) compounds. The mucopeptides are like those of Gram-negative bacteria (see Chapter 3.2B) and like the walls of many bacteria under some circumstances, the algal walls can be dissolved by the enzyme lysozyme to produce "protoplasts." (It is widely thought that bacteria and blue-green algae have evolved from a common ances- tral form.) An additional gelatinous sheath often surrounds the cell wall of the algae.

The photosynthetic apparatus of blue-green algae is somewhat more highly organized than that of photosynthetic bacteria. The pig- ments are present in flattened sacs called *lamellae* which are arranged in parallel array. In usual microscope preparations their membranes

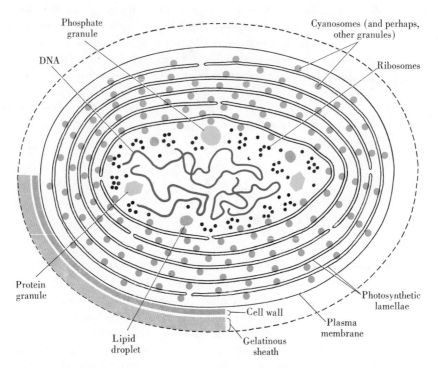

Fig. III-7 *Diagram of a blue-green alga cell. Based on the work of M. Lefort-Tran and others.*

show "unit" structure. The lamella closest to the periphery of the cell may be continuous with the plasma membrane.

Granules, approximately 400 Å in diameter are observed between the lamellae, with at least some apparently attached to the outer lamellar membrane surface. These granules may be seen by electron microscopy of thin sections and of freeze-etched (see Fig. II-12) preparations. The granules are often referred to as *cyanosomes* or *phycobilisomes*; recent studies suggest that they may be the sites of phycobilin pigments. The cells can be disrupted (using a device known as a "French press") and fractions prepared by density gradient centrifugation. When this is done, the cyanosomes are separated into a fraction found to contain phycobilins while the lamellar membranes are found in the fraction containing chlorophyll and other pigments (such as the yellow *carotenoids*). If the cells are fixed (in glutaraldehyde, Section 1.2.1) prior to disruption and centrifugation, then fragments are obtained which consist of membranes with cyanosomes still attached. Such cyanosome-membrane fragments retain photosynthetic functions that neither membranes nor cyanosomes possess alone. Recently, by negative staining procedures (Fig. II-37), the cyanosomes have been shown to consist of about 10 subunits, each about 130 Å in diameter. These subunits as well as other particles on the membranes are yet to be characterized chemically and functionally. (See p. 189.)

The cells divide by inward growth of plasma membrane and wall producing two cells of equal size as in bacteria. Mesosomes or similar structures (Chapter 3.2B) have yet to be observed and the mechanisms for separation of daughter "chromosomes" are not known.

Although they have no obvious specialized motion-related organelles such as cilia or flagella, blue-green algae are capable of movement. Gliding and rotatory motions have been observed whose mechanisms are unknown.

c h a p t e r **3.3**
PROTOZOA

Protozoa are single-celled, eucaryotic organisms with the same basic functions and with the same organelles as cells of higher organisms. Many different types occur (amebae, ciliates, flagellates, and so forth), each with particular characteristics associated with its own mode of life; unlike the cells of multicellular organisms, each protozoon must carry out all the functions required for both survival and duplication of the whole organism. For this reason, although they usually are thought of as low on the evolutionary scale, the protozoa include the most complicated and diversified types of known cells. Pinocytosis and phagocytosis are used by most protozoa in feeding. Amebae often have elaborate surface coats (see Fig. II-5) that adsorb material to be taken in by phagocytosis and pinocytosis, and many other protozoons have special "mouth" regions where food is ingested. The vacuoles and vesicles that form acquire lysosomal enzymes which digest the contents. Finally, egestion, or defecation, of indigestible residues occurs by fusion of the vacuoles membranes with the cell membrane.

Interesting structural specializations are found in various protozoon species. For example, some secrete a hard, extracellular shell containing calcium or silicon. Others possess *trichocysts*, cytoplasmic bodies that fuse with the plasma membrane and extrude filamentous structures used in defense and trapping of food. (Similar bodies found in *Hydra*, called nematocysts, will be discussed in Section 3.6.2). Many contain *contractile vacuoles*. In fresh water, the organisms are not in osmotic equilibrium with their environment, and they constantly tend to take in water. In bacteria and higher plants living in water, this tendency is counteracted by the presence of a rigid cell wall which prevents swelling. In protozoa, the water that comes in ultimately is removed by the contractile vacuoles which fuse with the plasma membrane and expel the water at the cell surface. In some organisms, for example, *Paramecium*, the vacuoles are large, round, membrane-delimited bodies surrounded by an elaborate arrangement of small vesicles and tubules that carry fluids to the vacuoles.

Some flagellate protozoa contain *kinetoplasts*, unique structures having sufficient DNA for detection by light microscope cytochemistry, but having as well the typical mitochondrial membrane configurations. As the cells multiply, these structures divide; their existence is one of the lines of evidence suggesting that duplication of mitochondria may depend on nucleic acids present in the organelles themselves.

In addition to the kinetoplasts, other specialized features of protozoa are particularly interesting for the understanding of cell organization and function. For example, a few species have giant "centrioles," rodlike structures tens of microns long that, like ordinary centrioles, are involved in formation of the cell division spindle. The limited information available suggests that the structure of the giant centrioles is quite different from ordinary centrioles, although their function is apparently similar. Their study may therefore shed light on centriole function.

Some protozoa can be conveniently grown as large populations in a medium of controlled and well-defined composition. Many are large enough for microsurgery.

3.3.1 *MACRONUCLEUS AND MICRONUCLEUS*

In ciliates and some other protozoa, two nuclei are regularly present. One of these often is quite large (Figs. III-8 and III-9), contains much DNA, and shows the presence of nucleoli. This is the *macronucleus*; it is considered to be the nucleus that actively participates in RNA production for cell metabolism. The smaller *micronucleus* contains much less DNA and is involved chiefly in sexual reproduction. During sexual reproduction, the macronucleus degenerates and is re-formed by alteration and growth of a micronucleus. Both nuclei normally divide when the cell does. In some species of protozoa, *amicronucleate* strains are known that function with only a macronucleus present. In other strains with both types of nuclei, removal of the macronucleus causes cell death even if a micronucleus is present.

The simplest explanation for these observations is that the macronucleus contains many copies of the chromosomes present in the micronucleus and represents an amplification mechanism whereby numerous replicates of DNA-contained genetic information are made available for such functions as RNA production. This might explain the observations made on macronuclear regeneration in the ciliate *Stentor*. As Fig. III-8 shows, the macronucleus of this organism has the form of a string of beads. All but one of the beads can be removed without killing the cell. Fragments of cells provided with a single "bead" can regenerate complete cells. If only one or two copies of

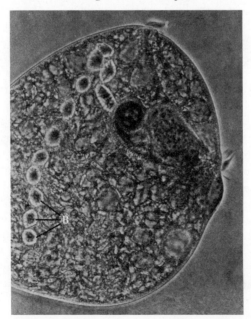

Fig. III-8 Portion of the protozoan, Stentor. The macronucleus appears as a chain of beads (B). The micronuclei are too small to be well demonstrated in this photograph. × 100. Courtesy of L. Margulis.

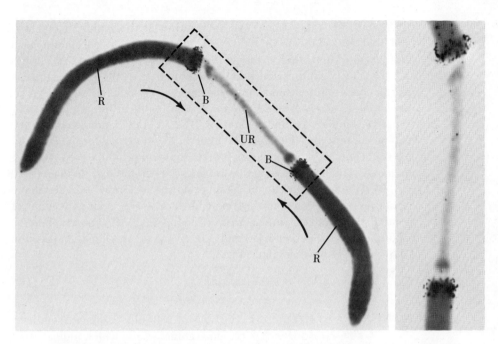

Fig. III-9 *Autoradiogram of the macronucleus of the ciliated protozoan,* Euplotes. *The cell had been exposed to tritiated thymidine for 20 minutes before fixation; the DNA synthesized during this interval is therefore radioactive. Grains are found only over two short lengths of the nucleus; these are the replication bands (B). The bands reflect a wave of replication passing toward the center of the nucleus (arrows); the wave has not yet reached the zone at UR. The chromatin in the region between the bands and the ends of the nucleus has already been replicated (R).* × *600; insert is an enlargement of the area in dotted lines.* × *1,000. Courtesy of D. Prescott.*

each gene were contained in the macronucleus, it would be difficult to see how a complete nucleus, capable of sustaining cell life, could be regenerated from a small fragment that did not contain all of the genes.

Interesting studies of macronuclear replication have been performed with the ciliate, *Euplotes* (Fig. III-9). The macronucleus of this species is a long, sausage-shaped structure. Replication of the contents occurs in a progressive fashion, starting at both ends and moving toward the center. The actual points of replication can be distinguished in the microscope as a pair of bands whose appearance differs from the rest of the chromatin. This provides a most convenient system for studying duplication of nuclear material. The chromatin in the bands appears to transform from clumps and patches into dispersed fibrils (individual chromosomes have not been distinguished). Autoradiography shows that DNA synthesis occurs at the bands (Fig. III-9). On the other hand,

RNA synthesis takes place everywhere but at the bands. Thus, it is held that no synthesis of new RNA takes place on chromatin as it duplicates. From other evidence, this seems true of other cell types as well. Newly synthesized proteins accumulate at the bands, suggesting that new chromosomal proteins become associated with DNA as it replicates; newly labeled protein is also found in the nucleus, away from the bands. As the amount of DNA in the bands doubles, the histone content also doubles. The total duplication time for the *Euplotes* macronucleus is several hours. After duplication of its contents, the nucleus is pinched in two across its long axis as the cell divides.

Macronuclei appear to be one device whereby the metabolic capacities of the nucleus are expanded to serve a large volume of metabolically active cytoplasm (see Section 3.3.2). Polyploidy (Section 4.3.6), polyteny (Section 4.4.4) and selective gene amplification (Section 4.4.5) probably have similar effects.

3.3.2   **NUCLEUS AND CYTOPLASM**   In Sections 1.2.4 and 2.2.3, we outlined autoradiographic results which indicate that synthesis of RNA occurs in the nucleus and that the RNA then passes to the cytoplasm. Some of the experiments were based on transplantation of the nuclei of amebae (with RNA that had been radioactively labeled) into nonradioactive host cells. Similar nuclear transplants have been made with the *proteins* labeled. Again, the host cell cytoplasm becomes radioactive but, unlike the RNA experiments, the host cell nucleus also soon shows many grains when studied by autoradiography. This suggests that, at least in amebae, proteins can move both from nucleus to cytoplasm and then back from cytoplasm to nucleus; this is in contrast to nuclear-produced RNA, whose movement is essentially unidirectional. Such bidirectional movements of proteins are being studied for their significance in reciprocal interactions of nucleus and cytoplasm. Other interesting experiments on the relations of nucleus and cytoplasm have explored the hypothesis that the initiation of cell division depends on the cell's reaching a certain size. One form of this hypothesis suggests that division mechanisms have evolved so that a rapidly growing cell tends to divide before it exceeds a critical volume. For a sphere, the surface is given by $4\pi r^2$, while the volume is $4/3\pi r^3$. For a given increase in r, the volume increases proportionally more than the surface, since cubes of numbers increase faster than squares. While cells are not usually spheres, this sort of relationship of surface increasing at a slower rate than volume is a general one; as cells get larger, there is a *relatively* smaller surface available for exchange of material with the environment. Similarly, a fixed amount of DNA services an increasing mass of cytoplasm.

If one repeatedly cuts off portions of cytoplasm from growing amebae and thus prevents them from attaining a certain size, the cells do not divide. This suggests that one set of factors controlling cell division is related to cell size. However, the relationship is not a simple one. The growth of amebae may be curtailed by placing the cells in a nutrient-poor medium. Division is slowed, but does occur despite the fact that the cells are smaller than normal. Thus, while growth and division may be under some common controls, these are flexible. Perhaps some key compound must reach a critical level before division occurs and the normal rate of production of this component is such that it usually accumulates to the appropriate level as the cell reaches a certain size. Or perhaps a much more subtle or complex control is responsible. This question will be discussed again in Section 4.2.3.

3.3.3 BASAL BODIES, CENTRIOLES, CILIA, AND FLAGELLA

In ciliated protozoa like *Paramecium*, the cell surface and the cortex (the cytoplasm just below the membrane) are highly organized (Fig. III-10). The basal bodies of the cilia are arranged in a precise, geometrical fashion. In a series of experiments, portions of the cortex were removed and reimplanted so that the basal body pattern was reversed and the cilia beat in the opposite direction from their neighbors. Such changed orientations persist through repeated cell duplications; some have been followed for 700 generations. This is one of a number of experiments that lend weight to the idea that much of the surface pattern

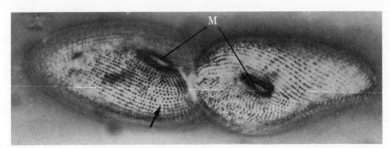

Fig. III-10 *A dividing* Paramecium *stained to show the basal bodies and related structures in the cell cortex (the cytoplasm just below the plasma membrane). The basal bodies are arranged in rows which appear as lines of granules (arrow); the two daughter cells have identical patterns of basal bodies. The "mouths" of the daughters are indicated by M. Courtesy of R. V. Dippel and T. Sonnenborn.*

depends for its replication on local determinants rather than on the nucleus. Electron microscopy of duplicating protozoa show that new basal bodies form in close association with the old ones, suggesting some mode of self-duplication (see Section 2.10.5).

The basal bodies of ciliated protozoons show clearly the presence of a "cartwheel" structure within the $9 + 0$ tubule pattern (see Fig. II-59). This is restricted to the end of the basal body opposite to the end attached to the cilium or flagellum. Observations on protozoa and other cells have raised the possibility that this structural polarity may reflect a functional polarity; perhaps one end of the organelle is involved in producing cilia or flagella, and the other in basal body duplication or growth. In basal body duplication in ciliated protozoa, a new procentriole (Sect. 2.10.5) forms near the cartwheel end of the old body. Cartwheels appear early in developing procentrioles.

Finally, many of the ciliates and other organisms show elaborate systems of microtubules and fibers associated with the basal bodies; often these appear as a network that seems to link the basal bodies together. It is possible that such systems coordinate the cilia of a cell, and maintain the wave of ciliary motion that passes along the surface (Section 2.10.3). A few microsurgical studies indicate that interruption of the connections between basal bodies can affect coordination. However, the results of other studies are equivocal, and the extent to which various factors participate in coordination remains to be established. One proposal suggests that the plasma membrane may conduct some coordinating impulses analogous to nerve impulses. Another maintains that much of ciliary coordination depends on simple, mechanical interaction resulting from the close proximity of cilia to one another and that no special coordinating devices are needed. As evidence for this last suggestion, several students of protozoa cite observations on spirochetes that are found attached in large numbers to some protozoa. Spirochetes are elongate procaryotic organisms which have fibers helically wound around their cell body and which are capable of motion somewhat similar to that of a flagellum. Despite the absence of a common membrane or other obvious communication device, the spirochetes attached to one protozoon move in coordinated waves. It even appears that coordinated spirochete motion can produce motion of the protozoon.

c h a p t e r **3.4**

EUCARYOTIC PLANT CELLS

The procaryotic blue-green algae (Chapter 3.2C) are usually considered to be the simplest of the plant cells. Most plant cells contain the usual organelles of eucaryotes. In contrast to animal cells, they also show the presence of a cell wall and of plastids. Study of plant cytology has been hampered by technical difficulties. Many of these have now been overcome and progress in understanding subcellular structure and function is becoming increasingly rapid.

3.4.1   *ALGAE*   These morphologically simple plants, some unicellular, others multicellular, are classified partly on the basis of their color, determined largely by the nature of the pigments in their plastids. In red algae, alcohol extraction of cells removes a red pigment (phycobilin) from chloroplast preparations; when the red color is removed, a set of granules 350 Å in diameter and studding the chloroplast sacs in these algae disappears. These granules are left intact when the chlorophyll is extracted. Thus the granules are probably the chief site of red pigment localization. The close association of this pigment with the chlorophyll-containing lamellae may permit transfer of light energy absorbed by phycobilin to the chlorophylls of the photosynthetic apparatus.

Figure III-11 shows the structure of a brown alga; different species of this group may contain one or two or many plastids per cell, and these possess the pigment fucoxanthin, in addition to chlorophyll. The plastids are membrane-bounded, and the internal sacs are arranged as long parallel sheets; no grana are present. In addition to the membranes, the plastids contain 25-Å-fibrils, probably DNA. A *pyrenoid*, where starch is stored and probably synthesized, and a *pyrenoid sac*, which also stores polysaccharides, are part of the plastid.

As Fig. III-11 indicates, algae contain the usual organelles of eucaryotes. The cristae of the mitochondria are tubular, as in many protozoa (see Fig. II-36). The endoplasmic reticulum is closely associated both structurally and functionally with the nuclear envelope and Golgi apparatus. In contrast to many higher plant cells, paired centrioles are visible.

Like protozoons, algae have been used for a variety of important observations and experiments. As mentioned previously (Section 2.7.4), the fact that some of the unicellular forms have a single chloroplast which divides in synchrony with the cell is one of the lines of evidence suggesting that plastids are capable of self-duplication. Much attention is being devoted to the cytoplasmic streaming, or *cyclosis*, characteristic of many plant cells (Figure II-65). A large central vacuole occupies much of the cell (Figure III-12), and the rest of the cytoplasm may move around the vacuole at rates of tens to hundreds of microns per minute. The large cells (millimeters to centimeters in length) of

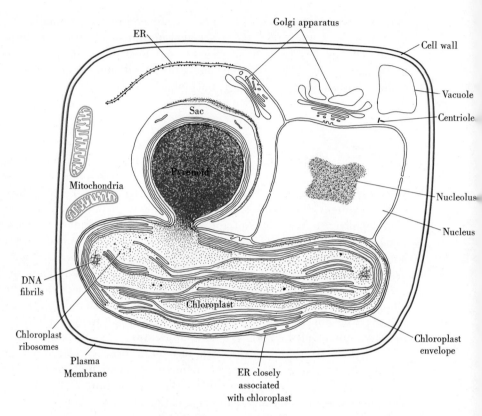

Fig. III-11   *The relationships among organelles in a hypothetical brown alga. Not included in this schematic diagram are free ribosomes, and several types of cytoplasmic granules. Note that the flat sacs within the plastid are arranged in extended parallel arrays rather than separated into grana and stroma systems as in higher plants (see Figs. II-43 and II-44). After G. B. Bouck.*

algae such as *Nitella* have been extensively used in studying this phenomenon; careful observations indicate that there is a stationary outer zone (the *ectoplasm*) located just below the cell wall. This contains most of the plastids and surrounds the inner moving zone, or *endoplasm*. A theory to explain cyclosis suggests that movement results from molecular interactions at the interface between ectoplasm and endoplasm; perhaps the interactions are analogous in some way to the interactions of thick and thin filaments during muscle contraction. Filaments present in the algal cytoplasm are being intensively studied in relation to this motion but we have already outlined some of the difficulties in interpreting the roles of microfilaments and microtubules in phenomena such as cyclosis (Section 2.11.4).

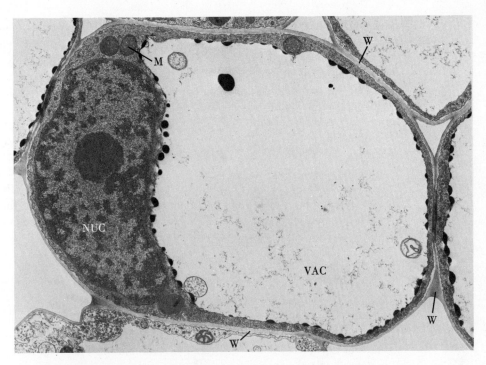

Fig. III-12 *A cell of the marsh plant,* Limonium. *It shows the major features of many mature plant cells. The vacuole occupies much of the volume while the nucleus and organelles such as mitochondria (M) are found in a thin layer of cytoplasm surrounding the vacuole. The cell is separated from its neighbors by a cell wall (W).* × *9,000. Courtesy of M. Ledbetter.*

3.4.2   **VACUOLES**   The cell shown in Fig. III-12 has features common to many cells of algae and of higher plants. The cytoplasm is disposed as a thin layer surrounding a large vacuole. The vacuole is surrounded by a distinct membrane or *tonoplast* that shows "unit membrane" structure in the electron microscope. A prominent wall separates the cell from its neighbors.

The number and size of vacuoles varies in different cells and during development. In mature cells, a single vacuole is often present and occupies up to 80 percent or more of the cell volume; at earlier developmental stages vacuoles occupy less of the cell and several small ones may be present. In the larger algae, the contents of single vacuoles may be sucked out by a glass tube and analyzed. These and other determinations have established that the vacuoles contain much water with high concentrations of inorganic salts, sugars, and other components. The concentrations of dissolved material may be considerably greater than those in the fluids absorbed by the plant from its environment. The tonoplast is semipermeable, and the high concentration of dissolved vacuole contents results in a tendency toward osmotic entry of water (Section 2.1.1). This generates a pressure balanced by the mechanical resistance of the cell wall, so that the cytoplasm is pushed firmly against the wall. During cell elongation, which may be very rapid (cells can elongate at rates of 20–75 μ per hr), the cell walls are somewhat elastic. They stretch under the pressure generated by water uptake in the vacuole; such elongation usually accompanies plant growth.

The vacuoles of many plants contain pigments such as the anthocyanins. The striking colors of petals, leaves, and fruit are due to such pigments; these colors are important in attracting insects and other organisms involved in pollination and seed dispersal.

3.4.3   ***WALLS***   Plant cell walls almost all contain cellulose, a polysaccharide made of glucose subunits. The cellulose molecules in the walls are in the form of multimolecular bundles or fibrils 100–250 Å thick and several microns long (Fig. III-13). A matrix of other material surrounds the cellulose and contains additional polysaccharides and compounds such as *lignin*, a complex polymer that imparts strength and rigidity to the cell wall and is characteristic of woody plants. Polysaccharides known as *pectins* help

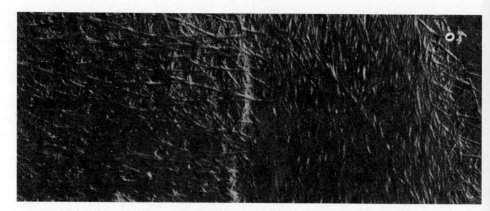

Fig. III-13   *A preparation of the wall surrounding a cell of a grass-like plant, the rush,* Juncus. *A shadowing technique has deposited a thin layer of metal on the specimen from an angle forming a shadow analogous to that formed with light. Many cellulose fibrils in the layer of the wall seen at the left of the figure run roughly perpendicular to the direction of the fibrils in the adjacent wall layer, seen at the right.* × 38,000. *Courtesy of A. L. Houwink and P. A. Roelfson.*

bind adjacent cells together; they are abundant in the *middle lamella*, a layer formed between two adjacent cell walls. Removal of the pectins by enzymatic or other means permits cells with their walls to fall apart from one another, although the walls maintain their shape. Waxy substances in the cell wall help to protect many plants from drying out in deserts or other environments.

Wall morphology and chemistry vary greatly among the different plants and plant tissues and are responsible for many major features of the tissues and for many of the useful qualities of woods, cotton, and so forth. Openings present in the walls of many cells of multicellular plants permit passage of fluids. The fluid-conducting system of higher plants is made partly of vessels composed of intercommunicating spaces bounded by complex cell walls left by dead cells. Cork owes its buoyancy to the closed and water-tight compartments bounded by the cell wall, which contain air-filled spaces left when the cells die. (Cell walls were the structures seen by Robert Hooke when, in 1665, he used the term *cell* in reporting his investigations of cork and piths of several plants.)

In many plant tissues, the cell walls are structures produced sequentially by the cell (Fig. III-14). The outermost layer is formed

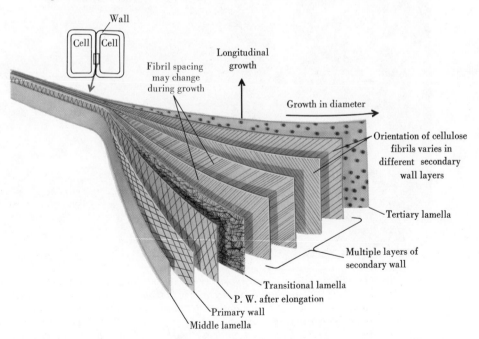

Fig. III-14 *Schematic representation of the various layers of the cell wall of a higher plant. After K. Muhlethaler. The chief orientation of the cellulose fibrils in the various layers is indicated. This orientation or the spacing of fibrils may change somewhat during growth of some plants.*

first. This *primary wall* is composed of fibrils running in many directions. The *secondary wall* may contain distinct layers. In each layer, the individual cellulose fibrils are oriented parallel to one another, but the fibrils of different layers are oriented at angles to one another (Fig. III-14). This layering contributes considerable strength to the wall; often adjacent layers have fibers at right angles to one another. A thin *tertiary* wall may also be present.

Little is known about how the cell secretes or controls the orientation of the material in its walls. Autoradiography indicates that early stages in glucose incorporation occur in the Golgi apparatus and

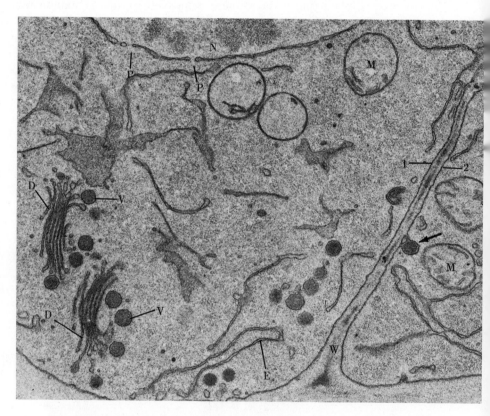

Fig. III-15 *Portions of two cells of corn root. The plasma membranes are seen at (1) and (2). (N) indicates a part of the nucleus and (P) nuclear pores. A mitochondrion is seen at (M) and endoplasmic reticulum at (E). The Golgi apparatus (D) consists of numerous stacks of saccules. The large Golgi vesicles (vacuoles; V) contain an electron-dense material; apparently they migrate to the cell surface, fuse with the plasma membrane (arrow), and contribute their content to the cell wall (W). Courtesy of W. G. Whaley, J. A. Kephart, and M. Dauwalder.*

Golgi vacuoles (Fig. III-15) participate in the formation of new walls during cell division (Fig. IV-17) and probably at other times as well. In some growing cells, the wall fibrils nearest the cell are oriented parallel to the microtubules found just below the plasma membrane, but no causal relationships between tubule patterns and cellulose fibril patterns have been firmly established. Formation of cellulose fibrils may occur at some distance from the cell surface, probably from smaller subunits secreted by the cell. There are hints of complex control mechanisms. For example, the initial cellulose molecules made by the cells that produce cotton fibers are of varying length; they contain up to 5000 linked glucose molecules. When secondary wall formation begins, the number of glucoses per cellulose increases to almost 15,000 and the lengths of the newly made cellulose molecules become more uniform.

3.4.4   ***PLASMO-***   Holes are often found in the cell walls
                   DESMATA   between adjacent cells. Frequently, within
                                   the holes, extensions of cytoplasm are present or adjacent cells show other signs of special relations with each other; these configurations are called *plasmodesmata*. As shown in Fig. III-16, endoplasmic reticulum often is closely associated with the cell surface at the points where plasmodesmata are present.

One line of thought holds that cytoplasmic continuity is often established by fusion of cells at the plasmodesmata as shown in Fig. III-16B. However, it has yet to be established that actual continuity of cytoplasm is the general case; there may be several different types of plasmodesmata. In either event, the structures provide a means for interaction between adjacent cells which are separated in other regions by thick cell walls; it is widely thought that material can pass from cell to cell through the plasmodesmata. The close association of endoplasmic reticulum may prove of considerable interest in this regard.

The simplest explanation for the origin of many plasmodesmata is by the persistence of points of continuity or close contact between dividing cells which become more widely separated at the other points along their surface. In many dividing cells, regions containing remnants of the spindle responsible for the earlier separation of chromosomes may be seen to maintain continuity between daughter cells for some time after division of the rest of the cytoplasm has been completed (see Fig. IV-17). Division of the cytoplasm is retarded at these regions, and the plasma membrane of one daughter cell thus retains continuity with the membrane of the other daughter.

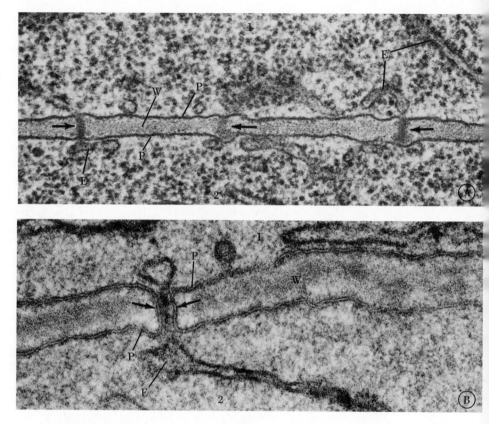

Fig. III-16 *Two types of* plasmodesmata. *A. Portions of two adjacent cells (1 and 2) of* Arabidopsis, *a plant of the mustard family. The cell wall is seen at (W) and the plasma membrane at (P). Arrows indicate plasmodesmata which, in this preparation, appear as dense regions that traverse the wall. Endoplasmic reticulum (E) is associated with the cell surfaces at the regions where the plasmodesmata are found. × 63,000. Courtesy of M. Ledbetter. **B**. A similar region showing two cells of corn. At the plasmodesma indicated by arrows, it can be seen that the plasma membranes of the two cells are continuous, forming a channel connecting the cytoplasm of one cell with the other. The plasma membranes show "unit" membrane structure. Courtesy of H. Mollenhauer.*

c h a p t e r **3.5**

ABSORPTIVE CELLS

**3.5.1   TISSUES; THE
          SMALL INTESTINE**   Figure III-17 diagrams a cross section of the vertebrate small intestine; like many other organs, this contains the four major tissue types formed in higher animals: epithelia, connective tissues,

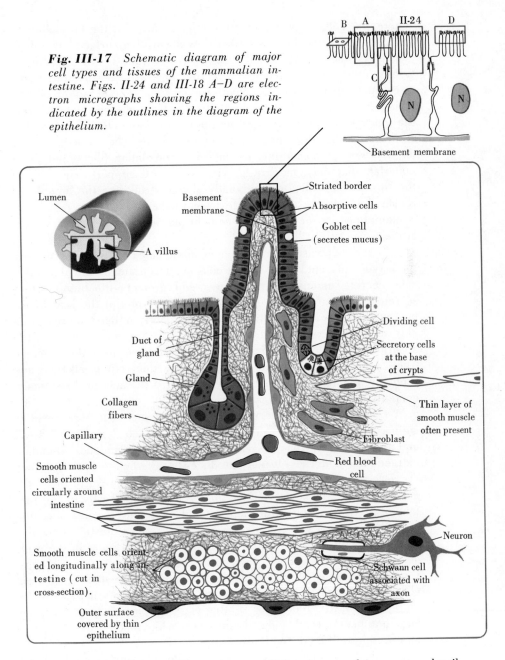

Fig. III-17 *Schematic diagram of major cell types and tissues of the mammalian intestine. Figs. II-24 and III-18 A–D are electron micrographs showing the regions indicated by the outlines in the diagram of the epithelium.*

muscle, and nerve. These tissues will be discussed in greater detail in subsequent chapters.

Many *epithelia* serve as covering tissues, at absorptive surfaces, in ducts, in skin, and elsewhere. They are continuous sheets of cells.

Where the cells play a protective role, as in skin, the epithelium is often several cells thick (stratified). At absorptive surfaces, the sheet usually is one cell thick. The absorptive surface of the intestine is thrown into a series of folds (*villi*) that increase the absorptive area in comparison to a simple smooth-walled tube. The organ is lined by a single layer of epithelial cells that selectively absorbs material from the *lumen*, the space enclosed by the organ. The cells are referred to as *columnar* because they are taller than wide. The epithelium rests on a *basement membrane*, a mat of extracellular fibers and other material, that may, in part, provide structural support. The cells at the tips of the villi are continually sloughed off into the lumen and replaced by others that migrate from further down. The population is maintained by the division of cells in the indentations (*crypts*) found at the base of the villi.

Gland cells also are classified as epithelial. The intestine contains three types of gland cells. *Goblet* cells scattered among the absorptive cells secrete mucus that lubricates and protects the lining. Other secretory cells are present in indentations (crypts) of the lumen surface at the base of the villi; the cells are poorly understood but probably secrete enzymes. Multicellular glands are present, well below the absorptive surface in the first part of the intestine, near the stomach. Their secretion passes into the lumen via ducts; it is alkaline and probably, among other functions it neutralizes the acid in the stomach contents that enter the intestine.

Connective tissues serve to bind other tissues together, providing support and a framework within which blood vessels, lymphatic vessels, and nerves course. They are composed of cells surrounded by abundant extracellular materials, usually produced by the cells with which they are associated. In fibrous connective tissues, widespread in the body, *fibroblasts* produce extracellular fibers, consisting mostly of the protein *collagen*. The fibers are surrounded by a matrix containing proteins and such polysaccharides as *hyaluronic acid*. Elastic connective tissues containing fibers of the protein *elastin* are found in the walls of arteries and elsewhere; such tissue provides resiliency to organs. Cartilage consists of cells surrounded by a stiff matrix of polysaccharides and proteins; bone cells are surrounded by calcium salts in an organic matrix. Blood also is often classified as a connective tissue; large numbers of red blood cells and fewer white blood cells are carried in a protein-rich fluid, the *plasma*.

The connective tissue of the intestine is largely of the fibrous kind. In it, blood capillaries and lymphatics are present; these transport oxygen to the epithelium and other layers and remove nutrients absorbed from the lumen, as well as CO_2 and other waste products.

There are three major types of *muscle tissue:* the voluntary *skeletal* muscle (Section 2.11.3) and the involuntary *cardiac* (heart muscle) and *smooth* muscle. The first two types show cross striations. In the intestine, a thin layer of smooth muscle cells underlies the surface epithelium and influences the overall shape of the epithelial layer. Separated from this by a thick region of connective tissues are two additional smooth muscle layers. In the inner one, the cells are oriented circularly around the intestine. In the outer, the cells have their long axes oriented longitudinally along the intestine. The presence of two outer layers of muscle cells with the cells in one layer oriented perpendicular to the cells in the other makes possible peristalsis and other complex movements of the intestine.

Nervous tissue consists of the impulse-conducting nerve cells (*neurons*) and certain associated cells called the *neuroglia* and *Schwann cells*. Some of the associated cells cover much of the surface of neurons and probably serve to modify neuronal activities.

One of the prominent groups of nerve cells in the intestine is disposed as a layer between the two outer muscle layers. These cells help to integrate the functions of the intestine with those of the rest of the body.

The outer layer of the intestine consists of a thin connective tissue layer within a covering of squamous (flat) epithelial cells. This outer epithelial layer is continuous with the *mesentery*, a thin sheet of tissue which attaches the intestine to the body wall.

3.5.2   JUNCTIONAL   Cells of epithelia are held together in sheets
STRUCTURES   or clusters by special junctional structures.
There are several types of such structures; the particular ones present vary in different epithelia. In the absorptive epithelium lining the intestine, there are regions near the lumen where the outermost layers of plasma membranes of adjacent cells fuse, obliterating the intercellular space that usually separates adjacent cells (Fig. III-18). These regions are known as *tight junctions* and occur in a continuous band around each cell. Some distance below them are *desmosomes*. In contrast to the tight junctions, desmosomes occur in localized areas, or patches, rather than as a continuous band. In desmosomes, the plasma membranes of the two cells are separated by the usual intercellular space of 100–200 Å or more. The extracellular material within the space may appear slightly denser than elsewhere, and sometimes it seems to be organized in closely associated layers. Filaments and dense material project from the plasma membrane for some distance into the cell cytoplasm.

In the intestine, and in some other tissues, the junctional structures near the lumen are known collectively as the *junctional complex*;

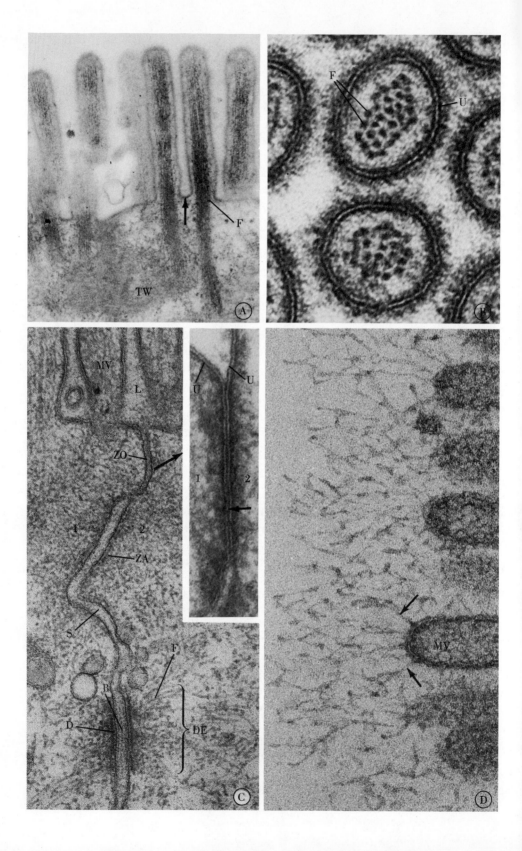

the tight junction is called a *zonula occludens* and the desmosome, a *macula adherens*. The region between desmosome and tight junction includes a *zonula adherens*; its structure sometimes is not as distinctive as the other regions but often resembles the structure of desmosomes; the two plasma membranes are separated by about 200 Å and a band of filamentous material, just below the plasma membrane, encircles each cell.

Desmosomes are probably involved in anchoring cells together in many epithelia; under experimental conditions (for example, shrinkage), in which cells tend to pull apart, they often remain attached to one another at desmosomes. Tight junctions are also widely found; they have somewhat different functions from desmosomes. It has been shown that tracer molecules — metals dispersed as fine particles — and proteins such as peroxidase (Section 2.1.4) do not penatrate into tight junctions, although they ordinarily pass readily through the spaces between cells. For this reason, it is believed that tight junctions can seal off a surface by preventing material from penetrating between the cells; in the intestine, kidney, and other absorptive sites, this makes movement through the cells themselves the only means of passage through the epithelium. Movement *through* cells is a more selective mechanism than movement *between* cells. In the latter, the size of a molecule as compared to the dimensions of the intercellular space is the chief limiting factor; in the former, many controlling devices determine movement (see Chapter 2.1).

◀ *Fig. III-18* *Electron micrographs of the regions of intestinal epithelial cells indicated on the diagram in Fig. III-17. See also Fig. II-24. **A**. The microvilli are finger-like extensions of the plasma membrane (arrow); each contains a core of filaments (F) that merges into a zone of filaments and amorphous material (the* terminal web, *TW) in the cytoplasm below the microvilli. × 50,000. Courtesy of J. D. McNabb. **B**. Cross-section of two microvilli. The plasma membrane appears as a "unit" membrane (U). The core filaments are seen transversely sectioned at F. × 275,000. Courtesy of T. M. Mukherjee. **C**. Junctional complex between two absorptive cells (1 and 2). (MV) is a microvillus on one cell and (L) is part of the lumen of the intestine. The desmosome (DE) and tight junction (zonula occludens, ZO) are seen. The insert at the right is an enlarged view of a tight junction (rat kidney) showing the very close approximation of the plasma membranes [each a "unit" membrane, (U)] found at such junctions (arrow). At regions other than the tight junction the cells are separated by an intercellular space (S). (ZA) indicates a zonula adherens (see text). At the desmosome, cytoplasmic filaments (F) radiate from a dense line (D) adjacent to the plasma membranes (the unit membrane appearance is barely evident at this magnification). An additional faint line bisects the intercellular space between the cells (B). × 96,000. Insert × 150,000. Courtesy of M. Farquhar and G. E. Palade. **D**. Tips of microvilli from a section specially prepared to show the polysaccharide-rich surface coat. The coat consists of fine filaments some of which may be seen attached to the plasma membrane (arrow). × 160,000. Courtesy of S. Ito.*

The tight junctions between some cells are referred to as *electrotonic junctions*. Such junctions are characterized by the fact that experimentally-induced electrical currents can readily flow through them from one cell to another; in the absence of the junctions, adjacent cells are insulated by their plasma membranes and associated material and thus do not readily pass currents to one another. Apparently, electrotonic junctions permit a rapid flow of ions and perhaps of other small molecules as well. (Ions are the carriers of biological electrical currents or *bioelectric currents*.) Many possibilities for roles of these structures are being explored; they are often thought to aid in the coordination of separate cells (see Section 3.9.2).

As studies have progressed, several new types of junctional structures have been described; their significance is being investigated. For example, there may be modified types of tight junctions which seal only part of the intracellular space; thus, some molecules can penetrate through these tight junctions. Particularly in insects, so-called "septate desmosomes" are found; in these, adjacent plasma membranes remain separate; fine bridges or *septa* of unknown nature appear to cross the intercellular space and may permit ionic communication between cells as just described for some tight junctions.

The junctions between the *endothelial* cells lining capillaries are of importance in controlling passage to and from the blood stream of molecules such as sugars or proteins, which cannot diffuse readily through the cells as water or gases probably can. In some tissues, such as brain tissue, adjacent endothelial cells are associated by tight junctions; in others, various less restrictive arrangements are found, and there are corresponding variations in the movement of material. For example, in the liver, most molecules can pass through readily through relatively large gaps in the endothelial lining of the modified capillaries, called *sinusoids* (see Fig. I-1). This presumably facilitates the extensive exchanges that occur between hepatocytes and blood stream. (An example of such an exchange is the passage of albumin, a major protein component of blood plasma, which is synthesized and secreted into the blood by hepatocytes.)

Of interest is the fact that pinocytosis vesicles, which form at one surface of a capillary and then move across the cell and fuse with the other surface (thus releasing their contents), are widely thought to contribute to the passage of molecules across some capillary walls (see Fig. II-4).

3.5.3 **ABSORPTIVE CELLS; MICROVILLI** All cells absorb small molecules through the plasma membrane, by diffusion or by active processes of the types discussed in Chapter 2.1. Many, and perhaps most, eucaryotic cells can also take up macromolecules by pinocytosis. A few cell types show specializations at the luminal surfaces of tubular organs that are clearly related to absorption. Two of these cell types will be considered here, one in the kidney tubule and another in the intestinal tract. Absorptive cells of the kidney epithelium are most interesting for the uptake and transport of protein, and those of intestinal epithelium for fat absorption and transport.

In both cell types, the absorptive area at the lumen surface is greatly increased by enormous numbers of microvilli, arranged in precise geometric array and collectively constituting the *brush border* or *striated border* (Figs. III-18 and III-19). In the intestine this surface increase, at the cellular level, is superimposed upon the increase of the lumen surface resulting from the folding of the epithelium into villi, so that an enormous absorptive area is present. It is considered that enzymes at or near the surface of the brush border hydrolyze various carbohydrates and other molecules; the surface is coated with filaments 25–50 Å thick that are especially prominent in man, cat, and bat. They are attached to the outer surface of the plasma membrane and appear to be rich in acid mucopolysaccharide. The chief role suggested for these filaments is as a filter that keeps large particles from approaching the plasma membrane. However, some enzymes of the brush border may also be associated with the filaments.

The plasma membrane of microvilli in both the intestine and kidney tubules contains cytochemically demonstrable phosphatases (Section 1.2.3) for which roles in transport have been postulated.

A terminological convention should be noted. In absorptive epithelia, the *base* of the cell is the region facing the capillaries, and the *free surface* or *apical region* is the opposite pole where absorption occurs. A comparable convention is used for exocrine secretory cells (Section 3.6.1); the apex of such cells is the zone where secretions are stored and released.

The occurrence of microvilli illustrates one evolutionary "solution" to the surface-volume "problem" raised in Section 3.3.2; the effective surface area is greatly increased by specialized cell shapes. Spherical shapes have minimal surface; alterations from the spherical increase the surface to volume ratio.

3.5.4   *PROTEIN ABSORPTION*   As Fig. III-19 outlines, kidney tubules receive a filtrate of blood from the modified capillaries that compose the glomerulus and the thin epithelial layer of the capsule; this filtrate is modified into urine by the tubule cells which reabsorb water, salts, sugars, and other components and pass many of them back into the blood in the capillaries present near the bases of the cells. A variety of tracer proteins have

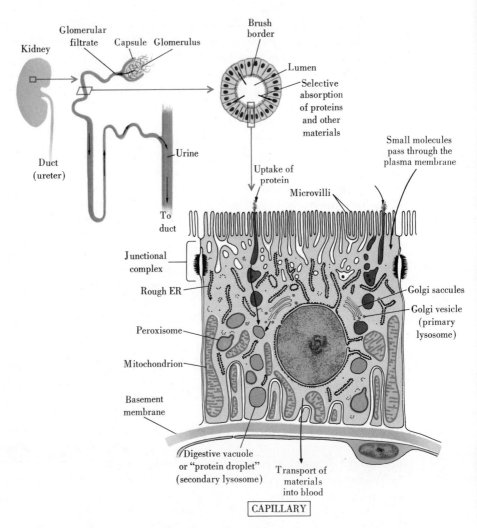

Fig. III-19 *Schematic representation of an absorptive cell of the rat kidney showing the path taken by protein absorbed from the lumen. Through selective absorption of some water, salts, proteins and other components the filtrate of blood that enters the tubule from the glomerulus (a special capillary system) is modified into urine. Many of the salts and small molecules absorbed by the tubule cells are returned to the blood in the capillaries at the base of the cells. The diagram stresses the uptake and fate of protein which appears to be digested within the tubule cells.*

been followed into kidney epithelium cells—mainly in the *proximal convolutions* the inital portions of the tubules, where much protein (especially in rats) normally is reabsorbed from the urinary fluid as it courses down the tubules. The tracers include radioactively labeled proteins detectable by autoradiography and the proteins ferritin and peroxidase that we have discussed previously (Section 2.1.4).

Proteins enter the cell in pinocytic vacuoles that form at the ends of canalicular structures projecting into the cell between the bases of the microvilli (Fig. III-19). These vacuoles merge to form larger *apical vacuoles* that move towards the base of the cell. Perhaps from Golgi vesicles, the apical vacuoles acquire lysosomal hydrolases, thus becoming secondary lysosomes. Proteins and other macromolecules are broken down within the vacuoles, or *protein droplets* as they often are called. The soluble digestive products are probably utilized by the cell.

3.5.5   TRIGLYCERIDE There is still some dispute regarding the
ABSORPTION manner in which triglycerides and other lipids enter the intestinal cell. Increasingly, it is felt that the few pinocytic vacuoles that are seen between the base of the microvilli play little, if any, role in the process. (Tracer experiments suggest that the vacuoles instead take up *protein* that eventually is digested in lysosomes). The most widely accepted theory holds that triglycerides are digested in the intestinal lumen outside the cell and diffuse into the cell at the microvillous surfaces, in the form of small molecules: monoglycerides, fatty acids and glycerol. These enter the smooth ER and are used to resynthesize triglycerides and perhaps other lipids (see Fig. II-24). The smooth ER is in two forms: an extensive network, a short distance below the microvilli, and numerous vesicles in which the triglyceride is transported toward the Golgi apparatus and toward the lateral cell borders. Most of the resynthesized triglyceride eventually enters relatively large spaces between adjacent cells, from which it is collected by capillaries of the lymphatic system which carry it to the bloodstream. The triglyceride apparently empties from the intestinal cells into extracellular spaces by fusion of the membranes delimiting the fat-containing vesicles with the plasma membrane. The fat in the extracellular spaces is in the form of *chylomicra* droplets. These contain some phospholipid and protein in addition to triglyceride. It is likely that the other components are added to the triglyceride in the ER vesicles; as Chapter 2.4 outlined, both triglycerides and phospholipids are thought to be made by smooth ER, and protein is made by rough ER. There are many continuities between rough and smooth ER, so that materials made in the rough portion could probably move readily into the smooth part.

It is interesting to note that after a fatty meal has been eaten, when a great deal of fat enters the cell, smooth ER becomes more abundant, possibly by loss of ribosomes from rough ER and by conversion of the cisternae and tubules into an interconnected meshwork and vesicles. This lends support to other evidence (Section 2.4.4) that the two systems, rough and smooth ER, may be interconvertible. After such a meal, the Golgi apparatus also contains some lipid resembling that seen in the ER; the significance of this remains to be determined.

3.5.6 DISTRIBUTION OF MITOCHONDRIA

In the kidney tubule cell (Fig. III-19), many elongate mitochondria are vertically aligned, in close relation to the plasma membrane at the base of the cell, near the capillaries. The plasma membrane shows deep infoldings. In addition, the lateral surfaces of adjacent cells fit together by protrusions from one cell fitting into indentations of the next; this *interdigitation* extends to the bases of the cells so that a thin section of one cell shows mitochondria apparently in membrane-bounded compartments that are actually extensions of neighboring cells not included in the section. The mitochondria have a great many cristae and high levels of oxidative enzyme activities, and they may be presumed to be producing much ATP. The abundance of mitochondria and the complex folding of the membrane are apparently devices that make energy available to a large area of cell surface which is actively transporting ions and other substances (reabsorbed from the tubules) between the cell and the blood.

In the intestinal cell, there are no basal interdigitations, and the mitochondria show no special orientation at the cell base. The apical mitochondria are more striking. They are very elongated and are oriented lengthwise in the cell, parallel to the ER strands that are also concentrated in the apical cytoplasm. Here, presumably, the mitochondria provide energy for the active processes involved in absorption and, perhaps, in lipid metabolism.

c h a p t e r **3.6**

SECRETORY CELLS

3.6.1 VARIETIES OF SECRETIONS

We have encountered several secretory cells in the chapters on ER and Golgi apparatus (Chapters 2.4, and 2.5). In pancreas (Figs. II-19, and II-20), pituitary (Fig. II-33), and intestinal glands (Fig. II-31), as well as in others, proteins are manufactured on the polysomes and are transported via the ER to the Golgi saccules (or saccule-derived vacuoles); here they are "condensed" into granules or viscous fluid and "packaged" into membrane-delimited vacuoles. These vacuoles subsequently open to the surface by fusion of the vacuole membrane and the plasma membrane, to discharge the secretion.

Cells that secrete protein in this way have well-developed ER and Golgi apparatus. The same is true of many cells that secrete polysaccharide-rich material, such as the mucus-secreting intestinal goblet cells and *chondroblasts;* the latter synthesize and secrete the mucopolysaccharide chondroitin sulfate, a major component of cartilage. In Chapter 2.5 we outlined evidence that steps in polysaccharide synthesis and "packaging" takes place in the Golgi apparatus, and we pointed out that in many (perhaps most) cells the secretion bodies formed by the Golgi apparatus are mixtures in which both carbohydrates and proteins are present in varying proportion and association.

There are some obvious differences among secretory cells. *Exocrine* glands release their secretions into special duct systems; the secretions of *endocrine* glands (chiefly hormones) directly enter the blood stream. Not all secretions are released by simple processes of membrane fusion; for example, in the *sebaceous glands* (which secrete the oils that coat the skin and hair) the cells fill with secretion then die and disintegrate, releasing their content. Section 2.8.2 outlines the complex path probably taken by the secretions of the thyroid gland. The gland first releases *thyroglobulin* to an extracellular storage lumen then apparently takes it back into the thyroid cells, digests it, and releases the digestion products to the blood stream. The synthesis of thyroglobulin is also an interesting process as Fig. III-20 outlines. Thyroglobulin is an *iodinated glycoprotein.* Attached to this protein are iodine atoms and short chains of carbohydrate molecules including sugars among which are *mannose, glucose,* and *galactose.* By autoradiography it can be shown that as usual, the protein part of the molecule is synthesized in the rough endoplasmic reticulum. The pattern of incorporation of carbohydrate differs with different precursors. If the cell is exposed to radioactive *mannose,* initial incorporation is in the ER whereas incorporation of radioactive *galactose* first occurs in the Golgi apparatus. Apparently, as the protein molecules are completed in the rough endoplasmic reticulum the synthesis of the attached carbohydrate chains begins; the initial portions of the chains contain mannose and glucose. Subsequently, the molecules reach the Golgi apparatus where galactose and other sugars are added, thus completing the carbohydrate chains. Iodine apparently is attached immediately after vesicles empty the glycoprotein into the lumen. Autoradiography shows that radioactive iodine (I^{125}) is rapidly incorporated into thyroglobulin near the cell surface. The enzyme responsible for the

A.

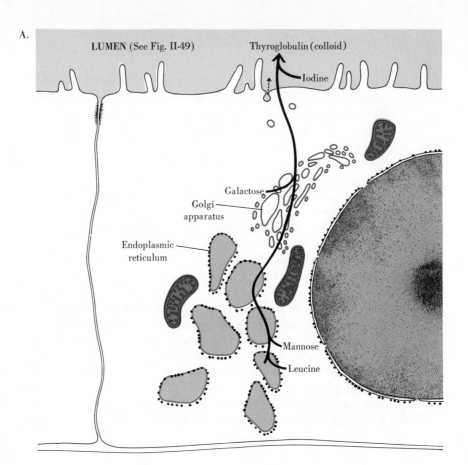

LUMEN (See Fig. II-49)

Thyroglobulin (colloid)

Iodine

Galactose

Golgi apparatus

Endoplasmic reticulum

Mannose

Leucine

B.

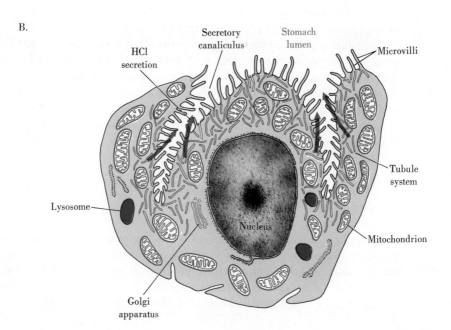

Secretory canaliculus

Stomach lumen

HCl secretion

Microvilli

Tubule system

Lysosome

Nucleus

Mitochondrion

Golgi apparatus

addition of iodine seems to be synthesized on the rough ER, transported in vesicles and then secreted into the lumen by the gland cells (Figs. II-23A, III-20A).

Important features of some interesting secretory cells are not yet entirely clarified. For example, in the antibody-forming *plasma* cells, immunoglobulins (Section 2.4.2) are synthesized in the rough ER and accumulate there (Fig. II-23) before release. Like thyroglobulin, immunoglobulins are glycoproteins to which the carbohydrate components are attached by sequential action of the ER and Golgi apparatus; this may be demonstrated by autoradiographic studies of cells of certain tumors with very high rates of immunoglobulin synthesis. It is not yet known how the organism is able to respond to the presence of foreign molecules by synthesizing antibodies (immunoglobulins) that *specifically* recognize these molecules and bind to them. How immunoglobulins are released from plasma cells is also not clear. Some observers believe that release involves fusion with the plasma membrane of vesicles that derive from ER or Golgi apparatus but lack special morphological features (such as a distinctive content) that would facilitate their identification by electron microscopy. Other suggested possibilities include fusion of portions of the ER to form large vacuoles that then fuse with the plasma membrane, disintegration of part of the cell or even direct passage of the protein molecules through the plasma membrane by unknown mechanisms.

Similar alternatives are under consideration for the release of steroid hormones synthesized in the smooth endoplasmic reticulum of the adrenal cortex and other organs.

The connective tissue components formed and secreted by fibroblasts include mucopolysaccharides, such as hyaluronic acid and the protein *tropocollagen.* Tropocollagen assembles (Section 4.1.3) to form the large extracellular collagen fibers that abound in many connective tissues. Autoradiographic studies with radioactive precursors of polysaccharides and proteins indicate the usual involvement of ER and Golgi apparatus. However, there is disagreement as to whether *all* of the protein to be secreted passes from the ER to the Golgi apparatus before secretion. The possibility remains open that at least some

Fig. III-20   *A. Synthesis and storage of thyroglobulin (after C. P. Leblond and coworkers). Details of protein release from the cell are incompletely understood; it is thought to involve fusion with the plasma membrane of vesicles derived from Golgi-associated membrane systems (perhaps from GERL; Sect. 2.4.5). An interesting possibility for vesicles transporting the iodinating enzyme is that some bud directly from the ER (Fig. II-23A) and fuse with the plasma membrane. The other phase of the thyroid secretory cycle, the reabsorption and digestion of thyroglobulin, is discussed on pp. 127–128 (see Fig. II-49). B. Schematic diagram of a parietal cell from the mammalian stomach. After S. Ito and R. C. Winchester.*

tropocollagen is released directly from the ER, perhaps by inclusion in vesicles that fuse with the plasma membrane.

Some secretory cells, such as those secreting ions, have relatively little ER and the Golgi apparatus is small. An important example is the *parietal cell* of the vertebrate stomach which secretes HCl into the stomach where the acid activates the digestive enzyme, *pepsin.* The cell surface exposed to the lumen is enormously enlarged by a great many microvilli and by deep infoldings into the cell (*canaliculi*) (Fig. III-20). *Active transport,* in which the cell expends energy (Section 2.1.3), is involved in moving hydrogen ions out of the cell into the lumen. It is hardly surprising, therefore, that parietal cells contain many stout mitochondria with numerous cristae and high levels of oxidative enzymes. Within the cytoplasm, numerous tubules delimited by smooth membranes are present. A widely held view is that ions are secreted into the tubules and probably are stored there, bound to macromolecules bearing charges opposite to the charges of the secreted ions. Connections of the tubules to the cell surface may provide a means for release of the ions to the canaliculi and thus, eventually, to the stomach lumen. The distinction between the tubule system and the scanty ER has recently been demonstrated; there are thickness differences in the membranes delimiting the two systems, and they show different staining reactions for specific phosphatases found in ER (and in the Golgi apparatus). The tubule system membrane is similar to the plasma membrane but thicker than that of ER (including the nuclear envelope). Golgi saccules, as in other cells (Fig. I-14 and I-15), show IDPase and TPPase activities but no such activities are seen in the tubule system. Finally, continuities between tubules and canaliculi, at least in certain phases of the secretory process, may be inferred from noting that when tissue is soaked in solutions of electron-opaque substances such as lanthanum salts (that cannot pass across the plasma membrane) the opaque material is later found in the tubule system, but not in the ER or Golgi apparatus.

The epithelial cells lining the swim bladders of some fish secrete oxygen and other gases into the bladder which gives buoyancy to the fish. Investigators do not all agree on how this occurs or on the roles of the epithelial cell organelles. The basal portions of the cells rest on a rich network of capillaries, and the cells attain a huge surface there by virtue of numerous deep infoldings of the plasma membrane. A current theory proposes that CO_2, glycolytically produced lactic acid (Section 1.3.1) and other substances are secreted from the cell across this surface into the blood. This induces changes in the blood that release gases from their bound form and make them available for passage into the bladder.

The mechanism by which gases are transported through the cells to the apical surface and then released to the space within the bladder in unknown. However, there are interesting cytoplasmic granules that probably play some secretory role related to the special gas-transporting function of the epithelium. They are prominent in the epithelial cells and often contain bubblelike structures (Fig. III-21). The granules apparently move to the apical cell surface and, by fusion of membranes, release their contents.

Currently, two major roles are proposed for the granules. Their content may be a material that provides a needed coating for the bladder. Granules of somewhat similar appearance are seen in lung cells and are thought to contribute to a special extracellular lining material (*surfactant*) important for lung function. Or, the granules may take part in gas transport. Do they contain enzymes similar to *catalase* and *peroxidase*, known to release free oxygen from hydrogen peroxide under certain conditions? Or, do the granules bind and concentrate gases from the hyaloplasm? Evaluation of these possibilities is especially difficult since there are no reliable methods for identifying gases by electron microscopy; the "bubbles," for example, might contain dilute fluids instead of gas.

The epithelial cells contain little ER, but the Golgi apparatus is extensive and many small vesicles are present. Perhaps the vesicles transport enzymes or other material to the granules.

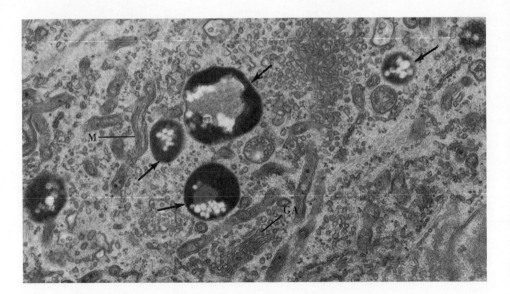

Fig. III-21 *Portion of a cell from the gas bladder of the fish,* Fundulus. *Near the Golgi apparatus (GA) bubble-like structures are seen within large granules (arrows). M indicates a mitochondrion. × 23,000. Courtesy of D. E. Copeland.*

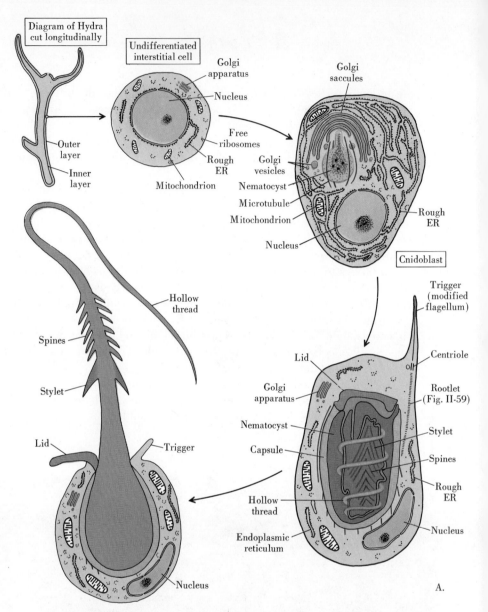

Fig. III-22 **A.** *Diagram illustrating the maturation of a nematocyst within a "cnidoblast". Based on studies of* Hydra *and other coelenterates by Slautterback, Westfall, Picken, Skaer and Lentz. As the nematocyst increases in size there is a great increase in extent of rough ER and size of Golgi saccules and vacuoles. It is after the recession in development of ER and Golgi apparatus that the nematocyst undergoes the major part of its development of complex structure.* **B.** *Developing nematocyst (NC) in a cnidoblast of* Hydra. *Oriented microtubules surround the nematocyst; in this section most are sectioned transversely (T). As indicated in A the nematocyst forms in association with the Golgi apparatus. Golgi saccules (G) and vesicles (V) are abundant. Much rough ER also is present (E). The arrow indicates a vesicle probably budding from the ER and contributing to the Golgi apparatus (or developing nematocyst; see Fig. II-19). × 30,000. Courtesy of D. B. Slautterback.*

3.6.2   ***CNIDOBLASTS*** An interesting modified secretory mecha-
              IN HYDRA nism has been described in the primitive
              animal, *Hydra*. The organism is made
essentially of two layers of epithelium. In the outer, there are special
cells (called *cnidoblasts*) that produce small projectiles, *nematocysts*,
that are shot out to pierce and paralyze prey.

The cnidoblasts develop from undifferentiated precursor cells,
the interstitial cells. The ER, sparse in the primitive cell, becomes
extensively developed in the maturing cnidoblast. Ribosomes, usually
lie free in the cytoplasm in the primitive cell, but are arranged on
the ER membranes during maturation (see Section 2.3.5). The Golgi
apparatus also becomes highly developed as the cell begins to secrete
the proteins that are stored in the *nematocyst*. The latter apparently
begins as a Golgi vacuole, small at first and then enlarging greatly.
Innumerable small vesicles develop from the much enlarged Golgi
saccules and fuse with the nematocyst. As the nematocyst enlarges,
its content becomes more electron dense (Fig. III-22). (A great many

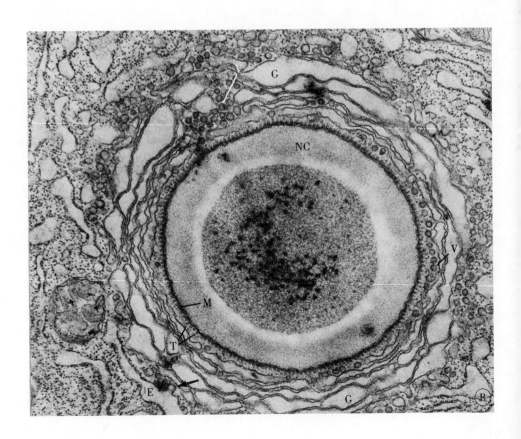

microtubules appear outside the nematocyst; they probably give rigidity to the area in which the secretory vacuole is rapidly enlarging.) When the nematocyst attains its maximum size, the ER and Golgi apparatus regress, breaking into vesicles which progressively diminish in number. Upon triggering, the nematocyst releases its contents to the extracellular environment (lower left diagram in Figure III-22A).

The striking development of ER and Golgi apparatus, followed by their virtual disappearance is but one of the interesting features of cnidoblasts. Another is the dramatic structural differentiation that occurs inside the nematocyst, *without apparent connections to other cell organelles.* (Speculation on how this might occur is found in Section 4.1.6.) Thus far, little attention has been given to the manner by which this differentiation occurs. Yet it seems likely to be a genetically determined process, involving a great many secretory proteins. In *Hydra*, there are four distinct types of nematocysts. In 100 species of coelenterates related to *Hydra*, 17 types of nematocyts have been described. Each type is characterized by a structure that is characteristic for the particular species. In one of the abundant nematocysts in *Hydra*, the complex structure includes a capsule (which apparently contains a variety of phosphatases), a lid, a coiled thread, large stylets, and smaller barbs of specific shapes (Fig. III-22).

The release of secretions from cells is under precise controls, but the mechanisms are not well understood. In *Hydra*, a *cnidocil* protrudes from the cnidoblast surface; in part, this is a modified cilium. Nematocyst release occurs when the cell is stimulated by the appropriate chemical and mechanical stimuli, such as might result from the presence of the small organisms used as food. The cnidocil is probably a mechanical receptor, and additional chemical receptors are presumably built into the plasma membrane. Release is also influenced by the primitive nervous system found in *Hydra*. In higher organisms, nervous and hormonal stimulation strongly influence the rates of release of different secretions, but how such extracellular influences can bring about appropriately controlled fusion of membranes surrounding secretion granules with the plasma membrane is not understood. One hypothesis currently being evaluated is the proposal that nervous and hormonal influences produce changes in the permeability of the plasma membrane that result in the influx of specific components (such as calcium ions); changes in the intracellular concentration of these components bring about release of secretion. Once granule release has occurred, some mechanism (perhaps resembling pinocytosis) must come into play to remove the membrane that has been added to the cell surface; cells stimulated to release many granules in a short period of time show only a transient increase in the area of surface membrane present at the cell poles where secretion is released.

chapter **3.7**

NERVE CELLS

Neurons include the longest cells in the body, ranging to several feet in length. They are the coordinating elements of an elaborately interconnected network. The network consists of sensory receptor cells feeding into sensory neurons, enormous numbers of connecting and integrating neuronal circuits, and motor neurons leading to effector organs such as muscle. A given neuron can interact, more or less directly, with hundreds of other cells.

Figure III-23 shows two of the major morphological types of neurons, *unipolar* and *multipolar*. Many unipolar neurons, such as vertebrate sensory neurons, are characterized by a single process that divides into two branches. One branch connects to a sensory receptor, and the other to other neurons via *synapses*. Multipolar neurons like the motor neurons of the spinal cord have numerous receptor processes (*dendrites*), which receive impulses at synapses with other neurons, and a single transmitter process (the *axon*), which carries impulses from the cell body to effectors. Both types of neurons are fundamentally similar in other respects.

3.7.1   **PERIKARYON**   The neuronal cell body, or *perikaryon*, con-
tains the cell nucleus and cytoplasmic organelles distributed around the nucleus in an arrangement that is roughly symmetrical (Fig. III-23). The Golgi apparatus is a well-developed network (Fig. I-15). Rough ER is extensive; it is concentrated in the so-called *Nissl substance*, patches (Fig. I-12) showing parallel rough cisternae plus many free ribosomes. Many lysosomes are present and tend to be concentrated near the Golgi apparatus. Mitochondria are numerous throughout the cell. In general, the cytological characteristics suggest extensive macromolecular synthesis: large nucleolus, many ribosomes, extensive rough endoplasmic reticulum, and large Golgi apparatus. This is to be expected from the role of the perikaryon as a synthetic center that supplies macromolecules and other material to the rest of the neuron.

The Golgi apparatus and associated endoplasmic reticulum are probably involved in part, in the formation of the numerous lysosomes. As in other cells, autoradiographic studies indicate that some protein made in the rough endoplasmic reticulum passes into the Golgi apparatus of neurons (Fig. II-32). In special *neurosecretory* neurons, found in the pituitary gland and elsewhere, the Golgi apparatus packages hormone granules which travel down the axons and which are released

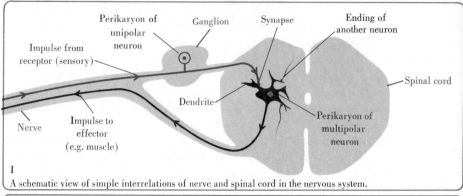

I

A schematic view of simple interrelations of nerve and spinal cord in the nervous system.

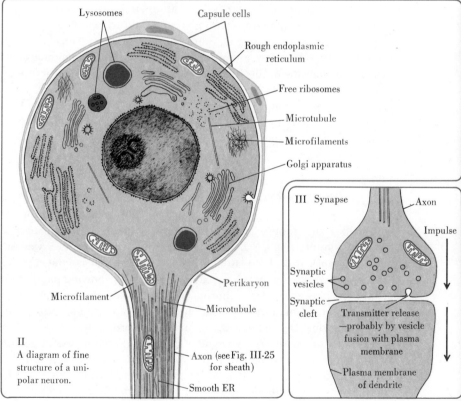

II
A diagram of fine structure of a unipolar neuron.

III Synapse

Fig. III-23 *Nerve cells.*

at the axon endings to enter extracellular spaces or the blood stream. In most neurons, the role of the Golgi apparatus is less clear, although at least some of the substances involved in transmission of nerve impulses from one cell to another appear to be packaged in membrane-delimited structures that originate in it: as discussed below, transmission is based on phenomena akin to secretion which occur at axon endings.

The abundance of lysosomes is of uncertain significance. Autophagia occurs in normal perikarya, but relatively infrequently. Pinocytosis vesicles are present at the surface of many neurons and protein uptake into lysosomes by these vesicles has been demonstrated in a few neurons. Yet, there is no reason to believe that either function is extensive enough to account for the numerous lysosomes. Many of the lysosomes of neurons are residual bodies. Neurons do not divide and many live as long as the organism, slowly accumulating residual bodies. As mentioned in Section 2.8.5, with time, lipofuscin ("aging pigment") accumulates within residual bodies.

The abundance of RNA in neuron perikarya has led to the speculation that learning and memory might be based on the coding of information in nucleic acids. Unfortunately, experimental tests of this idea are difficult and the results are equivocal. In some learning experiments in which animals are trained to perform specific acts, changes in the composition and amount of RNA are observed. However, these could reflect general effects on neuronal metabolism in which RNA is involved, as readily as the specific synthesis of a "memory" RNA; protein synthesis is thought to be involved in some phase of "memory storage." There also have been experiments in which some investigators believe they have transferred "memory" by injecting into one organism RNA purified from another. The more widely held theory of learning considers that new experiences are "stored," in the form of connections between neurons that are either newly formed after a novel experience or in some sense newly activated. There is indirect evidence that seems to support this theory, but fuller evaluation will require an understanding of the controls of nerve growth and function. The complexity of nervous organization in organisms capable of observable learning has thus far hampered direct testing of this theory; it is difficult to trace the neuronal circuits that might be involved.

3.7.2   *AXONS*   In contrast to the cell body, the axon contains neither Golgi apparatus nor rough ER. There are few ribosomes. On the other hand, microtubules and fine filaments are numerous; they are oriented along the long axis of the axon. There are some smooth-surfaced tubules and vesicles, probably including smooth ER. Mitochondria are frequent, lysosomes infrequent. The axon appears to be largely inactive in synthesis of protein or other macromolecules.

Studies of living nerves indicate extensive cytoplasmic flow in axons. Although it occurs in both directions, a net flow down the axon from the perikaryon results. Thus it appears that the perikaryon is continually manufacturing molecules needed for the maintenance and functioning of the axons. These molecules are transported down the axon. Even mitochondria and other organelles are believed by many to pass into the axon from the perikaryon. Different components move at rates ranging from one to many mm. per day, but the motive elements have not been identified. Such continual manufacture and transport of axoplasm is an enormous metabolic task for the perikaryon, especially since the cell body represents a small fraction of the total cell volume.

Since neurons do not divide and no reservoir of undifferentiated "precursor" cells exists, if the perikaryon is destroyed, the neuron is not replaced. However, if the axons are cut, only the portion no longer attached to the perikaryon degenerates; the remainder is capable of regenerating. Under appropriate conditions, neurons of the peripheral nervous system (nerves and ganglia) can re-establish some of the original connections and so restore neuron function. Proper conditions for full regeneration and re-establishment of the connections of neurons of the mammalian central nervous system (brain and spinal cord) have not been found.

3.7.3   **THE NERVE** The most readily observable feature of the **IMPULSE** passage of a nerve impulse along an axon is a change in the potential difference that exists across the plasma membrane; this change (*action potential*) moves like a wave along the axon surface. Current theory holds that the underlying mechanism is based on the flow of ions accompanying a wave of permeability changes that passes down the axon membrane. In resting state, K^+ ion concentration is maintained high and Na^+ concentration low in the axon (as compared with the extracellular space) by an energy-requiring active transport mechanism (Section 2.1.3). A somewhat oversimple but useful view of the consequences of this asymmetric distribution of ions is as follows: It can be shown in model experimental systems and by theoretical treatment that when such asymmetries are established across a membrane, an electrical potential will result if some of the ions can diffuse through the membrane more rapidly then others. K^+ can pass across the axon membrane much more rapidly than can Na^+. (What is being considered here is the passive movement down concentration gradients that continually tends to restore equal concentrations on both sides of the membrane and is counteracted by the active transport.) The tendency for K^+ to leave the axon more rapidly than Na^+ enters produces, at equilibrium, a net relative deficiency of positive charges on the inside of the membrane. In consequence the inside of the cell membrane is at an electrically negative potential as compared with the outside; this is referred to as the *resting potential*. Impulse conduction by a given region of the axon is based on the following sequence (Fig. III-24):

(1) A localized permeability change permits the rapid influx of Na^+. This influx is associated with changes in the potential across the membrane; eventually the inside becomes positive with respect to the outside.

(2) A second set of permeability changes restricts the movement

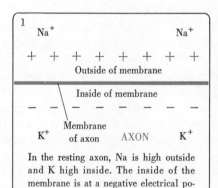

1 In the resting axon, Na is high outside and K high inside. The inside of the membrane is at a negative electrical potential compared to the outside.

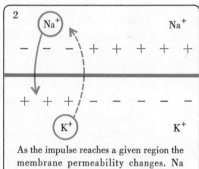

2 As the impulse reaches a given region the membrane permeability changes. Na enters and the potential difference reverses. Changes in permeability to K follow.

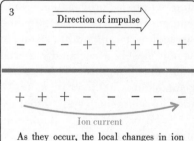

3 As they occur, the local changes in ion concentration and in potential lead to longitudinal currents of ions (or equivalent phenomena) which trigger permeability changes in adjacent regions of the membrane.

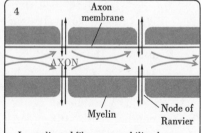

4 In myelinated fibers permeability changes take place only at the nodes where the insulating myelin sheath is interrupted. Between nodes longitudinal current flow carries the impulse which thus "jumps" from node to node.

Fig. III-24 *Outline of nerve impulse propagation. Currents that flow outside the axon accompanying the currents inside are not shown.*

of Na$^+$ but permits the rapid efflux of K$^+$. This is associated with eventual restoration of the original negativity of the inside.

(3) Propagation of the impulse to adjacent axon regions occurs. This results from the fact that the local changes of (1) and (2) produce an axon region that differs in ion concentration and electrical potential from adjacent regions. A "flow of electrical current" occurs between this region and the adjacent regions. The flow takes place both within the axon and outside it and is either an actual longitudinal movement of ions or some equivalent process tending to restore uniformity along the axon. It results in initiation of the cycle of permeability changes [(1) and (2)] in membrane regions adjacent to the local region under consideration. Thus, the impulse moves along the membrane. (Exactly how a longitudinal ion flow can trigger the cycle is not known. However, part of the answer probably concerns the fact that the flow decreases the potential difference across the membrane, and that decreases in potential difference can trigger nerve impulses.)

(4) The intracellular low Na—high K condition is restored, relatively slowly, by a "sodium pump" which produces active pumping out of Na$^+$ and reciprocal influx of K$^+$. This can be relatively slow since only a small percentage of the ions move in a single impulse; the very large initial asymmetry in Na and K distributions can permit the passage of many impulses before pumping becomes absolutely essential to further functioning. Normally the cycle of changes in permeability at a given axon region is completed in a few thousandths of a second. The impulse passes down the axons of different nerves at rates of tenths to tens of meters per second.

Only in the last step related to impulse propagation (Step 4 above) does the cell actively expend energy. In the living cell the mitochondria probably provide energy for the sodium pump, but the axon membrane itself (perhaps together with some closely associated material) seems to be solely responsible for conduction of impulses. This has been nicely demonstrated by using the giant axons, 500 μ in diameter, found in squid. Virtually all of the intracellular organelles and other cytoplasm can be removed from such axons and replaced by suitable solutions of ions in which the Na$^+$ concentration is kept low and the K$^+$ concentration high; the preparations still can conduct impulses quite efficiently. The nature of changes of the axon membrane that are reponsible for passage of impulses has yet to be determined. Very probably, rapid realignment or changes in shape of membrane molecules produce the observed permeability changes. The identity of these molecules and the mechanisms whereby they change are being actively sought. One proposal is that membrane molecules contain charged regions which change their alignment in response to changes in electrical fields, such as those occuring when an impulse passes down the axon; another proposal is that specific chemical reactions take place.

3.7.4   *MYELIN*   In vertebrate nerves, two major types of axons are found: *myelinated* and *unmyelinated*. In both, *Schwann cells* (or *neuroglial* cells in the central nervous system) are closely associated with the axon. In the unmyelinated type, the axon occupies a pocket formed by the indentation of the Schwann cells (Fig. III-25). The Schwann cells of the myelinated axons wrap repeatedly around the axon, surrounding it with many layers of their plasma membranes. The multilayered membrane system constitutes myelin (Fig. III-25); it is part of the Schwann cell. The rest of the cytoplasm and the nucleus remains at the outer surface of the multiple layers of plasma membrane. Each Schwann cell contributes one segment of myelin to the sheath that covers most of the length of the axon. The segments of sheath contributed by two Schwann cells abut, but do not fuse, at *nodes of Ranvier.* The lipoprotein of which myelin is made is a good electrical insulator. Thus, a myelinated fiber is surrounded by a layer of insulating material that is interrupted at the nodes. This modifies the conduction of the nerve impulse. The changes in membrane permeability responsible for conduction are thought to take place only at the nodes (Fig III-24). This kind of conduction, by "jumping" from node to node is called *saltatory transmission* (*saltation* refers to jumping); each node may be thought of as amplifying the impulse and passing it down the axon to the next node. The permeability change at one node results in local concentration changes and sets off a flow of ionic current within the axon. This it turn triggers the permeability change at the next node. The "jumping" is responsible for the more rapid impulse conduction by myelinated nerves, as compared with unmyelinated nerves; the latter lack the extensive insulation that makes saltatory transmission possible.

3.7.5   *SYNAPSES*   The impulse travels from one neuron to another through synapses. Transmission from cell to cell, in a few cases, appears to involve junctions of the electrotonic junction type (Section 3.5.2). More commonly, however, such direct transmission is precluded by the absence of the requisite special junctions. Instead chemical transmission takes place. *Acetylcholine* and *noradrenalin* (a compound related closely to adrenalin) are among the known chemical transmitters; different classes of neurons utilize different transmitters. A synapse is diagrammed in Fig. III-23. The numerous vesicles are generally believed to hold neurotransmitter

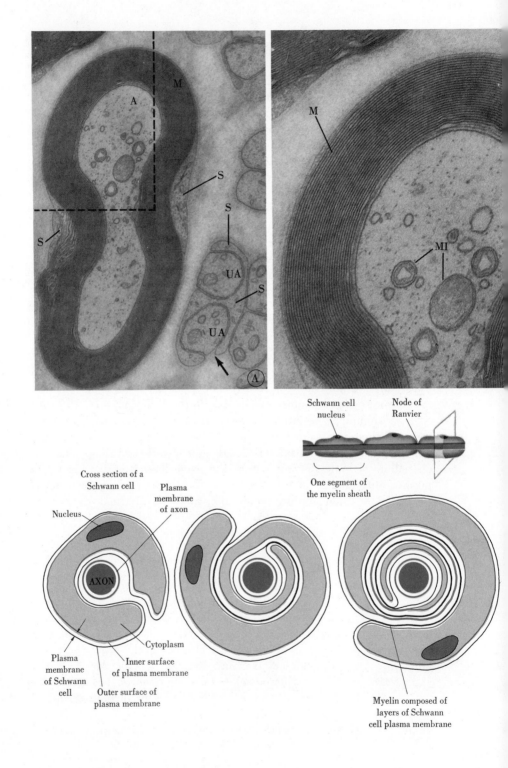

Schwann cell nucleus

Node of Ranvier

One segment of the myelin sheath

Cross section of a Schwann cell

Plasma membrane of axon

Nucleus

AXON

Cytoplasm

Plasma membrane of Schwann cell

Inner surface of plasma membrane

Outer surface of plasma membrane

Myelin composed of layers of Schwann cell plasma membrane

substances. When the impulse reaches the synapse, the vesicles apparently fuse with the membrane and release the transmitter into the extracellular "synaptic space." The transmitter interacts with receptor sites on the plasma membrane of the "receiving" cell (the *post-synaptic* plasma membrane), and these sites mediate initiation of an impulse; the initiation mechanism is not known but probably is based on changes in membrane permeability. Similar phenomena probably operate both for neuron-to-neuron transmission and for neuronal stimulation of other cells such as muscle cells. An enzyme, *acetylcholinesterase*, is present in synapses of neurons transmitting by acetylcholine. This enzyme can destroy acetylcholine very soon after it is released. This apparently gives an important measure of control to the system: if the acetylcholine remained intact in the synapse, it would continue to stimulate impulses in the receiving cell. Thus, for example, one nerve impulse might set off a long series of twitches in a muscle rather than the single one that is usually obtained. Axon endings that release noradrenalin also can absorb this compound; probably this is important in control and conservation mechanisms.

The adrenalin-related neurotransmitters in synapses include material in the form of membrane-delimited granules. Granules of very similar appearance are seen in the axon and perikaryon; they appear to form in the Golgi apparatus and are presumably transported down the axon to the synapse. The possibilities that some transmitters are synthesized or "packaged" in the axon or in the enlarged axon endings found at synapses are actively being studied. The granules just discussed probably contain enzymes involved in noradrenalin synthesis and storage. Enzymes that can synthesize acetylcholine are abundant at appropriate axon endings.

Fig. III-25   *Myelin. **A.** Electron micrographs of a portion of sciatic nerve (guinea pig) sectioned transversely. A myelinated axon (A) and several unmyelinated axons (UA) are shown. (S) indicates Schwann cell cytoplasm. A narrow space continuous with the extracellular space outside the Schwann cell (arrow) separates unmyelinated axons from the Schwann cells that surround them; the arrangment often resembles that shown in the lower left diagram. Myelin (M; enlarged at right) is of multiple membrane layers. Mitochondria in the axon are seen at MI. × 20,000; insert × 50,000. Courtesy of H. Webster.* ***B.*** *Schematic diagram of myelin formation by the establishment of a multilayered spiral-like pattern of Schwann cell plasma membrane. The cytoplasmic space and extracellular space originally separating membrane layers are obliterated so that the final pattern consists of layers where the original external surfaces of the membrane are closely apposed, alternating with layers where the original cytoplasmic surfaces are closely apposed. In mammals the myelin segment contributed by a single Schwann cell (this corresponds to the distance between successive nodes) may cover a length of several hundred microns of axon.*

c h a p t e r **3.8**
SENSORY CELLS

A wide variety of sensory receptor cells occurs throughout the body. Each is specialized so that it responds to a given type of stimulus, initiating a nerve impulse. Some respond to mechanical pressure, others to specific classes of chemicals. One of the most interesting responds to light. This is the *retinal rod cell*, capable of mediating vision at low illumination levels; another similar set of retinal cells, the *cones*, functions chiefly in high illumination and color vision. These photoreceptive cells line the retina and they make synaptic contact with connecting cells that synapse with neurons of the optic nerve.

3.8.1   ***THE RETINAL*** The rod cell (Fig. III-26) consists of a cell
ROD body, with nucleus, numerous mitochondria
and other organelles, connected by a cyto-
plasmic bridge or stalk to an *outer segment* specialized for light reception. At the base of the connection, a basal body with centriole-like
structure is found, and within the cytoplasmic stalk, an arrangement
of microtubules characteristic of cilia is present. From this appearance
and from developmental studies, it has been concluded that the outer
segment of the rod is a greatly modified derivative of a cilium. Other
cilia of sensory type are known, but often the modifications are not as
great as in the rod. Sensory cilia often lack the central pair of microtubules characteristic of most cilia and flagella. (See Section 3.6.2
for another example.) They are spoken of as "9 + 0" cilia, in contrast
to the "9 + 2" cilia discussed earlier (Chapter 2.10).

The rod outer segment is a cylindrical body, delimited by a plasma
membrane, containing several hundred flattened sacs stacked on top
of one another (Fig. III-26). The membrane-bounded sacs contain
molecules of the visual pigment, *rhodopsin*. The sacs apparently originate as plasma membrane infoldings which then pinch off. (The chief
morphological difference between cone cells and rods is the persistence of continuity of the stacked sacs with the cone cells' plasma
membranes).

The arrangement of the photoreactive material in stacked membranes provides a large surface where light can be absorbed. Studies
with polarizing microscopes have shown that the rhodopsin molecules
within the membrane exist in orientations that provide extraordinarily
high efficiency of light absorption. When it absorbs light, the rhodopsin
molecule splits into *retinene*, related to vitamin A, and *opsin*, a protein.

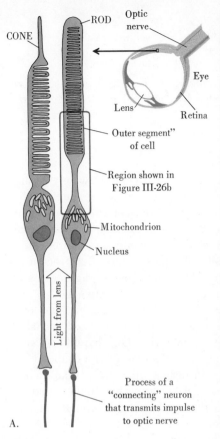

CONE

ROD

Optic nerve

Eye

Lens

Retina

Outer segment" of cell

Region shown in Figure III-26b

Mitochondrion

Nucleus

Light from lens

Process of a "connecting" neuron that transmits impulse to optic nerve

A.

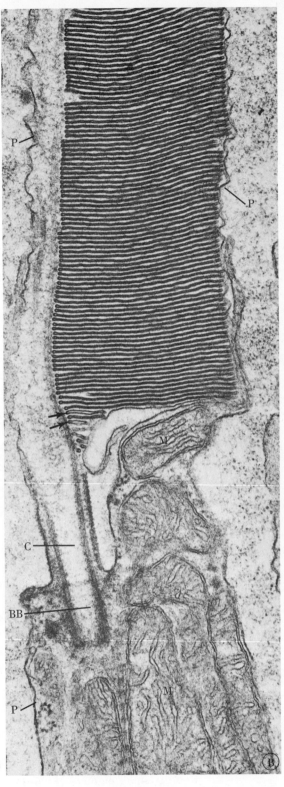

P

P

M

C

BB

P

M

B

Fig. III-26 *Retinal receptor cells of vertebrates.* **A.** *Schematic diagram of a portion of the retina. The receptor cells have their light sensitive portions directed towards the back of the eye. As indicated in the diagram, the stacked membrane systems in the cones remain connected to the plasma membrane.* **B.** *An electron micrograph of a portion of a rod cell like that outlined in* **A.** *The basal body (BB) and cilium (C) that connect the inner and outer segments of the cell are seen in longitudinal section. Many mitochondria (M) are present near this connection. P indicates the plasma membrane. The outer segment contains the stacked membrane systems (lamellae) in which the photoreceptive pigments are located. These appear to form as sacs (arrows) which eventually flatten, obliterating the space within. Courtesy of D. W. Fawcett.*

The appearance of one or a few molecules of altered photosensitive molecule can set off a nerve impulse by mechanisms that are as yet unknown.

3.8.2   ***LIGHT***   The structure of eyes differs widely among
        RECEPTORS IN   organisms. In contrast to the single light-
        INVERTEBRATES   sensitive retina of the vertebrate eye, many
                  invertebrates (for example, crustaceans
and insects) have *compound eyes* made of repeating photoreceptive units, each with its own set of cells. In these units, light reception centers at the *rhabdomeres* which are densely packed aggregates of microvilli protruding from the surfaces of the receptive cells. The microvilli, like the retinal rod sacs, are oriented with their long axes perpendicular to the surface from which light enters. They probably provide a large area of light-receptive material. One hypothesis suggests that light-induced changes in the molecules of the rhabdomere membrane lead directly to changes in permeability to inorganic ions; since the microvilli are part of the cell surface, this could initiate a nervelike impulse spreading across the surface and leading ultimately to transmission to connecting neurons.

Some invertebrates possess light-sensitive cells outside of their eyes. For example, crayfish have light receptors on their abdomens which apparently serve as protective devices against predators; the response to altered illumination leads to rapid movement of the organism. One of the cells that has been tentatively identified as an abdominal light receptor is shown in Fig. III-27. The cell contains many membranes arranged in a whorl which presumably provide extensive membrane surfaces, comparable to those found in other light receptors and in sensory cells generally.

Many unicellular organisms possess special light absorbing *eyespot* or *stigma* regions. These are probably parts of systems for orienting the organisms with respect to light; such orientation may be especially important in photosynthetic forms. Sometimes the eyespots are discrete zones within chloroplasts (see Fig. II-43). In other cases they are separate structures. Often they are found near a basal body of a flagellum. The pigment in the eyespot regions may be in the form of large spherical bodies.

The role of light-absorbing pigments in eyespots may be quite different from that played by photosensitive pigments in receptors of higher organisms. For example, an interesting explanation has been put forth for the light responses of the flagellate *Euglena*. It is suggested that the small pigmented eyespot near the flagella is so positioned that

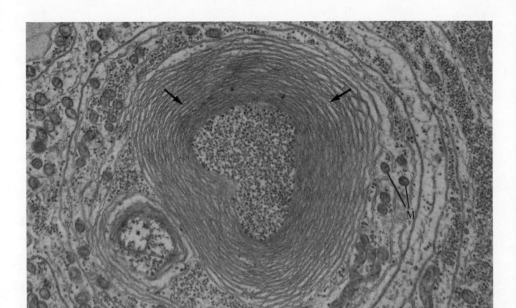

Fig. III-27   *Portion of a cell thought to be an abdominal light receptor in the crayfish. The extensive system of concentric membranes found in the cytoplasm is indicated by arrows. Mitochondria are seen at M. × 15,000. Courtesy of A. J. D. DeLorenzo.*

when the organism is oriented at certain angles to the light, a shadow of the eyespot is cast on a special swelling located at the base of one flagellum. The swelling is presumed to be light sensitive and to respond to the alterations in light intensity (resulting from different shadow positions) by controlling flagellar motion and thus the movement of the organism.

c h a p t e r **3.9**
MUSCLE CELLS

Skeletal muscle is made of bundles of large multinucleate cell masses called *fibers* (see Fig. II-63). Both smooth and cardiac muscles are made of separate uninucleate cells; smooth muscle is composed of bundles or sheets of cells (Fig. III-17), cardiac muscle of a branched network of cells joined end to end.

The contraction of all three major types of muscles—skeletal, cardiac, and smooth—is dependent upon the presence of the proteins, actin and myosin. As discussed in Section 2.11.3, it is the regular arrangement of actin- and myosin-containing filaments that is responsible for the *striations* of cardiac and skeletal muscle. The sliding of filaments past one another is the basis of contraction of these types of muscle.

Relatively little is known about the mechanism of smooth muscle contraction. Longitudinally arranged filaments abound in the cytoplasm, but do not show the ordered arrangement seen in striated muscle. No discrete myofibrils and no well-differentiated systems of alternating thick and thin filaments are present.

3.9.1 STRUCTURE AND FUNCTION OF SKELETAL MUSCLE

A skeletal muscle fiber (Figs. II-63 and III-28) is bounded by a *sarcolemma*, the plasma membrane with an overlying layer of extracellular material. Numerous peripherally located nuclei are present in each fiber. Golgi apparatus, rough ER, and ribosomes are scanty and are concentrated near the nuclei. Myofibrils occupy most of the cytoplasm. Closely adjacent to the myofibrils are mitochondria; their close topographic relation to the contractile material results in the efficient transfer of ATP. Glycogen, in the form of particles 200–300 Å in diameter, is scattered through the cytoplasm, providing a storage supply of carbohydrates.

A membrane-bounded tubule is associated with each sarcomere. The tubules constitute the *T system*. Marker molecules such as ferritin and peroxidase can be seen to pass rapidly from the extracellular medium into the T system. This confirms the finding by electron microscopy that the tubule system is continuous with the plasma membrane and is therefore open to extracellular fluids. The smooth ER of striated muscle is called the *sarcoplasmic reticulum*. Sarcoplasmic reticulum is closely associated with each T-system tubule and forms a network between tubules (Fig. III-28). Although the sarcoplasmic reticulum comes into close contact with the tubules, the reticulum and T system are not continuous with one another; extracellular peroxidase and other markers do not pass into the reticulum.

The elaborate arrangement of the T system and sarcoplasmic reticulum membranes is thought to coordinate the contraction of the fiber. Nerve impulses set off impulses in the plasma membrane of the muscle fiber that are probably similar in mechanism to the nerve impulses. One hypothesis to explain coupling of the electrical impulse to the contraction of the muscle fibrils proposes that the impulse passes into the T system, and that this results in release of calcium ions from

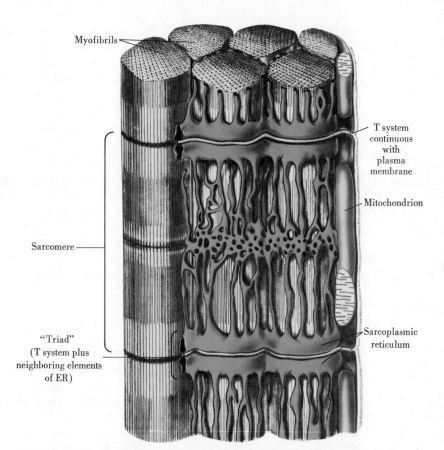

Myofibrils

T system
continuous
with
plasma
membrane

Mitochondrion

Sarcomere

Sarcoplasmic
reticulum

"Triad"
(T system plus
neighboring elements
of ER)

Fig. III-28 *The T-system and sarcoplasmic reticulum of a frog muscle. The conduction of an impulse by the plasma membrane at the fiber surface depends on mechanisms comparable to those occurring at the axon membrane (Fig. III-24). The possibility is being explored that conduction within the T-tubules occurs by flow of ionic current somewhat comparable to the longitudinal flows within the axon. The mechanism of coupling between T-systems events and changes in the sarcoplasmic reticulum is incompletely understood. Diagram modified from L. D. Peachey and W. Bloom and D. W. Fawcett.*

the sarcoplasmic reticulum into the myofibrils. This, in turn, is thought to activate the ATPase of myosin, which splits ATP and initiates contraction. Following contraction, the calcium is reabsorbed by the sarcoplasmic reticulum. This theory is based in part on observations that calcium can help to activate myosin ATPase activity and that isolated microsome fractions (Section 1.2B) containing portions of the sarcoplasmic reticulum are known to bind calcium and concentrate it from the medium. ATPase activity, possibly related to calcium transport, is also found in the sarcoplasmic reticulum, as shown by studies of isolated membrane fractions and by microscope cytochemistry (Section 1.2.3). In the absence of something like the T system, it is difficult to see how an external plasma membrane change could produce virtually simultaneous response of all the sarcomeres on all the myofibrils throughout the mass of a muscle fiber many microns in diameter; if "trigger" substances had to diffuse in from the outer surface of the fiber, the outer fibrils would contract far in advance of the ones located at the center of the fiber. The T system penetrates throughout the fiber, so that even the sarcomeres of fibrils at the center can be affected directly by the impulse at the plasma membrane. Impulse conduction by the membrane is a much more rapid process than diffusion; molecules move by diffusion at rates measured in microns per second while impulses are transmitted at rates of meters per second. Muscles of different organisms often contain somewhat different arrangements of the T system and sarcoplasmic reticulum, but the basic function is probably the same in all striated muscles.

3.9.2 COORDINATION OF CARDIAC AND SMOOTH MUSCLE Sarcoplasmic reticulum and a T system are also found associated with the myofibrils of cardiac muscle. However, in this muscle type and in smooth muscle, an additional problem in coordination results from the fact that both these muscle types consist of groups of separate cells, each relatively small in comparison to a skeletal muscle fiber. In both smooth and cardiac muscle, many tight junctions between cells are observed. It has been assumed that these junctions coordinate adjacent cells by providing a pathway for coordinating impulses (Section 3.5.2). An impulse triggered by a nerve ending on one cell could spread, via tight junctions, to other cells.

c h a p t e r **3.10**
GAMETES

The details of cellular mechanisms related to sexual reproduction vary considerably among different organisms. However, in virtually all sexually reproducing eucaryotes, two features are constant. At some point in the life cycle, two special cell divisions occur; together, the two divisions constitute *meiosis*. The meiotic divisions result in the halving of the number of chromosomes and, thus, halving of the amount of DNA per nucleus. At a subsequent stage, two cells with the halved chromosome numbers resulting from meiosis, fuse to form a zygote. The nuclei also fuse to produce a single nucleus combining the chromosomes from the two cells. The details of chromosome behavior and the genetic consequences of these processes will be considered in Chapter 4.3. For the moment, it should be noted that zygote formation usually involves the fusion of cells and combination of chromosomes from two different parent organisms. This results in a zygote nucleus with a new genetic combination.

The cells that fuse and contribute the chromosomes of zygotes are generally referred to as *gametes*. Sometimes, as in many unicellular organisms and in some lower plants and animals, male and female gametes do not differ much in structure. In some cases, fusion is partial and temporary. Thus, during conjugation in *Paramecium*, two micronuclei are present in each cell and one from each cell passes into the other cell through a cytoplasmic bridge that forms as a transient structure. In most higher organisms, male and female gametes are distinctly different, and gametes fuse completely (fertilization). Usually, though not invariably, the male gametes are smaller and motile; cilia or flagella are present on most male gametes of multicellular animals and lower plants, but some species have ameboid gametes. In flowering plants, pollen grains germinate to form a pollen tube that grows through the female structures of the flower and ultimately brings the male gamete nuclei into the ovary where the eggs are located.

In this chapter, we will discuss the gametes of higher animals which exemplify a high degree of gamete specialization.

3.10.1 **NUCLEI** The differences between sperm and egg are related to their different roles in zygote formation. Transmission of chromosomes from the male parent depends on specialized features of sperm cells. Egg cells carry the chromosomes from the female parent and also contain the reserve material used in the early development of the embryo. The gametes form from *gonial* cells (spermatogonia or oögonia). These divide to produce a population of *spermatocytes* or *oöcytes*. Each spermatocyte undergoes the two cell divisions of meiosis to produce four *spermatids*, which start off as more or less round cells but differentiate into mature, elongate *sperm* (Fig. III-32). Oöcytes usually undergo extensive growth before or during meiosis; in many animals the meiotic divisions are not completed until after fertilization and usually only one of the four division products gives rise to a mature *ovum* (the others degenerate).

The *nuclei of mature sperm* are small and the chromatin is densely packed and entirely inactive in RNA synthesis. In some animals (for example, fish) the histones (Section 2.2.2) are replaced by *protamines*, very basic proteins unusually rich in the amino acid, *arginine*.

Often unusual lamellar or tubular patterns of chromatin distribution are seen (Fig. III-29). In other cases, the chromatin is so densely packed that it appears to be a homogeneous mass. Some sperm nuclei are elongate; in extreme cases, they may be more than ten times longer than wide. In such nuclei, the chromosomes are often lined up as coiled fibers, with long axes running along the length of the head.

The *nuclei of growing oöcytes* and ova are quite large and the chromatin is dispersed within to such an extent that often it can not be readily seen in the microscope even after use of highly specific stains for DNA. The chromosomes may assume a special "lampbrush" configuration and sometimes a great many nucleoli are present; these features will be discussed in Section 4.4.5. The large oöcyte nuclei engage in intensive RNA production, supporting their own cytoplasmic growth and providing RNA that is stored and used in later embryonic development.

3.10.2 **CYTOPLASM** Sperm of higher animals are elongated, highly modified cells (Fig. III-32); widths of a few microns and lengths of 50–100 μ are not uncommon. Most of the length is accounted for by the flagellum. The *head* which contains the nucleus may make up only 5–10 percent of the length (Fig. III-32). The front of the head contains an *acrosome*. This large, membrane-bounded structure is produced by the Golgi apparatus as the spermatid matures. Lysosomal enzymes have recently been shown to be present in

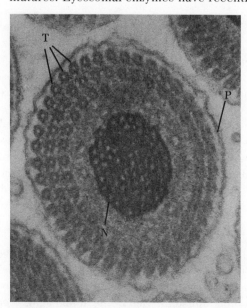

Fig. III-29 *Cross section of the head of a sperm from the insect, Steatococcus. The head has an elongate cylindrical shape; the many microtubules oriented longitudinally (T) are presumed to contribute to the establishment or maintenance of this shape. In the nucleus (N) the chromatin is arranged as dense material surrounding less dense tubular regions which appear in section as light circles. P indicates the plasma membrane.* × *100,000. Courtesy of M. Moses.*

the acrosome of some sperm, leading to the suggestion that acrosomes are highly modified lysosomes.

It is also during spermatid maturation that the flagella grow in association with the centrioles. As the flagella form, the centrioles become situated at the base of the nucleus. The flagella show the usual $9 + 2$ tubule pattern, often accompanied by additional fibrous structures that run alongside the tubules (Fig. III-32). In the *midpiece*, the region just behind the head of the sperm, the mitochondria are aggregated, near the flagellum. Quite often the mitochondria are in ordered arrays; sometimes they form a spiral ribbon that twists around the flagellum for some distance. Presumably this facilitates transmission of ATP used in sperm movement. The flagellum projects beyond the midpiece as the *tail*.

The entire sperm is covered by a plasma membrane, but aside from the acrosome, mitochondria, centrioles, and flagellum, little cytoplasm is present. Golgi apparatus, ER, ribosomes, and other cytoplasmic components are sloughed in a cytoplasmic bud that forms as the spermatid matures into a sperm and eventually separates from the sperm and disintegrates.

In contrast to sperm, ova are often enormous cells with diameters ranging from roughly 100 μ in many mammals to over 1 mm in some invertebrates and amphibians; occasionally they are even larger, as in reptiles and birds. Their abundant cytoplasm accumulates during oöcyte growth. The cytoplasm contains the usual organelles, plus distinctive *yolk bodies* which may occupy much or most of the cytoplasmic volume (Fig. III-30). Yolk contains varying proportions of stored lipids, carbohydrates, and proteins, used later in development. Numerous ribosomes some bound, many free (Section 2.3.5) are present in egg cytoplasm. These are used by cells of the early embryo; in some organisms, new ribosomes are not synthesized by the embryo until relatively late in development. (Thus far this has been studied only in a few organisms.) The polysomes of mature but unfertilized eggs are inactive in protein synthesis. They are activated after fertilization.

The details of yolk formation in the oöcyte varies from species to species. Yolk may accumulate in Golgi vacuoles, endoplasmic reticulum (Fig. III-30), or occasionally in mitochondria. In at least some species, considerable amounts of material are also taken up by pinocytosis and contribute to the yolk (Figs. III-30 and III-31). Some of this material may come from follicle ("nurse") cells that surround many growing oöcytes; much material probably comes from the blood. The pinocytosis vesicles carrying extracellular material fuse with developing yolk bodies.

The cytoplasm just below the egg surface is often organized as a gel-like cortex. In many species, special cortical granules or fluid-

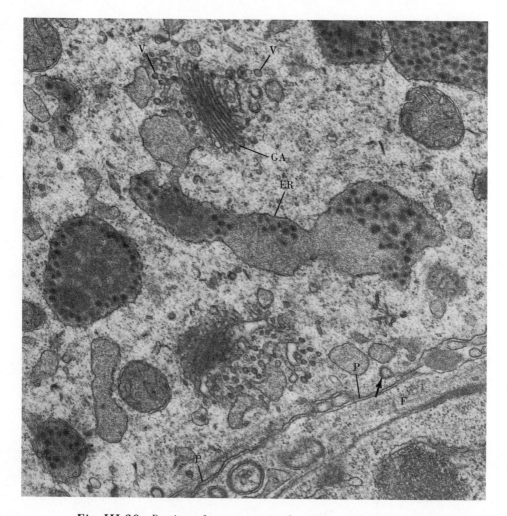

Fig. III-30 *Portion of an oöcyte in the spider crab, Libinia. Part of one follicle cell (many surround the developing egg) is seen at (F). Yolk accumulates as electron-dense material in cisternae of ER. Dilated regions appear to separate from the ER as large yolk spheres. Vesicles, some showing "coats" (arrow; see Section 2.1.4) appear to form at the plasma membrane. These coated pinocytosis vesicles and other vesicles (V) associated with the Golgi apparatus (GA) are considered as probable sources of material added to the yolk. Presumably the vesicle membranes fuse with the ER-derived membrane of the yolk sphere. × 50,000. Courtesy of G. Hinsch and V. Cone.*

containing vacuoles accumulate there; the role of these granules in fertilization is discussed below.

The egg often shows asymmetric distribution of cytoplasmic components. Of particular interest are the special cytoplasmic regions

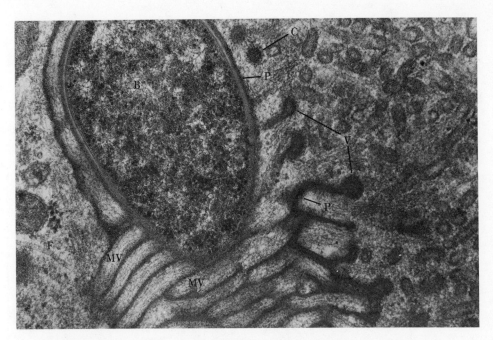

Fig. III-31 *Portion of the surface of an oöcyte of the cockroach, Peri-planeta. A follicle cell is seen at (F). Electron-dense material is present in the extracellular space. The oöcyte plasma membrane (P) is folded into a complex pattern of microvilli (MV) at the base of which pinocytosis vesicles form; some seen at (V) are still attached to the plasma membrane. The vesicles are "coated" (see the one at C). (B) indicates one of the bacteria that are often found asso-ciated with the developing oöcytes of this species; perhaps the bacteria are symbionts. × 45,000. Courtesy of E. Anderson.*

in some eggs that contain substances involved in differentiation of specific cell types during development (Section 4.4.3).

A variety of extracellular coats, sometimes in multiple layers and usually rich in polysaccharides, surround the eggs of many species. In some cases, a thin surface layer may be formed by the egg itself, but additional thick coats often are contributed by ducts down which the eggs pass after release from the ovary, or by follicle cells.

3.10.3   FERTILIZATION   Although the details of fertilization also vary from species to species, in many in-stances a general pattern is followed (Fig. III-32). The acrosome of the sperm probably plays an important role in sperm penetration through the exterior coats of the egg. On contact with the egg coats, the acro-

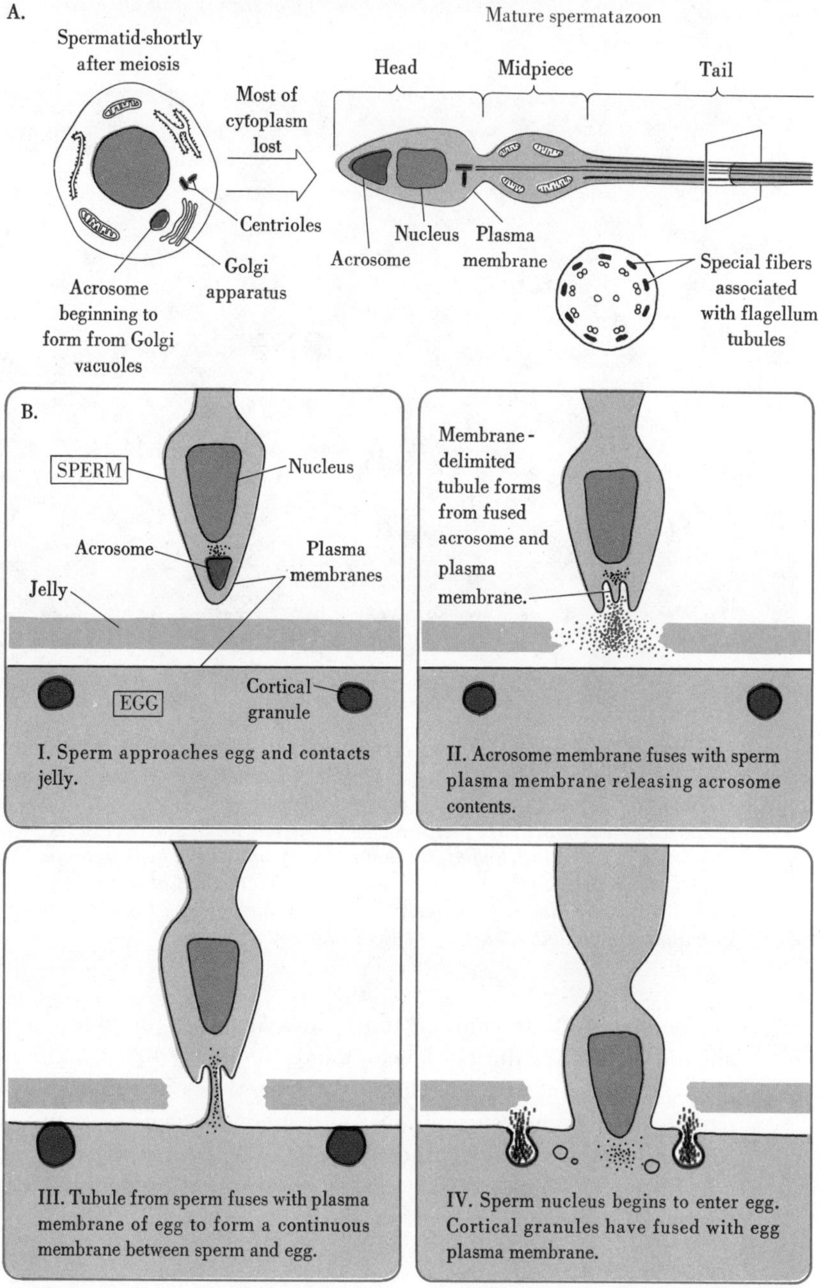

A. Schematic diagram comparing a spermatozoon (of a higher animal) with the spermatid from which it differentiates. The arrangement of mitochondria, the acrosome, centrioles and other organelles varies in detail in different animals. **B.** Outline of fertilization based primarily on studies of invertebrates. After L. Colwin and A. Colwin.

Fig. III-32

some membrane fuses with the plasma membrane of the sperm, releasing the acrosome contents; this resembles the release of secretion from a secretory cell (Chapters 2.5 and 3.6). Acrosomal enzymes of a class known as *hyaluronidases* probably break down polysaccharides in the egg coat; *proteases* that break down the proteins of the coats are also present. After the release of the acrosome contents, the acrosome membrane (which is now continous with the *sperm* plasma membrane) forms one or more tubular projections that contact the *egg* plasma membrane and soon fuse with it. This establishes continuity between egg and sperm. The sperm nucleus thus can enter the egg without exposure to the extracellular environment.

As sperm and egg fuse, the egg undergoes changes. One rapid alteration, of unknown character, prevents fertilization by additional sperm. This barrier is soon reinforced in many eggs by the formation of a *fertilization membrane* that surrounds the egg. This is not a unit membrane, but rather a thick layer, probably of proteins and polysaccharides. The egg cortical granules contribute to the fertilization membrane. Fusion of their delimiting membranes with the plasma membrane releases the granule contents to the extracellular surface. Granule release occurs in a wave that proceeds around the egg from the point of sperm fusion.

The role played by the sperm in bringing about these changes in the egg is still unclear. Even before the sperm nucleus enters the egg, a small amount of material from the region just behind the acrosome (Fig. III-32) passes into the egg. It is speculated that this, or perhaps material in the sperm plasma membrane or the acrosome membrane, normally initiates changes in the egg. In a few species, the egg develops parthenogenetically (without sperm); eggs of other species can be induced to do so by pricking with a needle or by treatment with certain chemicals. Thus the effect of the sperm in activating the egg may be relatively nonspecific.

In many species, not only the sperm nucleus but also the mitochondria, centrioles, and flagellum enter the egg. The sperm centrioles are believed to function in the division of fertilized eggs, though probably not in all species. The fate of the mitochondria is not clear, although in the eggs of some animals they have been seen to degenerate, as does the flagellum. The sperm nucleus and egg nucleus either fuse to form a single zygote nucleus, or, in some cases, remain separate until development begins. In the latter instance, the behavior of the maternal and paternal chromosomes is coordinated to achieve normal chromosome separation during the first division of the fertilized egg (Chapter 4.2), and a single nucleus forms in each daughter cell.

chapter *3.11*

CELLS IN CULTURE

An increasing variety of cell types isolated from multicellular organisms has been grown in culture. Many can be grown on glass in single layers that are well suited for microscopic study of living cells (Fig. I-10). Some can be grown suspended in liquid media to produce huge numbers of cells that can be readily used for biochemical studies. Some cultures survive a few days or a few weeks; others have been carried for a great many years. If provided with nutritional requirements (amino acids, vitamins, glucose or some other sugars, salts, and certain proteins), one cell can divide to produce a cell line which can continue division after division without limit. Most of these long-term cultures result from cells not obviously specialized in morphology and function. Some, however, are cultures of cells that do retain a number of the specialized features they have in the intact organism, such as the synthesis of particular macromolecules. In early work, media used for culture were based on nutrient-rich extracts from natural sources such as coconut milk or blood serum; these are complex mixtures with many components, some poorly understood. Small amounts of such extracts are still added to most culture media to provide some necessary factors of unknown nature, but increasingly the bulk of the medium is an artificial mixture of known components.

By culture techniques plant and animal cells, including mammalian cells, can be grown almost as microorganisms isolated from influences of the cells elsewhere in the body and in an environment that can be readily manipulated. Essentially uniform cell populations may be derived by isolating a single cell and permitting it to divide repeatedly, to produce a *clone*. Cloning and other techniques that have been so successful in elucidating the biochemistry and molecular biology of bacteria are being applied to cultured cells from multicellular organisms. An example is shown in an experiment (Fig. III-33) which uses HeLa cells, a widely used type of cultured cell (from a human cancer) that has been propagated for several decades. The experiment indicates that large RNA molecules of nucleoli are probably the precursors of the smaller RNA's of ribosomal subunits.

Feedback inhibition of metabolic pathways (Section 3.2.5) and induction of enzyme synthesis by specific compounds in the growth medium have been observed in cultured cells. These effects are of particular interest for the eventual understanding of how levels of specific enzymes and other products of different cell types of a multicellular organism are established. Experiments of these types, like the

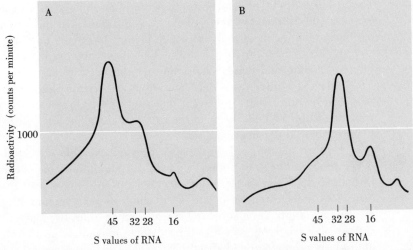

Fig. III-33 *RNA synthesis in HeLa cells. The cells were exposed for 25 minutes to radioactive RNA precursor (H³-uridine). **A.** At this time a sample was taken from which RNA was extracted and centrifuged on a density gradient (see Chapter 1.2B) to separate RNA molecules of different size classes (measured here in terms of S: see Section 2.3.2). **B.** RNA extracted from cells exposed to radioactive precursor then grown for an additional 20 minutes in nonradioactive medium in which the drug Actinomycin D was included. This drug prevents the transcription of new RNA molecules on DNA templates. Thus, changes in distribution of radioactivity during the 20 minute exposure reflect modifications of molecules made before the cells were placed in the drug-containing medium rather than synthesis of new RNA molecules. Comparison of **B** with **A** indicates that there are fewer radioactive 45S molecules and more 16–18S and 32S molecules. This is interpreted as meaning that 45S molecules give rise to 16–18S and 32S molecules; probably a given 45S RNA molecule is fragmented to produce one 16–18S and one 32S molecule. The 16–18S and 32S RNAs are precursors of the RNAs in the small and large ribosomal subunits. From J. E. Darnell and J. Warner.*

one in Figure III-33 are quite difficult to carry out in intact organisms. Many of them require large uniform cell populations into which a radioactive label or other material can be readily introduced, and from which samples for analysis can be taken conveniently and rapidly at successive intervals. Using, for example, the liver of intact animals involves waiting until the material injected into the blood stream reaches the liver. Also, injected material is greatly diluted in the blood and a large part of it enters organs other than the liver. There may be serious problems in obtaining successive samples; if, as is customary, several animals are used for different time periods, it is often impossible to be sure that the animals are sufficiently similar to one another in nutrition, genetics and so on, to permit reliable comparison. If the agents injected are drugs, hormones, or similar compounds, the direct effects on the liver may be difficult to distinguish, for example, from secondary effects in which the agent has acted upon another organ (such as a gland) which in turn has acted upon the liver.

Cell cultures provide a useful tool in *pathology*, for evaluating effects of abnormal conditions or of specific harmful agents on cells. They may also facilitate identification of biochemical abnormalities or other features that appear in diseased cells. Observations on the transformation of cells from normal to malignant in cell cultures will be described in the next chapter.

Culture techniques have also proven valuable for studies of development. The interactions of cells in populations can be readily investigated. When mixtures of several cell types are grown in a culture, *like* cells often associate with one another (Section 2.1.6), indicating an ability of cells to "recognize" one another. In many cultures, cells of a given type grown on glass slides move and proliferate only until their number and distribution reaches the point where most of the glass surface is covered by a single layer of cells in close contact with one another. At this point, *contact inhibition* occurs; movement of cells slows or stops, and cell growth and division also are often inhibited. If the cells are separated after such inhibition, they generally resume active movement and proliferation. It has been suggested that inhibition of movement, cell growth, and cell division all stem from some basic cell response to contact with other cells. At present this suggestion appears oversimplified; different factors may contribute to inhibition of different activities. The strength with which cells of a given type adhere to glass and to one another sometimes appears to parallel the sensitivity of the cell type to inhibition of movement; presumably strong cell-to-cell adherence inhibits movement when the cells are crowded together, and thus continually encounter one another as they move. Production by cells of inhibitory substances or depletion of other substances needed for growth have been proposed as mechanisms that might help explain inhibition of growth and division; as cell number and crowding increases, the concentrations of such substances might pass critical thresholds. When extensive contact take places, it is possible that the cell can transfer such substances to one another with greater efficiency. Eventually, study of the control of cell growth and movement in cultures should contribute to understanding of controls of cell number and distribution in the organism. Of much interest is the fact that many cultured cancer cells are less sensitive to contact inhibition of movement, growth, and division than are normal cells (Figure III-35).

Interactions involving embryonic differentiation of specialized cell types are also being studied. Often the presence of one tissue in culture, or of some of its products, is required for the differentiation of a second tissue. The cell interactions involved in differentiation in cultures appear to involve fairly general effects (such as the clustering of cells in groups as a prelude to differentiation) and also specific interactions of different cell types, presumably involving specific cell products.

Cultures of certain embryonic cells have yielded particularly interesting results. Single, uninucleate, embryonic muscle cells will divide repeatedly; they will then fuse to form multinucleate skeletal muscle fibers. This strengthens the proposal that the presence of several to many nuclei in one skeletal muscle fiber results from fusion of separate cells, rather than from repeated division of a nucleus without division of cytoplasm.

In Section 4.2.3, experiments will be discussed relating to the question, "Is a given cell capable of synthesizing specialized products and simultaneously carrying out the metabolic processes responsible for growth and cell division?" Cultured cartilage cells are useful for such experiments, because they can be induced to divide rapidly and repeatedly in culture and can also make a well-defined and easily detected set of specific products (notably the polysaccharide, chondroitin sulfate).

Cultures of plant cells have proven useful for many investigations. For example, masses of unspecialized cells (*callus* tissue) grown in culture can be used to investigate possible influences of "plant hormones" or other agents in development. Structures resembling roots or buds can be induced to form by treatment with appropriate levels of *auxins* (affecting cell growth) and *cytokinins* (affecting cell division). Cells from mature carrots have been grown in culture for prolonged periods; small clusters of cells have produced complete new plants. (Some believe that a single cultured cell can give rise to an entire carrot.) The usual interpretation of such experiments is that a given plant cell is not irreversibly fixed in differentiation. Normally the cells of the mature carrot from which the culture was begun would have restricted functions; after growth and division in cultures however, they can produce all the cell types of a plant. Other experiments emphasizing the plasticity of some plant cells have been done with tobacco tumor tissue; single tumor cells can be isolated from cultures and grown to form multicellular masses. When these masses are grafted to tobacco plants, they appear able to form normal tissues. Such reversal of the cellular changes leading to tumor formation is being intensively studied for its implications for possible control of tumors.

A recently developed experimental technique can induce fusion of cultured cells. If cells of different morphological or functional types (even from different species) are grown together, fusion occurs among a few of them spontaneously. Addition to the cultures of certain viruses (rendered noninfective by exposure to ultraviolet light) greatly increases the frequency of fusion. Interesting experiments on the interaction of nucleus and cytoplasm and on genetic control mechanisms have used this technique. In one experiment outlined in Section 4.4.3, a metabolically active cell was fused with a metabolically inactive one to form a *heterocaryon*, a cell with two separate nuclei derived from different sources. The inactive nucleus shows dramatic responses to its new cytoplasmic environment.

If, in fused cells, the nuclei also fuse, the product is a *somatic hybrid*. In many cases, such cells can divide and be perpetuated in culture as more or less stable types. Hybrids have been produced by fusion of the cultured pigment-producing cells from hamsters with the nonpigmented cells of mice. The hybrids do not synthesize pigment (melanin), suggesting that the chromosomes of the mouse cells carry genes for factors that repress pigment synthesis; since pigment production is based on the genes in the hamster chromosomes, this experiment, in effect, has produced control of one set of genes by another. Detailed study of the underlying mechanism should contribute much to the understanding of controls of metabolism.

chapter **3.12**

CANCER CELLS

This chapter ends with the consideration of a most important form of cell abnormality. The basis (or bases) of cancer remains unknown. But optimism regarding its conquest is generated by current progress toward understanding the genetic machinery of cells. In a manner not understood, permanent changes in heredity of cells can upset the mechanisms that restrict their division and that keep them in their normal places. When cells divide to form abnormal large masses, they are called *tumors. Benign* tumors remain as discrete masses. *Malignant* tumors (cancers) generally *metastasize:* cells spread to distant parts of the body, through the blood or lymph, and there they take root and grow into new cancer masses.

The chief sources of cancer cells for morphological and biochemical studies have been the transplantable tumors of mammals. In transplanting tumors, a small piece or a suspension of tumor cells is injected under the skin or into the leg muscle or body cavity. Before the tumor has grown to a size that kills the animal (from five days to several months, depending upon the growth rate of the tumor), a bit is again transplanted. Some transplantable tumors are obtained in *ascitic* form. These grow in the body cavity where they elicit the production of fluid (called *ascites*); the cell-laden ascites can be removed from the body cavity as a convenient source of malignant cells. For the study of molecular biology of cancer, growth of malignant cells *in culture* has proved of great value.

Ultimately, all studies are directed towards understanding the nature and cause of the changed heredity. Many, but not all, cancer researchers believe that despite their diversity (animal and human cancers vary enormously in etiology (developmental history), histology, and biology), the fundamental nature of the transformation to the malignant state is essentially the same in many, and perhaps all, cancers.

Tumorlike growths are also found in plants. One type, called *crown-gall*, arises when specific types of bacteria enter wounds.

3.12.1   **TRANSPLANT-**   Cancers of a wide variety of organs have
              ABLE   been studied, but none more extensively
         HEPATOMAS   than tumors of the liver, or *hepatomas*.
                                  This is because the liver is so well known
biochemically and morphologically, and because a broad spectrum of hepatomas, varying in growth rate, are available for investigation.

Initially, much of the work on the biochemistry of cancer was done with rapidly growing tumors of the liver and other organs. However, many of the findings, both morphological and biochemical, seem chiefly to be secondary features, manifestations of the rapid rate of division of the cells; they shed little light on the nature of the underlying transformation from normalcy to malignancy. Thus, attention is focusing on the more slowly growing hepatomas; it is thought that analysis of differences between these tumors and the normal liver may more readily lead to understanding the crucial nature of the transformation to malignancy. Perhaps the most firmly established feature of malignant cells had been their maintaining a high rate of glycolysis (Section 1.3.1), even in the presence of oxygen (aerobic glycolysis). Analysis of the slowly growing hepatomas, however, showed some to have low rates of aerobic glycolysis, similar to that of normal liver.

Both rapidly growing and slowly growing hepatomas posses organelles like those of normal cells, although the number and arrangement of the organelles may differ somewhat from normal. The rapidly growing *Novikoff hepatoma* differs from the slowly growing *Morris hepatomas* in lacking peroxisomes (perhaps it was derived initially from nonhepatocyte cells of the liver, cells that lack peroxisomes), possessing much less endoplasmic reticulum, having many more "free" polyribosomes, smaller mitochondria, and a smaller Golgi apparatus (Fig. III-34). In general, the cytology of the slow-growing Morris hepatomas is quite like that of normal hepatocytes, but it is not identical. For example, differences from normal hepatocytes are evident in the morphology and enzyme activities of the plasma membrane of Morris hepatoma cells: the microvilli are less regularly arranged and there is a different distribution of some phosphatases among different regions of the cell surface. It is possible that this reflects other, more subtle important differences of the surfaces of normal and malignant cells. Such surface differences could result in abnormal cell-to-cell interactions; some investigators believe that cancer cells generally differ from normal in their surface interactions with one another.

A number of other differences have been detected between normal and malignant hepatocytes, but their significance for malignancy is a matter of speculation. The fact that tumor cells may retain many of the specialized features of the tissues from which they derive suggests that the transformation to malignancy may be based on fairly

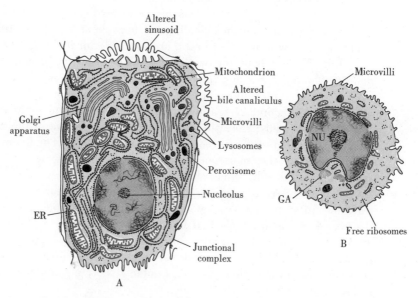

Fig. III-34 *Diagram of the cells of two transplatable hepatomas of rat.* **A.** *A slowly-growing ("differentiated") Morris hepatoma and* **B** *a rapidly growing ("undifferentiated") Novikoff hepatoma. Comparison with Figures I-2 to I-6 will confirm that the differentiated hepatoma cells resemble hepatocytes in many respects far more closely than do the undifferentiated hepatoma cells. Yet, even the differentiated type differ from hepatocytes in cell-surface features, decreased amounts of ER and in the size and nature of the lysosomes.*

subtle changes. One theory suggests that abnormalities in polysomes or in their attachment to the ER may underlie the observed failure of certain enzymes (for example, an enzyme involved in metabolism of the amino acid tryptophan) to increase in hepatomas when substrate (tryptophan) is added to the diet—such induction like (Section 3.2.5) increase does occur in normal liver. This has led to speculation that all cancers, whatever their diverse nature, may show a defect in control of enzyme synthesis and turnover. Some investigators maintain that in malignant cells, nucleolar RNA's are abnormal in composition; others that differences exist in the chromosomal proteins that combine with DNA. But no suggestion seems more likely, at least for many cancers, than the possibility of alterations in the DNA itself, including the incorporation of foreign DNA of viral origin. Observations on *carcinogenesis*— the process by which normal cells are converted into cancer cells and, ultimately, into cancers—both in animals and in cultured cells, point to this suggestion.

3.12.2 **CARCINOGEN-** Agents that produce cancers are called *car-*
ESIS IN ANIMALS *cinogens.* A great many chemical carcinogens are known, in animals and man. There are also physical carcinogens, such as X-ray and ultraviolet radiation, and biological carcinogens, mostly viruses. About a dozen *oncogenic* (carcinogenic) viruses are known in animals (Fig. III-3). They include DNA viruses, such as polyoma virus, and RNA viruses, such as mouse leukemia virus. Thus far oncogenic viruses have not been demonstrated in man, with the possible exception of so-called *Burkitt's lymphoma.* There is a widespread belief that leukemia, and possibly other human malignancies, will prove to be virus induced. Viruses are believed to induce some tumors in plants.

Some chemical carcinogens or products of their metabolism are known to combine with specific proteins of the cell. An increasing number have been shown, at least in the test tube, to affect nucleic acids, especially DNA. Similarly, the effects of X-rays and of the oncogenic viruses that have been studied most closely can best be explained in terms af action on DNA. A widely discussed theory suggests that the DNA of some oncogenic viruses becomes incorporated into the malignant cell's chromosome, as is known to occur with some bacteriophages (Section 3.1.2). A change in the cell's genetic constitution, or *genome,* may well be the common pathway to malignancy in many instances, however different the initiating factor or "cause" may be in each particular case (see Section 3.12.3).

Such genetic changes are apt to be much more subtle than the abnormalities in numbers (aneuploidy; Section 4.3.6) and shapes of

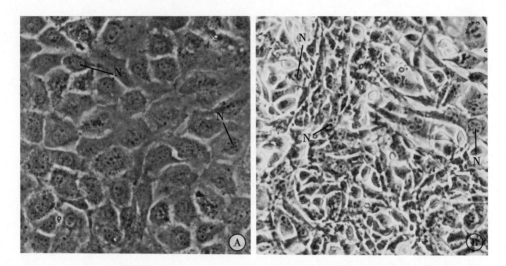

Fig. III-35 *Cells of a strain of cultured mouse fibroblasts photographed through a phase-contrast microscope. Nuclei of a few of the cells are indicated by (N). The cells in **A** are normal cultured cells that have formed a single layer (monolayer) of flattened cells on the glass surface. The cells in **B** have been infected with polyoma virus and thereby "transformed". They no longer form a monolayer but overlap in multiple layers, clumps and irregular arrays. Courtesy of R. Dulbecco.*

chromosomes that are seen in many, but not all, cancer cells. The chromosomal abnormalities are likely to be an effect rather than a cause of the malignant transformation; they may reflect cell division "errors" engendered by a very high rate of division. Only in one instance, a type of chronic leukemia in man, is a particular chromosome change (absence or reduction in size of one or two of the chromosomes), consistently found; in this case, there *might* be a causal connection between the chromosomal abnormality and the malignancy.

3.12.3 CARCINOGENESIS IN CULTURED CELLS

Important ideas concerning the subtle changes in the cell genome come from studies on cultured cells. Most work on carcinogenesis in cultured cells centers on oncogenic viruses. These are used because some viruses are exceedingly simple, and because the large fund of information on the molecular biology of viral infection aids in the interpretation of findings.

Polyoma virus is one of the simplest known viruses. It produces cancers in a variety of tissues when inoculated into newborn mice, rats, or hamsters. When embryonic fibroblasts are infected with virus, some of the cells are transformed. Their morphology then resembles that of rapidly dividing cells, and when transplanted into an animal, the cells multiply and produce cancers. Transformed cells continue growing in culture after the normal cells cease to do so, and their growth is not restricted by contact inhibition (Chapter 3.11). Unlike many normal cells, most malignant cells grow over each other and form clumps of unoriented cells several layers thick (Fig. III-35). One important observation on polyoma and some other viruses is that, once the cell has transformed, the virus is no longer detectable as a separate infective particle. However, the presence of viral DNA in the host cell (probably in its nucleus), is indicated by several facts. RNA that hybridizes with viral DNA (Section 2.2.3) and some virus-specific proteins have been identified in transformed cells. When fusions are induced (Chapter 3.11) between certain transformed cells the virus "reappears"; that is, separate viruses are again detectable.

With polyoma and other viruses, as few as 4–8 viral genes appear to be involved in a cell's transformation to malignancy. These are such relatively simple systems that they are expected to contribute much to rapid progress in understanding the basic nature of cell transformation to malignancy. A recent finding of great interest is that oncogenic RNA viruses contain a unique enzyme that can generate DNA copies of the viral RNA, "reversing" the usual transcription.

FURTHER READING

Austin, C. R., *The Ultrastructure of Fertilization*. New York: Holt, Rinehart and Winston, Inc., 1968. 196 pp.

Dulbecco, R., "The induction of cancer by viruses," *Scientific American*, April 1967, 216(4):28.

Eccles, J., "The synapse," *Scientific American*, Jan. 1965, 212(1):56

Ephrussi, B. and M. C. Weiss, "Hybrid somatic cells," *Scientific American*, April 1969, 220(4):26.

Ham, A. W., *Histology*, 5th ed. Philadelphia: Lippincott, 1965. 1041 pp. and W. Bloom and D. W. Fawcett, *A Textbook of Histology*, 9th ed. Philadelphia: Saunders, 1968, 858 pp. Outstanding textbooks of mammalian histology.

Katz, B. W., *Nerve Muscle and Synapse*. New York:McGraw Hill, 1966, 193 pp. An account of bioelectric phenomena.

Ledbetter, M. C., and Porter, K. R., *An Atlas of Plant Structure*. New York: Springer Co., (in press, 1970).

Porter, K. R. and M. A. Bonneville, *Fine Structure of Cells and Tissues*, 3d ed. Philadelphia: Lea and Febiger, 1968. 196 pp. A collection of electron micrographs of animal cell types with explanatory material.

Porter, K. R. and C. Franzini-Armstrong, "The sarcoplasmic reticulum," *Scientific American*, March 1965, 212(3):72.

Ross, R., "Wound healing," *Scientific American*, June 1969, 220(6):40.

Sharon, N., "The bacterial cell wall," *Scientific American*, May 1969, 220(5):92.

Smith, D. S. *Insect Cells: Their Structure and Function*. Edinburgh: Oliver and Boyd, 1968. 372 pp. A collection of electron micrographs of insect cell types with explanatory material and discussions of key features.

DUPLICATION AND DIVERGENCE: CONSTANCY AND CHANGE

Duplication of even the simplest procaryote involves far more than replication of DNA. Cell growth and division are integrated processes based on coordinated activities of virtually all cell components. Mechanisms operate for duplication of all cell structures and for separation of the cell into two daughters each with its share (in general, approximately equal) of the structures and macromolecules of the parent.

In bacteria such as *E. coli* and presumably in other procaryotes, only a single chromosome is present per nuclear region. In some bacteria, the mesosomes appear to function in separating the two daughter chromosomes produced by DNA replication (Section 3.2.4).

In eucaryotic cells, there is much more DNA than in procaryotic cells, and it is "packaged" in several or in many chromosomes. The chromosomes, unlike those of bacteria, contain much protein and other non-DNA material. In almost all eucaryotic cells, the chromosomes are duplicated and separated into daughter cells by the process of *mitosis*, which involves the formation of a temporary intracellular structure, the *spindle*.

Mechanisms for synthesizing nucleic acids, proteins, and other macromolecules are now reasonably well analyzed. Organelle duplication has only recently begun to be understood. As has been previously indicated, mitochondria, plastids, and probably centrioles are capable of some form of self-duplication. Other structures, such as flagella and ribosomes, are duplicated in a less direct manner. But whether an organelle can reproduce itself or not, a key part of the analysis of its formation is an understanding of the manner by which the appropriate structure is assembled from macromolecule building blocks. Part of the answer may lie in the process of *self-assembly*, the spontaneous organization of molecules into specific multimolecular configurations. Some macromolecules possess built-in features that automatically produce correct association with other macromolecules, so that a given three-dimensional structure results from properties inherent in the molecules of which it is composed. No special, external, "structure-specifying" system is needed to impress three-dimensional ordering on groups of macromolecules in a manner analogous to the way in which mRNA controls the amino acid sequence of proteins. If such structure-specifying systems are involved at one level or another, they themselves must ultimately derive their structure from a self-duplicating system (presumably containing nucleic acids) that essentially specifies its own structure. Otherwise one must postulate an endless chain of systems specifying the structure of other systems.

The evidence for the operation of self-assembly is strongest for relatively simple structures, such as some viruses and protein fibers. These systems have been dissociated into their component macromolecules or subunits and then reconstructed by mixing the components *in vitro* (in the test tube). Even the simplest *in vitro* self-assembly system requires careful control of the composition of the medium in which the subunits are suspended. Ions such as calcium (Ca^{2+}) or magnesium (Mg^{2+}) can have dramatic effects on structures; for example, the two subunits of isolated ribosomes separate if the magnesium concentration in the suspension medium is reduced below a certain level. The two positive charges of calcium or magnesium can bind two separate negative charges and, in so doing, hold separate molecules or regions of molecules in fixed relations to one another. Probably many other effects of charged ions and molecules also operate in the control of structure. For example, two positively charged molecules will tend to remain separate due to the fact that similar charges repel one another. But, if each complexes with negatively charged ions, the net charge of the complex will be zero and the molecules can more closely approach each other. For assembly of structures *in vivo* (in the living cell), mechanisms must operate to collect appropriate amounts of components at the proper point, to control levels of ions, acidity, ATP, and other factors, and probably to ensure the presence of enzymes required to speed assembly.

The degree to which self-assembly operates in the formation of complex cell organelles, particularly those delimited by membranes, is not known. Experiments have been outlined previously in which mitochondrial and bacterial enzyme systems have been reconstructed from their components (Section 2.6.5 and 3.2.4). Experiments like these suggest that parts of organelles may self-assemble. However, it is unlikely that all the molecules of a chloroplast or mitochondrion, mixed in correct proportions in a medium of proper composition, would produce a *complete* organelle. Mitochondria and plastids grow and divide; the relationship between organelle growth and structure assembly has only begun to be explored, but it seems likely that as an organelle grows the previously existing structure may aid in orienting newly added molecules. In the next chapter we will outline experiments dealing with the formation of reasonably complex structures; the experiments provide hope that the duplication of organelles will soon be understood in molecular terms.

Duplication leads to similar cells, but cells also undergo change. Change occures on three time scales: the very rapid physiological changes that have been discussed at several points in the preceding parts, cell differentiation during embryonic development, and the slower processes of evolutionary change.

The cell diversification that occurs in embryonic development results from complex patterns of interaction between nucleus and cytoplasm and between different groups of cells. When an egg is fertilized, a precise program of events is set in motion, a program that depends on factors many of which are only beginning to be appreciated. An important part of the story is the formation of different messenger RNA's in different cells as different genes are freed from the repression that keeps them inactive. Interesting cytological changes are associated with this gene activation.

It is self-evident that basic cellular processes, such as photosynthesis or mitosis, have an evolutionary history: at some point in evolution they arose from simpler processes. Their history is known only in vaguest outline, but some research and much speculation have been devoted to its elucidation.

Evolutionary changes are based on the natural selection of variant organisms resulting from mutations of DNA. Mutated DNA is usually expressed as altered proteins which effect changes in cellular metabolism. A change in any major step of metabolism tends to affect many distantly connected metabolic processes due to the interweaving of pathways. Variation is greatly increased by sexual reproduction which generates new combinations of genes. In most eucaryotic organisms, sexual reproduction depends on *meiosis*, a special pair of cell divisions.

The evolution of cells involves the evolution of chromosomes, as evidenced by differences in the number and shape of chromosomes in different species. Cytoplasmic factors also are involved; some cytoplasmic organelles possess genetic information that is expressed partly independently of the nuclear genes. Mutations of cytoplasmic genetic factors can occur and can result in permanent alterations of cell characteristics.

c h a p t e r **4.1**

MACROMOLECULES AND MICROSCOPIC STRUCTURE

4.1.1   **SELF-ASSEMBLY**   If a protozoon or a tissue culture cell is
            OF MEMBRANE   cut or torn with a microneedle, rapid healing
            STRUCTURE   occurs unless the opening created is large
                            enough to permit extensive loss of cell
contents. The simplest explanation for healing is that the plasma membrane can grow and repair itself rapidly, using components already present in the cytoplasm.

When artificial mixtures of proteins and lipids, or mixtures of components isolated from cellular membranes are placed in proper solvents, they often form arrays resembling unit membranes (Section 2.1.1). The phospholipids play a central role in this; they can spontaneously assume an ordered arrangement (Fig. IV-1) in which hydrophilic

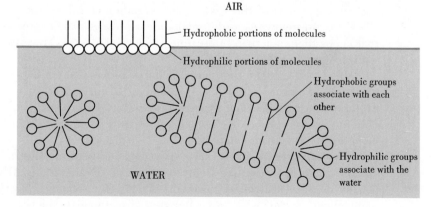

Fig. IV-1 *Polar lipid molecules associate spontaneously in ordered arrays when placed in water and at air water interfaces.*

regions (Section 2.1.2) associate with the water and hydrophobic regions associate with one another. This creates layered structures in which each layer is two lipid molecules thick. When different proportions of various other lipids (for example, cholesterol) are added to the phospholipids and proteins, unit membranes of different thicknesses form. Variation of lipid or protein content may be a partial explanation of the different thicknesses of membranes of different organelles. In addition, some mixtures of lipids and other molecules form structures that resemble the hypothetical globular or polygonal membrane sub-units discussed in Section 2.1.2.

These observations, though not conclusive, raise the possibility that the basic structure of cellular membranes may form by self-assembly (or by relatively simple addition of molecules to pre-existing membranes). How enzymes become arranged with specific geometry on the membrane remains unknown, and we cannot yet account for functional differences among different membranes. Perhaps different enzymes assemble in complex groups before they associate with lipids and become incorporated in membranes.

4.1.2   SELF-ASSEMBLY OF PROTEIN STRUCTURES   Proteins have specific three-dimensional conformations (Fig. IV-2) which are central to enzyme activity and other properties. The folding of the chain into a specific three-dimensional structure will be strongly influenced by the *primary structure* (the amino acid sequence). For example, if sulfhydryl (—SH) groups on the amino acid cysteine are present at two points on the chain, they may react to form a disulfide bond (S—S). This bond will hold the two portions of the chain in a fixed relation to one another (Fig. IV-2). Similarly, interactions between other amino acids may form *noncovalent* stabilizing bonds such as the ionic bonds resulting from attraction between the oppositely charged NH_3^+ and COO^- groups, hydrogen bonds (Section 2.2.1), and complex interactions (hydrophobic bonds) that promote associations between hydrophobic groups (Sections 2.1.2 and 4.1.1). Most of these bonds are easily broken and re-formed, in contrast, for example, to the bonds that hold amino acids together within the polypeptide chain (Fig. II-14). The latter are *covalent* bonds based on sharing of electrons by two atoms. Generally, drastic chemical treatment or enzymatic digestion is needed to break covalent bonds, whereas noncovalent bonds are responsive to mild treatments such as changes in the ion concentration of the medium.

Experiments have suggested that some proteins of given primary structures may fold spontaneously into the correct three-dimensional conformation. Enzymes have been unfolded (*denatured*), with consequent loss of function, by exposure to agents such as *urea* which breaks hydrogen and hydrophobic bonds. When the denaturing agents are removed, some enzymes are capable of returning spontaneously to a conformation indistinguishable, in terms of enzymatic activity, from the original. For other proteins, conditions for such reversal of denaturation have not yet been found and it is possible that they fold by more complex processes. Perhaps, during protein synthesis, the completed portion of a polypeptide detaches from the ribosomes and begins to fold while the rest of the chain is being completed. Sequential folding in this way, with the part synthesized earliest folding partly independently of portions synthesized later could lead to complex three-dimensional patterns. When the protein is denatured in the test tube and allowed to re-form its structure, the original distinction between early-and late-synthesized ends does not exist and it may fold in a different manner.

Many proteins have subunits joined together in specific manners. For example, hemoglobin consists of four subunits, each a folded polypeptide chain. If the hemoglobin subunits are dissociated from one another, then mixed in a test tube, they readily reconstitute the original tetrapartite structure. Apparently the proper amino acid groups are present at the surfaces of the folded polypeptide chains to promote association of subunits in specific patterns (Fig IV-2).

These observations indicate that the information encoded in nucleic acids, by specifying the amino sequences of polypeptide chains, can contribute directly to determination of three-dimensional protein structures.

4.1.3 FILAMENTS AND FIBERS

The muscle proteins actin and myosin exist in the striated muscle cell as distinct filaments containing linked molecules of actin (thin filaments) or myosin (thick filaments) (Section 2.11.3). In the test tube, the molecules can be dissociated from one another and then reassociated to reconstitute filaments. The reconstituted filaments from myosin resemble the thick filaments of striated muscle; those from actin resumble the thin filaments. However, there are differences between reconstituted filaments and *native filaments* (those found in muscle). For example, in most vertebrate muscles, the thin filaments are uniformly about 1 μ in length; reconstituted filaments are of variable length, extending up to 10 μ or more. Reconstituted filaments consist of two helically wound chains, in which each chain is composed of apparently identical globular subunits about 50 Å in diameter; each subunit is an actin molecule. Similar chains are present in the native

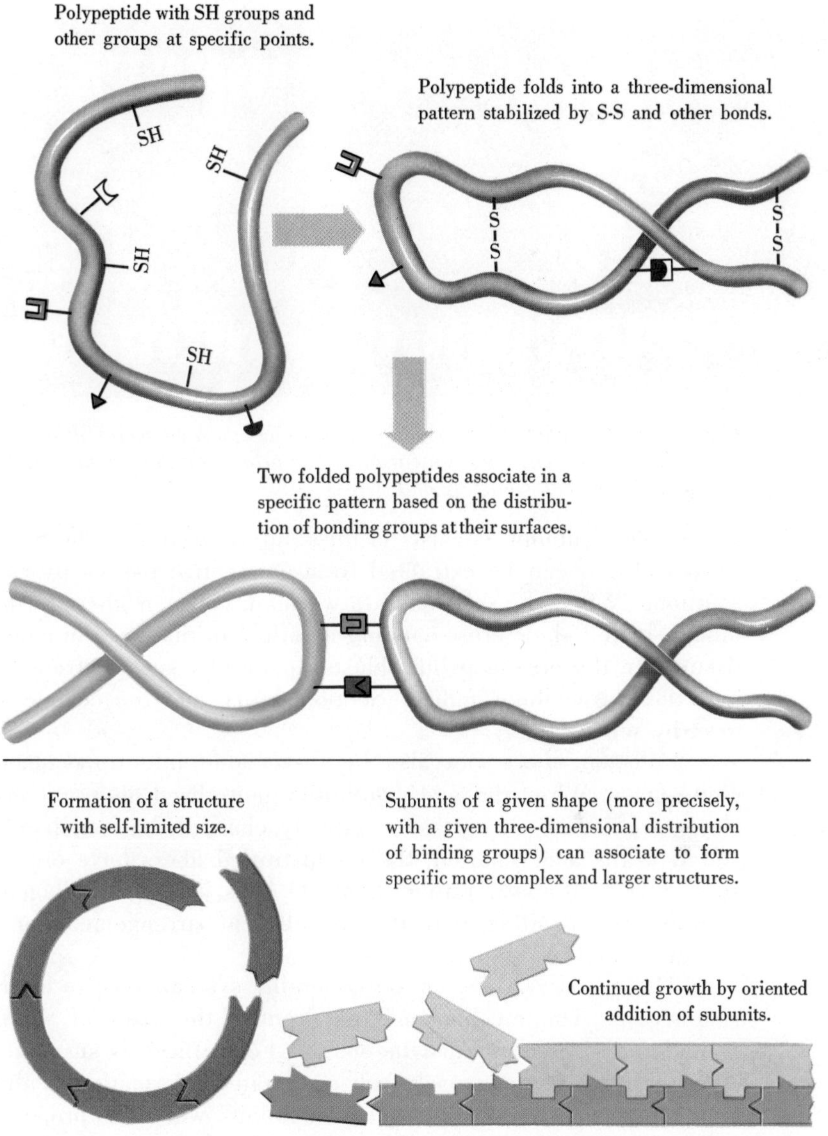

Polypeptide with SH groups and
other groups at specific points.

Polypeptide folds into a three-dimensional
pattern stabilized by S-S and other bonds.

Two folded polypeptides associate in a
specific pattern based on the distribution of bonding groups at their surfaces.

Formation of a structure
with self-limited size.

Subunits of a given shape (more precisely,
with a given three-dimensional distribution
of binding groups) can associate to form
specific more complex and larger structures.

Continued growth by oriented
addition of subunits.

Fig. IV-2 *The self-assembly of macromolecules to form complex shapes and arrays.*

filaments (see Fig. II-63), but there are several hundred subunits in native filaments and a thousand or more in reconstituted filaments.

Collagen, the major protein of connective tissue, appears in the electron microscope as thick fibers with distinct cross-bandings reflecting orderly arrays of constituent molecules (Fig. IV-3). Tropocollagen,

Fig. IV-3 *Collagen fibers shadowed with evaporated metal atoms to bring out their three-dimensional appearance. The prominent pattern of cross-banding is evident. × 50,000. Courtesy of J. Gross.*

the collagen subunit, consists of three intertwined polypeptide chains. Tropocollagen can be extracted from connective tissues by cold salt solutions. When the solutions are warmed, collagen fibers form spontaneously and show cross-banding identical to that seen in connective tissues. In the organism, fibroblasts apparently secrete tropocollagen into the extracellular milieu (Section 3.6.1), and the collagen fibers form by self-assambly.

Collagen fibers can also be dissociated into tropocollagen by dilute acids. When the acid solution is neutralized fibers reconstitute However, if ATP or certain negatively charged macromolecules are added to the acid solution, the reconstituted fibers have cross-bands spaced 2800 Å apart, rather than 700 Å as in native collagen. This altered cross-banding indicates a different arrangement of tropocollagen molecules.

These observations on self-assembly systems lead to interesting conclusions. The mechanisms that control the sizes of some self-assembling structures must be sought. For actin, it is suspected that other proteins in the muscle cell cooperate in formation of thin filaments to terminate the addition of subunits when the proper length is reached; one model suggests that native thin filaments contain a protein called *tropomyosin* in addition to the actin. The width of collagen fibers in connective tissue varies from place to place in the body. The differences may be due to variations in the rate of release of tropocollagen resulting in different rates of fiber formation, or to variation in the concentration of ions or other substances that interact with collagen.

Also, it appears that the same set of subunits may aggregate in different patterns. Conditions needed to make tropocollagen form fibers that do not have the 700 Å spacing are very rarely found in the body. However, collagen fibers themselves are the subunits of larger aggregates, and these aggregates are of varying form. In tendons, collagen fibers are grouped in thick bundles, while in some other connective tissues they form large flat layers. Probably interaction of collagen with other connective tissue components influences the form of the aggregates.

4.1.4 MICROTUBULES; BACTERIAL FLAGELLA

A number of proteins that do not normally form microtubules in cells (abnormal hemoglobin and a few enzyme proteins) can form tubular structures *in vitro*. Some of these tubules are similar in dimensions to microtubules. This suggests that microtubule formation may depend partly on fairly general properties of protein molecules with a roughly globular shape. The formation of the microtubules of the cell is controlled by unknown factors. But, the microtubules of isolated sea urchin sperm flagella can be dissociated into small, soluble protein subunits. Mixtures of these subunits (plus some larger material from fragmented tubules) can be induced to reassociate into microtubules by careful adjustment of the concentrations of subunits and inorganic ions such as K^+. This correlates well with the findings previously mentioned (Sections 2.10.4 and 2.11.1), that microtubule-rich cell structures can often be broken down and re-formed relatively easily. Experiments on test-tube assembly of microtubule structures from purified proteins may aid in the analysis of the formation of organelles such as the spindle or cilia; such *in vitro* experiments should provide insight into crucial control mechanisms.

Bacterial flagella (Section 3.2.4) can be dispersed into separate molecules of the protein flagellin. In the presence of short pieces broken from intact flagella, the flagellin molecules rapidly reconstitute long fibers similar to flagella. The fragments serve as "seeds" for the growth of the fibers; flagellin molecules attach to them in oriented fashion.

The flagella are relatively straight in some bacterial strains and wavier in others. If long fragments are used to "seed" growth, then fragments from "wavy" flagella combined with dispersed flagellin molecules from "straight" flagella produce reconstituted fibers that are wavy. If both fragments and flagellin are from "straight" flagella, reconstitution results in straight fibers. These experiments suggest that some cell structures can grow by adding additional subunits in oriented fashion. Most likely, binding sites are present that control the addition of subunits (Fig. IV-2).

The existence of orienting binding sites is also suggested by an experiment on the protozoon *Tetrahymena*. This experiment is especially interesting because it involves the formation of a more complicated structure by adding one type of subunit to an apparently simpler structure made of a different type of subunit. The central tubules and arms of the peripheral tubules (Fig. II-57) can be selectively extracted from cilia. The extracts contain a protein, *dynein*, which has ATPase activity. When dynein is added again to the cilia from which it was prepared, the arms of the peripheral tubules reappear and the cilia regain the ability to split ATP. (Reappearance of the central tubules has not yet been observed.) There must be appropriate binding sites for dynein built into the peripheral tubules which orient the dynein as projecting arms. The ATPase activity may be important in providing energy for the movement of the cilia.

4.1.5 **VIRUSES** Those *spherical* viruses with simple polyhedral shapes are thought to form by the assembly of protein subunits constructed in such a way that they assume automatically the proper three-dimensional array. For simple closed surfaces such as the sphere, it is not difficult to conceive of subunit shapes that will associate in a structure of fixed size and shape (Fig. IV-2). Note that in such a case, the control of size may not be a problem; no sites are left for additional subunits. This is a *self-limiting* structure.

The cylindrical tobacco mosaic virus can be dissociated into RNA and separated protein subunits. The proteins can reassemble alone to form cylinders of two alternative types, only one of which has the subunits arranged as in the native virus (Fig. IV-4). If RNA and protein subunits are mixed under the proper conditions, viruses are reconstituted. Both nucleic acid and protein subunits have the same arrangements as in native viruses.

Different mutants of the bacterial virus called T_4, which has a complex morphology, make incomplete viruses when they replicate in bacteria. Some fail to make heads; others cannot make tails or parts of tails. If test-tube mixtures of incomplete viruses are made so that all the parts are present, complete viruses may form (Fig. IV-4). However, there are restrictions. In the systems so far studied, for example, tail filaments will not attach to tails unless a head is attached (at the other end).

Two of these findings are of particular interest. In the tobacco mosaic virus experiments, the RNA not only carries the genetic information for the structure of individual protein subunits, but the nucleic acid also appears to act as a core or scaffold that interacts with the

A.

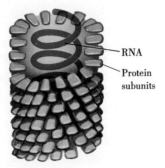

Intact virus — consists of RNA surrounded by helical array of protein subunits. The RNA and protein subunits can be separated and their behavior studied in the test tube.

RNA

Protein subunits

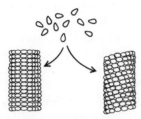

Isolated protein subunits can aggregate in two arrays, helical and non-helical. The helical array is similar to the intact virus.

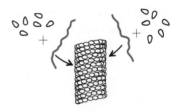

Isolated protein subunits mixed with isolated RNA reconstitutes the virus with the protein subunits arranged as in the native virus.

B.

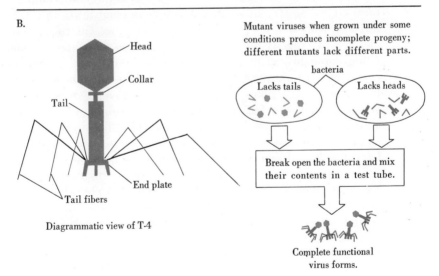

Head

Collar

Tail

Tail fibers

End plate

Diagrammatic view of T-4

Mutant viruses when grown under some conditions produce incomplete progeny; different mutants lack different parts.

bacteria

Lacks tails

Lacks heads

Break open the bacteria and mix their contents in a test tube.

Complete functional virus forms.

Fig. IV-4 *Viral reassembly in the test tube (see Fig. III-1 for electron micrographs of the viruses). **A.** Experiments with tobacco mosaic virus. **B.** Experiments with T-4 bacteriophage. From the work of Frankel-Conrat, Kellenberger, Wood, Edgar and others.*

protein subunits in promoting association into one of several alternative configurations. In addition, the RNA is of fixed length and probably determines the length of the assembled structure. In a sense, the RNA may be operating as a system that is both self-duplicating and can also specify the three-dimensional ordering of other molecules (as discussed in the Introduction to this Part). However, the final disposition in the virus of the RNA and proteins can probably best be thought of as reflecting mutal interaction in which the RNA "helps" align the protein molecules and the interaction with the proteins contributes to determining the helical arrangement of the RNA.

The T_4 observations indicate that the assembly of complex structures may involve *sequences* of assembly. The experiments have led to the proposal that in normal virus formation, heads, tails, and tail filaments, each are assembled from their components as separate structures. Completed heads and tails associate spontaneously and once this has occurred tail filaments are attached. An interesting finding is that enzymes may be required for certain steps in attachment of the filaments. Since they can catalyse the formation of specific bonds between specific groups, the participation of enzymes adds a level of complexity and of possible control. The formation of blood clots is based on the assembly of fibers made of the protein *fibrin* in a series of steps, some enzyme-mediated and some apparently spontaneous.

In terms of assembly mechanisms, T_4 is the most complex structure studied so far, although it is simple in comparison to a mitochondrion or chloroplast. At least 40 different genes control the structure of T_4. Some specify the amino acid sequences of the protein subunits. Others probably affect assembly in less direct ways, for example, by specifying the enzymes that are involved in attachment of one part to another. The genes that control formation of more complicated structures may number in the hundreds. Assembly of structures such as the complex organelles of eucaryotic cells almost certainly requires control of size, use of scaffolds, specified sequences of assembly (perhaps resulting from genetically specified sequences of synthesis of components), ordered growth of preexisting structures and probably the involvement of enzymes to make specific links between components and to speed processes of association. The studies discussed in this section have made the assembly of structures much less mysterious and have provided invaluable guidelines to the directions in which further investigations might proceed.

The overall conclusion is that self-assembly, modified by relatively simple mechanisms, is of importance in the formation of complex structures by cells. Some progress in studies of the assembly of ribosomes will be outlined in Chapter 5.1.

4.1.6   ***NEMATOCYSTS***   Earlier (Section 3.6.2) it was shown that the
         OF HYDRA   nematocysts of *Hydra* have a complicated
structure which arises inside a vacuole
originating from the Golgi saccules. A wide variety of proteins and
other substances is probably present inside the vacuole. From these,
the capsule, lid, coiled thread, stylets, and barbs are fashioned. All
these have a morphology that is specific for a given species; therefore, they must be under genetic control. The structures develop at a
time when the endoplasmic reticulum and Golgi apparatus are rapidly
regressing, suggesting that protein synthesis has been turned off and
that little, if anything, is being added to the vacuole.

An interesting possibility is that nematocyst structures within
the vacuole form like the T_4 virus — by the separate assembly of individual parts, with the presence of binding sites on each part for the
attachment to other parts, and by the sequential addition of parts to
form a whole structure. Perhaps enzymes present in the vacuole are
involved. This kind of assembly process has not been demonstrated
for nematocysts and the possibility cannot be excluded that more complex processes are involved. However, it would be interesting to attempt
isolation of individual nematocysts at a time when protein synthesis by
the cnidoblast is turned off and when visible differentiation of nematocyst structures has not yet progressed. If the notion of T_4-like assembly
within a vacuole containing all necessary components has any merit,
differentiation of structure might well proceed in isolated nematocysts.
The next experimental steps would be the extraction of nematocyst
proteins and attempts to reconstitute the complex structures in the
test tube.

c h a p t e r **4.2**

CELL DIVISION IN EUCARYOTES

4.2.1   ***CHROMOSOME***   That genes are arranged in linear order on
   CONSTANCY AND   chromosomes was one of the great dis-
        MITOSIS   coveries of this century. It opened the route
to a series of studies that have led in recent
years to the analysis of heredity in terms of DNA base sequences. Some
major conclusions emerging from these studies are relevant to the
discussion here:

(1) In dividing and nondividing cells, the nuclear DNA is located
in the chromosomes. When division is not occurring, the chromosomes

are usually uncoiled and the DNA participates in controlling cell structure and function. During division, the chromosomes coil into compact structures whose behavior accounts for transmission of DNA to daughter cells. Chromosomes of a dividing plant cell are shown in Fig. IV-5. Note that the chromosomes are of different lengths and that recognizable specialized regions are visible; *kinetochores* (also called *centromeres*) are present in all chromosomes; they function in chromosome movement during cell division. *Secondary constrictions* are present only on a few chromosomes; they include the sites of nucleolar organizers. Each chromosome is divided longitudinally into two identical *chromatids*; this is characteristic of cells soon to enter division.

(2) Figure IV-6 shows the human *karyotype*, (chromosome complement of man). The chromosomes of one cell have been arranged to demonstrate the fact that they occur in pairs of similar (*homologous*) chromosomes, referred to as *homologous pairs*. Cells in which the chromosomes are present in such pairs are referred to as *diploid*. The pairs may differ considerably from one another in sizes of chromosomes and in position of kinetochores. Although unicellular organisms and many lower plants differ from this pattern (see Section 4.3.1), most cells of usual multicellular animals or of higher plants contain identical

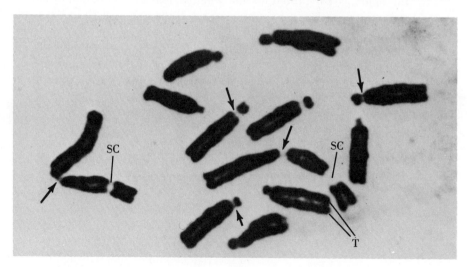

Fig. IV-5   *The twelve chromosomes (6 pairs) of the bean,* Vicia, *during mitotic metaphase (see Fig. IV-7). Each consists of two chromatids (T) and each has a kinetochore (arrows) seen as a depression or lightly-stained region of the chromosome. The chromosomes differ in length and position of the kinetochore. Two of the chromosomes, members of one pair, show a second "gap" along their lengths, the* secondary constrictions (SC). *These are the nucleolar organizer sites, the places where the nucleolus is attached when it forms (see Fig. II-17). × 2,000. Courtesy of J. E. Trosko and S. Wolff.*

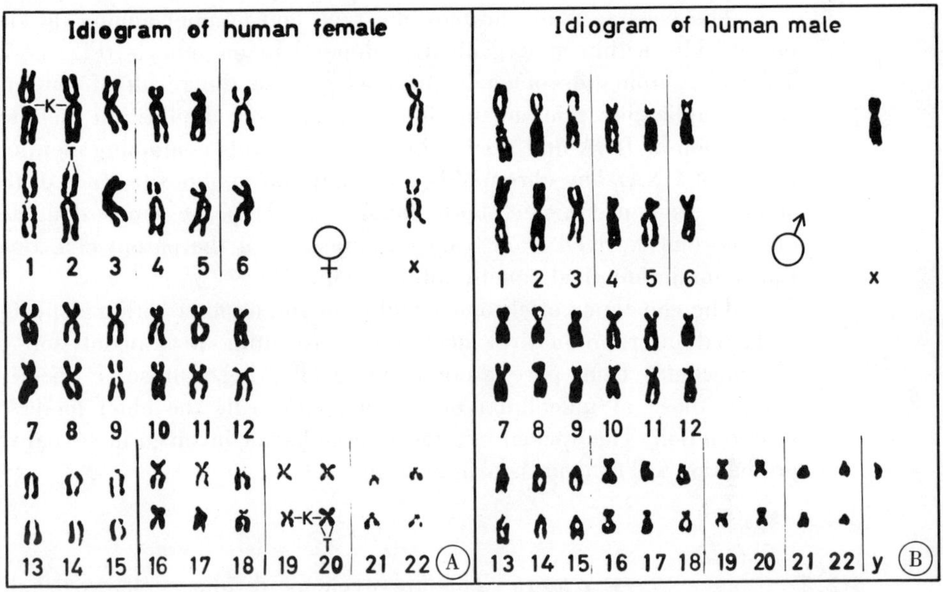

Fig. IV-6   *Metaphase chromosomes (Fig. IV-7) from human cells. The cells have been spread on a slide, stained and photographed. The chromosomes have been cut from the photographs and arranged as an "idiogram" showing the 22 pairs of "autosomes" and the two "sex chromosomes" (in females, two X chromosomes, in males, one X and a smaller Y chromosome). (K) indicates kinetochores and (T), chromatids. × 2,000. Courtesy of T. Puck and J. H. Tijo.*

diploid sets of chromosomes. The major exceptions are the mature sex cells which are *haploid*; only one chromosome of each pair is present. Also, especially in plants, some cells may contain twice (or greater) the normal total number of chromosomes with corresponding numbers of extra identical copies of each chromosome.

(3) Most organisms of a given species have the same karyotype. Different species differ in chromosome number, which varies from two to several hundred, and in chromosome morphology (lengths, positions of kinetochores, and so forth.)

(4) A particular chromosome in a given species carries a particular set of genes controlling characteristics of that species, and nonhomologous chromosomes carry different genes. (Differences between homologous chromosomes will be discussed in Section 4.3.1.) A gene may be thought of as a DNA segment usually with the information necessary to specify via mRNA, the amino acid sequence of a particular polypeptide chain or protein molecule. (However, some genes produce RNA's which are not translated into protein, for example, the tRNA's.)

The constancy of chromosome type and number among the different cells within an organism is dependent on mitosis (Fig. IV-7). The two chromatids of each chromosome are the result of chromosome duplication prior to division. Chromosome duplication includes replication of DNA and results in two chromatids containing identical copies of DNA. The chromatids are separated to opposite poles of the spindle; in consequence, both daughter cells possess one daughter chromosome derived from each chromosome of the parent cell. Both thus contain identical genetic information.

The constancy of chromosomal type and number within a species is based on reproductive mechanisms resulting in transmission of chromosomes from parent organism to offspring. For some species, mainly those of unicellular organisms, mitosis is the chief mode of reproduction. The special chromosome behavior involved in sexual reproduction will be discussed in Section 4.3.1.

4.2.2   **THE CELL LIFE CYCLE**   The life cycle of dividing cells is usually divided into (1) *interphase* and (2) the mitotic division stages, *prophase*, *metaphase*, *anaphase*, and *telophase* (Fig. IV-7). Interphase was originally given this name because it is the stage between successive mitotic divisions and usually shows few dramatic chromosomal changes that are readily recognizable in the microscope. It was once regarded as a resting stage. However, interphase is now known to be the period of intense cellular metabolic activity; cells usually spend most of their time in interphase.

Cytochemical and autoradiographic evidence indicates that DNA synthesis (replication) in preparation for division occurs during the interphase prior to division. In interphase, a period of DNA synthesis (S), a pre-S period (G_1), and a post-S predivision period (G_2) (Fig. IV-8) are recognized; these are distinguishable in experiments like the following:

If tissue culture cells are exposed very briefly to tritiated thymidine (Section 1.2.4), and then grown in nonradioactive medium, only those cells actively synthesizing DNA during the brief exposure will incorporate label. If the population is studied by autoradiography at intervals thereafter, it is observed that the first cells to enter division

Fig. IV-7 *Mitosis in an animal cell. The behavior of chromosomes is similar* ▶ *in virtually all eucaryotes, plants and animals. Some of the other organelles vary in behavior as outlined in the text. See Fig. IV-17 for one of the special features of plant divisions. Once the two chromatids have separated after metaphase, it is a matter of terminological convenience as to when they should be referred to as daughter* chromosomes.

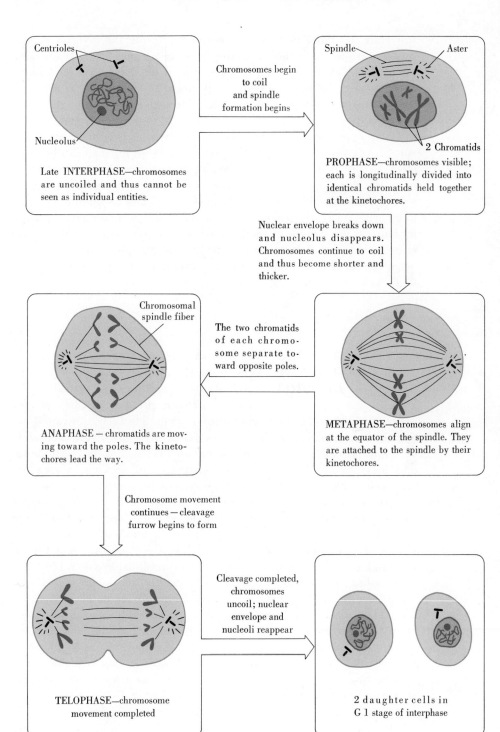

Centrioles

Nucleolus

Late INTERPHASE—chromosomes are uncoiled and thus cannot be seen as individual entities.

Chromosomes begin to coil and spindle formation begins

Spindle   Aster

2 Chromatids

PROPHASE—chromosomes visible; each is longitudinally divided into identical chromatids held together at the kinetochores.

Nuclear envelope breaks down and nucleolus disappears. Chromosomes continue to coil and thus become shorter and thicker.

Chromosomal spindle fiber

The two chromatids of each chromosome separate toward opposite poles.

ANAPHASE — chromatids are moving toward the poles. The kinetochores lead the way.

METAPHASE—chromosomes align at the equator of the spindle. They are attached to the spindle by their kinetochores.

Chromosome movement continues — cleavage furrow begins to form

Cleavage completed, chromosomes uncoil; nuclear envelope and nucleoli reappear

TELOPHASE—chromosome movement completed

2 daughter cells in G 1 stage of interphase

will show no label in their chromosomes during the division stages. These were in G_2 at the time label was given. They had already completed DNA synthesis and so incorporated no radioactive thymidine. After a time, the division figures will begin to show label as the cells that were synthesizing DNA when label was given pass through G_2 and go on to divide. The time between exposure to tritiated thymidine and the *first* appearance of labelled cells in division is a measure of G_2, since it represents the time taken by cells almost finished synthesizing DNA to proceed to division. Similar approaches are used to determine the duration of the other interphase stages. For example, under some circumstances, the percentage of cells found by autoradiography to incorporate H^3-thymidine during a very brief exposure gives a rough measure of the length of the S phase as compared to the other cycle stages; this is so because the probability of any given cell being in S when the label is presented depends directly on the percentage of the total cycle represented by S.

4.2.3   ***RATES AND*** Rates of division vary considerably among
          CONTROLS OF cell types (Fig. IV-8). The rates are de-
          DIVISION pendent on conditions of growth. For
          example, *Ameba proteus* at 23°C divides
every 36–40 hr but if the temperature is lowered to 17°C, this rate slows to one division every 48–55 hr. Few eucaryote cells divide as rapidly as bacterial cells; a bacterial population may double in cell number and mass every 15–20 min. Early cleavages of developing eggs are relatively rapid, probably because no growth occurs between divisions; the egg is separated into smaller and smaller cells. Such cleavages may occur at rates of one per hour to several per hour; the fastest are found in some insect eggs where only the nuclei divide in the earliest stages of development (see Section 4.4.3). Cells of plant root tips or mammalian cells in culture double in mass and divide every 12–24 hr. Most other cells divide much less frequently and some highly specialized cells, such as mature neurons, do not divide at all.

The relations between division rates and cell specialization are often summarized by statements like the following: The more specialized a cell becomes in its morphology and metabolism, the less likely it is to divide. To some extent this is applicable to cell types of multicellular organisms, but it cannot be regarded as a hard and fast rule. It is true that some highly specialized cells rarely divide. Some organs and tissues maintain populations of relatively unspecialized cells which divide relatively rapidly; some of the division products differentiate,

	Interphase	Mitosis (M)
Sea urchin egg cleavage		
at 2 cell stage	0.25-0.5 hours	0.5 hours
at 200 cell stage	2	0.5
Plant root meristems	16-30	1-3
Cultured animal cells	12-24	0.5-2
Mouse intestinal epithelium	12-24	1

	G_1	S	G_2
Mouse intestinal epithelium	9	7	1-5
Cultured mouse fibroblasts	6	8	5
Bean root meristems	9-12	6-8	4-8
Ameba proteus	undetectable probably no G_1	3-6	30

Fig. IV-8   *Approximate rates of division and durations of cell cycle stages for some rapidly dividing cells. For a given cell type, G1 is the most variable period. Non-dividing cells are sometimes thought of as remaining in a modified G-1 called G_0. Data from the original observations of Howard and Pelc, from Mazia and from others.*

others remain as a reservoir for future division. Examples of this behavior are found in the intestinal epithelium, in the bone marrow cells that give rise to blood, and in the *meristems* near the tips of plant roots. In these tissues, continual division is required to replace dead or lost cells (intestine and marrow) or to provide for continued growth (roots). However, it is not to be concluded that rapidly dividing cells are in some sense necessarily "simpler" than slowly dividing cells. For example, many unicellular organisms such as protozoa are extremely specialized and complex, but still can divide rapidly. Cells that rarely divide when they are surrounded by other cells in a multicellular tissue may divide at a rapid pace when separated and grown in tissue culture. Cartilage-producing cells from embryonic chicks, when grown in culture, show interesting behavior in this regard. While in the chick, they may be seen (by autoradiography) to incorporate radioactive sulfur compounds into the polysaccharide chondroitin sulfate, used in cartilage synthesis. On isolation in culture, they show a drastic decline in incorporation of cartilage precursors, but much incorporation of tritiated thymidine. This indicates replication of DNA leading to division. After dividing for a while, the cells in culture form clusters, stop incorporating thymidine, and resume incorporation of cartilage precursors. It appears that preparation for division and extensive synthesis of specialized products do not occur simultaneously in the same cartilage cell. However, a given cell can alternate between a dividing state and a specialized state. (See also p. 240).

This experiment should not be interpreted as indicating that cells never prepare to divide and simultaneously to synthesize specialized products. Rather, like other observations, it emphasizes that cell growth, specialization, and division must be under complex controls which produce different behavior for different cell types and, probably for a given cell type under varying conditions. These controls result, for example, in the fact that many cells which undergo repeated divisions almost precisely double their mass during the interphase preceding each division. As mentioned above, however, others (such as the cells in early divisions of developing embryos) may not grow at all between divisions. Perhaps the most important thing to be borne in mind is that *specialized* is a vague word. It is used to designate cells that produce a unique product, such as muscle proteins, or those that have a special architecture, such as neurons. However, while every cell has divided at some stage in its history, it is not absurd to think, for example, of the plant meristems mentioned above as groups of cells specialized for continued rapid division.

Knowledge of the factors that control division rates might, for example, help to explain why most cancer cells are able to divide much more rapidly than the normal cells from which they derive. However, it is not clear as yet whether division results from an activation or triggering process, such as the accumulation of a critical amount of some enzyme or other component, or whether division is initiated by the removal of inhibitory factors. In discussing protozoa (Section 3.3.2) experiments on amebae were mentioned, exploring the possibility that division was triggered when a certain cell size was attained. The conclusion was reached that while cells of a given type normally may tend to divide at a specific size, the relations of cell size to the onset of division are not simple. If 70 percent of a rat's liver is removed, the hepatocytes (which normally divide infrequently) begin to divide at rates faster than rat cancer cells until the liver has regained its original mass. (Some other tissues also show similar responses to wounding or removal of portions.) The onset of the wave of division may result from metabolic changes in the organism that are related to the surgically reduced liver mass. Perhaps some stimulator of cell division is produced. There are hints that the blood stream of an animal with a regenerating liver carries factors that can induce division in the liver of a normal animal. Such division may be observed when the circulatory systems of two animals (one with a normal liver, the other with a regenerating liver) are made continuous. On the other hand, hepatocytes might normally synthesize a substance that inhibits cell division within the liver. When liver mass is reduced, the amount of inhibitor may fall below a threshhold of effectiveness, to be restored only when the organ size is restored. Possibly this sort of mechanism contributes to the control of organ size in embryonic development.

Experiments may show that looking for *the* inhibitor or *the* activator will not be a fruitful approach in studying the control of division. Since many organelles and diverse activities are involved in division, several possible control points probably exist in a given cell type, and somewhat different controls might operate in different cell types.

Division of cells may be reversibly prevented by conditions such as exposure to abrupt temperature changes or to certain metabolic inhibitors. These treatments prevent cells from passing through one or another critical stage in the cycle necessary for initiation or completion of mitosis. If a population of cells is treated for a time period long enough to permit most cells to reach the same stage and to stop there, and if the conditions preventing division are then reversed, then synchronous division often occurs. Over 90 percent of the cells may divide at the same time. Normally, dividing cell populations are asynchronous; different cells divide at different times (early cleavages of a developing embryo being among the few exceptions). The availability of methods for artificial synchronization is an important asset for biochemical studies of division stages since these techniques provide homogeneous populations for analysis. It should be possible to investigate the accumulation and utilization during the division cycle of components such as ATP, specific enzymes, or spindle proteins and thus, perhaps, to identify key control points.

Another promising approach is the study of the interaction of nucleus and cytoplasm by the direct experimental alteration of the cytoplasmic environment surrounding a given nucleus. In three different situations [adult toad brain nuclei transplanted into toad eggs; nuclei transplanted from amebae in G_2 to amebae in S; and hybrids formed by fusing HeLa cells with chicken red blood cells (Section 4.4.3)], DNA synthesis is induced in nuclei that ordinarily do not synthesize DNA, by exposure to cytoplasm of cells preparing for division. Efforts are under way to identify the specific activating factors that are involved: Are enzymes transferred from the cytoplasm to the nucleus? Is there some critical cytoplasmic concentration of specific metabolites in dividing cells, or are inhibitors absent? Hopefully, such questions will be answerable in specific molecular terms and insight will be gained into the remarkable coordination of nuclear and cytoplasmic events during division.

4.2.4 ***CHROMOSOME*** DNA replicates by a *semiconservative*
DUPLICATION mechanism; that is, the double helix separates into its two polynucleotide strands,
each strand remains intact and acquires a new complementary partner
that is formed by the alignment of nucleotides along the old strand
(by base pairing) and by the enzymatic linking of the nucleotides into a
polynucleotide chain (Fig. II-7). In *E. coli* and other bacteria, the
chromosome is probably one circular molecule of DNA. Replication
begins at one point of the circle and proceeds around the circle by
separation of the two DNA strands and formation of new partners. In
higher organisms, the chromosomes are larger, more complex, and more
numerous. They consist of DNA, histone, and nonhistone proteins;
in many cases, RNA is also present. However, in some respects the
duplication of chromosomes parallels the replication of DNA, as in
bacteria. Note: as we use the terms here, a *molecule* of DNA refers to
a *double helix*, composed of *two polynucleotide strands* (Fig. II-7).

When cells of higher organisms are exposed for a few minutes
to tritiated thymidine and then grown in a nonradioactive medium and
studied by autoradiography, not all the chromosomes of a given cell
are invariably labeled. Those that are labeled show radioactivity at
several discrete, and often quite distant, points along their length.
These findings indicate (1) that different chromosomes may duplicate
at different times during interphase (only those that are duplicating
during the brief exposure to label become radioactive); and (2) that
several regions replicate simultaneously on a given chromosome
(if only one replication point per chromosome were operating, only one
site would be labeled). Of interest is the fact that, the sequence of
DNA replication in different chromosomes and chromosome portions
appears to be the same in successive divisions of a given cell type.

Once the DNA is replicated, how is it apportioned between
daughter chromosomes? An experiment designed to answer this question is outlined in Fig. IV-9. The key finding is that the distribution of
label over chromosomes after the second duplication is different from
that at the first duplication following label administration. This is interpreted as indicating that, as in bacteria, the DNA of a duplicating
chromosome is distributed between the products of duplication (the
chromatids in Fig. IV-9) as if it were in the form of one long DNA

Fig. IV-9 *J. H. Taylor's experiment on chromosome duplication. Taylor* ▶
used the drug, colchicine *to surpress cytoplasmic division (Section 4.2.7)
without preventing chromosome duplication; chromatids separate but in the
absence of a spindle, they remain in the same cell. This provides a convenient
method for determining the number of times the chromosome set being studied
has duplicated during the experiment; each duplication results in doubling of
the number of chromosomes per cell. Red strands are radioactive.*

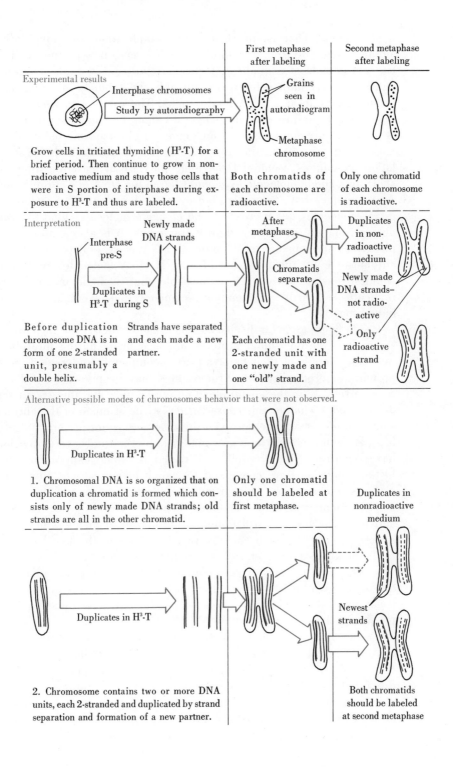

	First metaphase after labeling	Second metaphase after labeling

Experimental results

Interphase chromosomes

Study by autoradiography

Grains seen in autoradiogram

Metaphase chromosome

Grow cells in tritiated thymidine (H³-T) for a brief period. Then continue to grow in non-radioactive medium and study those cells that were in S portion of interphase during exposure to H³-T and thus are labeled.

Both chromatids of each chromosome are radioactive.

Only one chromatid of each chromosome is radioactive.

Interpretation

Interphase pre-S

Newly made DNA strands

After metaphase

Duplicates in H³-T during S

Chromatids separate

Duplicates in non-radioactive medium

Newly made DNA strands—not radioactive

Only radioactive strand

Before duplication chromosome DNA is in form of one 2-stranded unit, presumably a double helix.

Strands have separated and each made a new partner.

Each chromatid has one 2-stranded unit with one newly made and one "old" strand.

Alternative possible modes of chromosomes behavior that were not observed.

Duplicates in H³-T

1. Chromosomal DNA is so organized that on duplication a chromatid is formed which consists only of newly made DNA strands; old strands are all in the other chromatid.

Only one chromatid should be labeled at first metaphase.

Duplicates in nonradioactive medium

Duplicates in H³-T

Newest strands

2. Chromosome contains two or more DNA units, each 2-stranded and duplicated by strand separation and formation of a new partner.

Both chromatids should be labeled at second metaphase

double helix. That is, in chromosome duplication, two longitudinal subunits separate from one another, each forms a new partner, and the two chromatids of a duplicated chromosome consist each of one old and one newly built subunit. Presumably the subunits are equivalent to single strands in a DNA double helix. The words *equivalent to* are important. From present knowledge, it cannot be concluded that there is actually only one long DNA molecule per chromosome before duplication and per chromatid after duplication. Using even the best isolation techniques, isolated DNA molecules are at least ten times too short for a single molecule to account for all the DNA of a chromosome. As techniques have improved, however, the length of DNA isolated as intact molecules has increased; the elongate molecules tend to break during isolation, producing shorter fragments. It is also simpler to explain the simultaneous replication of DNA at several points along a chromosome by the presence of several DNA molecules replicating simultaneously; however, it is possible that a single molecule can replicate as partly independent segments that eventually become linked together.

The findings so far discussed concern only the DNA of the chromosome. New chromosomal proteins are also made during interphase, but many details of their behavior are incompletely understood. Some proteins of preduplication chromosomes may be preserved and they may participate in the postduplication chromatids. But there is current debate about whether this *conservation* is true of much of the chromosomes' protein and whether even the most persistent protein molecule has not been replaced by a new molecule by the time several divisions have occurred. The possibility is raised by some experiments that histone molecules are conserved for much longer periods than other chromosome proteins. One group of investigators reports that radioactivity incorporated into histones during a brief exposure to radioactive amino acids is still found in chromosomal histones many cell generations later. (Such conservation is of course the usual case for DNA, in which a given strand is potentially immortal.)

The total nuclear protein content doubles during interphase but the sites of synthesis, nucleus or cytoplasm, remain to be learned. Most, if not all, of the doubling of chromosomal histones takes place in the S period. At least in some cultured cells such as HeLa cells (Chapter 3.11), the use of radioactive amino acids and cell fractionation at successive intervals permit the tentative conclusion that histones made in the cytoplasm enter the nucleus; they are partly, if not entirely, responsible for the increase in histone amount. In sum, the information available is very fragmentary, and much investigation must be done on chromosomal proteins before an understanding of chromosome duplication is possible. If chromosomal proteins are made in the cytoplasm, how they enter the nucleus and become incorporated into the chromosomes has yet to be elucidated. Systems like the nuclei of protozoa discussed in Section 3.3.1 may prove of considerable use in further study of these problems.

4.2.5   ***CHROMOSOME*** A chromosome that measures, at mitotic
            STRUCTURE metaphase, 5–10 μ in length and a micron
            or less in width may contain an amount of DNA that would be several millimeters to several centimeters in length, if stretched out in one straight double helix. "Packaging" of this DNA in chromosomes involves extensive coiling and folding. This is readily evident through light microscopy; chromosomes during cell division are seen to consist of coiled fibers (Fig. IV-10).

On occasion, by light microscopy, longitudinal subdivisions are seen in chromatids. This suggests the presence within the chromatid of two or more subunits, seemingly much larger and more complex than the two strands of a DNA helix. The nature of such subunits is not clear, but it is conceivable that the observed subdivisions are artifacts of one sort or another. However, such observations and some other indirect evidence have contributed to a long debate between those who believe that the chromosome is "single-stranded" and those who believe that it is "multistranded." The work *strand* in this context

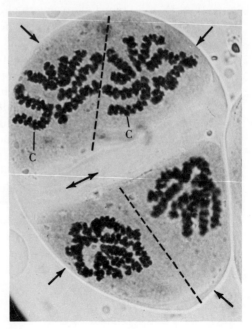

Fig. IV-10 *A microsporocyte of the plant,* Trillium. *It has been fixed at the end of anaphase (second meiotic division) as the chromosomes (C) begin to uncoil. The two meiotic divisions (Fig. IV-18) produce four cells that give rise to male gamete nuclei. In the present case, the first division took place along the plane of the double-headed arrow; the second would have been along the dotted lines. The single-headed arrows indicate cell borders.* × 1,000. *Courtesy of A. Sparrow.*

refers not to the complementary halves of a DNA molecule, but rather to some hypothetical basic unit of chromosome structure. A single-stranded chromosome would contain one longitudinal structural unit that presumably is equivalent to a DNA double helix. A multistranded chromosome might contain several such units lying side by side. Some alternative models are diagrammed in Fig. IV-11. Indirect evidence already outlined (Section 4.2.4) suggests the presence of a single functional unit equivalent to a DNA molecule in a preduplication chromosome or in a postduplication chromatid. It is not yet possible for electron microscopy definitely to associate such a functional unit with a structural unit and thus settle the "single-stranded" versus "multistranded" issue at the morphological level. When isolated whole (unsectioned) chromosomes are examined in the electron microscope, they appear to be masses of long fibers, as large as 200–300 Å in diameter (Fig. IV-12). No delimiting membrane is present. The fibers often show looping, as would be expected if there were actually one very long fiber folded into a compact chromosome. But technical difficulties in examining thick structures in the electron microscope and in minimizing the breakage of fibers and other alterations during preparation have hampered these studies; the presence of a few distinct fibers rather than one cannot be unequivocally ruled out.

Sections of chromosomes show numerous fine fibers (*fibrils*), grouped in a variety of bundles; in diameter they are as small as 20–30 Å, the dimensions of a DNA molecule or DNA-protein complex (Fig. IV-13). Only a short length of a given fibril is included in a section so that it is impossible to tell whether the chromosome is of one or a few coiled fibers, of many independent fibrils, or of a complex, interconnected network. Attempts at serial-section reconstruction (Section 1.2.1) have been of limited value owing to the small dimensions of the fibrils and to their high degree of coiling.

Many efforts have been made to dissect chromosome structure by chemical and enzymatic procedures. Proteins may be removed from isolated chromosomes by digestion with proteolytic enzymes; when this is done, fibers normally 200–300 Å thick are replaced by much thinner fibers. Some feel that this may indicate that the 200–300-Å-fiber is a bundle of several DNA molecules, bound together by protein. But proponents of the "single-strand" view are convinced that the 200–300-Å-fiber is seen to contain only a single DNA molecule when protein is removed; normally, this DNA molecule is extensively coiled and cannot be seen because of the presence of much chromosome protein.

Some chromosomes are fragmented into shorter pieces, when treated with the enzyme DNAase (Section 4.4.5) or when cells are grown in solutions of abnormal metabolites that interfere with DNA replication. This may indicate that the structural continuity of some if not all chromosomes depends on DNA; if DNA polynucleotide chains are interrupted, chromosome structure is interrupted.

Many models of chromosome structure have been proposed to explain the variety of observations (Fig. IV-11), but no one is universally accepted. One model pictures a chromosomes as consisting of a continuous non-DNA backbone along which DNA molecules are attached at intervals. Experiments just cited indicate that this arrangement is unlikely, since it predicts that breaking DNA molecules might make the chromosome thinner but should not interrupt chromosome continuity. Some models propose that the chromosome is actually made of one elongate DNA molecule or that it consists of several molecules linked end-to-end, so that they behave as one. Because of their simplicity and because they agree with much of the available cytological and genetic information, these models are widely favored. More complicated structures, however, cannot yet be ruled out. The arrangement of chromosomal proteins is unknown, although the histones appear to be bound closely to the DNA and some of the histone molecules may run along the double helix (following the course of the polynucleotide chains).

The architecture of the eucaryote chromosome is one of the outstanding unsolved problems of cell structure. When it is finally understood, it may bring quick resolution to problems such as the nature and control of chromosome coiling mechanisms, the arrangement of replication and transcription enzymes, and the structure and functioning of

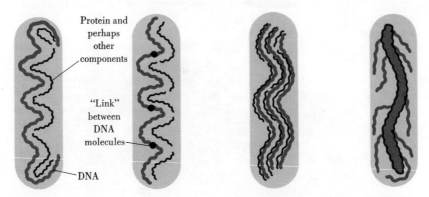

Fig. IV-11 *Schematic sketches of several hypothetical models of chromosome structure: **1**. One elongate, coiled DNA molecule (double helix) with protein and other material running alongside; **2**. Several DNA molecules linked together end to end by unknown materials; **3**. Several to many DNA molecules arranged longitudinally in parallel strands or coiled as in a rope; **4**. A non-DNA backbone (for example, of protein) to which DNA molecules are attached at intervals. In all cases, the material in the actual chromosomes would be far more extensively folded and coiled.*

specialized chromosome regions such as the kinetochore. Although kinetochores can often be readily recognized by light microscopy (Fig. IV-5), it is only recently that the electron microscope has shown that their structure differs characteristically from the rest of the chromosome (Fig. IV-13). Before long, it may be possible to describe such differences in molecular terms.

4.2.6   SPINDLES, The precision of chromosome behavior in
KINETOCHORES mitosis is related to the *spindle*. This highly
AND CENTRIOLES organized, fairly rigid, gel-like region of cytoplasm has a fibrous appearance in the light
microscope. Three types of spindle fibers are often distinguishable by light microscopy: *continuous* ones that pass from pole to pole; *chromosomal* fibers that attach to the chromosome kinetochores; and *interzonal* fibers that are present between two groups of chromatids as they separate at anaphase (Figs. IV-7 and IV-17). The interzonal fibers probably include the continuous fibers that persist after the chromatids separate under the influence of the chromosomal fibers.

In animal cells, spindles generally form in association with the centrioles (Fig. IV-7). At each pole, an aster is often present; in the light microscope, asters appear as roughly spherical arrays of fibers oriented radially around the centrioles and merging with the spindle. Centrioles and asters have not been seen in many plant cells.

Observations with the polarizing microscope resolved a long

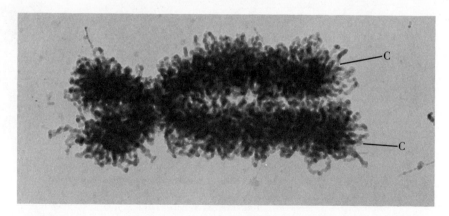

Fig. IV-12 *A human chromosome isolated at the metaphase stage of mitosis and viewed without sectioning. The two chromatids are readily visible (C); they are attached to one another at the kinetochore. Since very few ends of fibers are seen, it has been argued that each chromatid consists of one long looped and coiled fiber. Approx. × 50,000. Courtesy of E. J. DuPraw.*

debate among light microscopists as to the reality of spindle fibers. The presence of oriented molecules, in a fiberlike organization, was shown in the spindle of living cells (Fig IV-14). In the electron microscope, spindles (and asters) show numerous microtubules (Fig. IV-13), in addition to ribosomes and vesicles (some of which probably derive from the endoplasmic reticulum). The fibers of light microscopy are shown to contain groups of microtubules; it is probable that some of

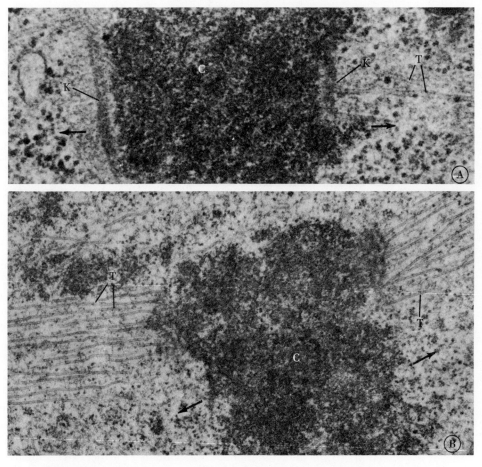

Fig. IV-13 *Chromosomes during mitosis as seen in sectioned material. A. Portion of a chromosome (C) from a cultured fibroblast (Chinese hamster). Kinetochore regions (K) are seen as special bands of dense material pointing towards the spindle poles (directions indicated by arrows). The structure of the remainder of the chromosome is difficult to interpret; it includes fine fibers sectioned at various angles with only a small portion of a given fiber region included in any given thin section. × 80,000. B. Portion of a chromosome in a cultured cell from the Tasmanian Wallaby (Potorus). Spindle microtubules are indicated by T. Courtesy of B. Brinkley and E. Stubblefield.*

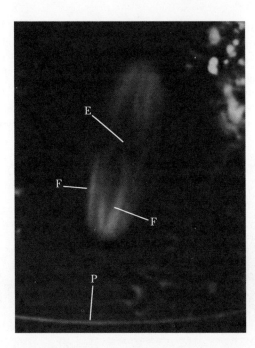

Fig. IV-14 The spindle of a living *egg of the worm,* Chaetopterus, *photographed through a polarizing microscope (Section 1.2.2). The equator, where the metaphase chromosomes (not visible in this photograph) are located, is indicated by E. Clear indication of the fibrous organization of the spindle may be seen at F. The plasma membrane is present at P. × 1,500. Courtesy of S. Inoue.*

the tubules are continuous from pole to pole, whereas others terminate at the chromosomes.

The *mitotic apparatus* (spindles, centrioles, asters, and chromosomes) can be isolated as a unit from developing eggs (Fig. IV-15). The isolated apparatus maintains enough of its original morphology that chromosomes remain attached to it. However, it has not yet been possible to induce normal movement of chromosomes on isolated spindles; probably alterations and loss of material occur during isolation. The isolated preparations have been very useful in studies of spindle chemistry. A portion of the protein isolated from the spindle resembles the proteins of other microtubule-rich structures. Biochemical studies have confirmed the cytochemical finding that SH groups and S—S bonds are important in the maintenance of spindle structure. They have shown in addition that other relatively weak bonds, such as those involved in three-dimensional protein structure (Section 4.1.2), also function in the spindle. Further, by comparing the proteins isolated from spindles with the proteins isolated from developing egg cells, it has been demonstrated that spindle proteins are present in the cells in large amounts long before the spindle forms. Thus, spindle formation probably is based in part on assembly (perhaps self-assembly) of previously made subunits. In agreement with this, the spindle can undergo reversible dissolution relatively easily (Section 4.2.7).

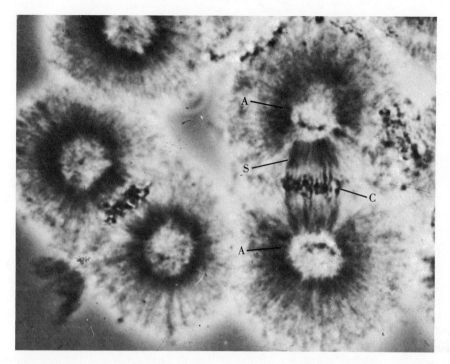

Fig. IV-15 *Phase-contrast micrograph of the mitotic apparatus isolated from sea urchin eggs. The spindle (S), asters (A) and attached chromosomes (C) are visible. Courtesy of D. Mazia.*

    In most cells with centrioles, the spindle forms between the centriole pairs as they separate (in prophase). Centriole duplication occurs well in advance and may begin as early as telophase of the prior division. In interphase, as DNA synthesis progresses, each centriole of the initial pair is accompanied by a new daughter centriole that apparently forms from a procentriole oriented at right angles to the parent (Section 2.10.5). Generally, the nuclear envelope breaks down in prophase, as the chromosomes become associated with the spindle.

    However this pattern is not universal. Many plant cells have no obvious structures corresponding to centrioles. In some protozoa and other cells, the spindle forms within the nucleus, and the nuclear envelope does not break down. In other cases, even though the envelope remains, the chromosomes attach to an extranuclear spindle, in a manner not understood. In a few fungi in which no centrioles are present, an intranuclear spindle forms between special thickened regions of the nuclear envelope. In some protozoa, even if the nucleus is removed from the cell, a spindle of continuous fibers forms between the centrioles. On the other hand, in some insect and plant cells,

spindle fibers are seen to form initially in association with the kinetochores of the chromosomes.

Were only a single cell type studied, one or another structure
would be assigned a primary role in controlling spindle formation.
Usually it is the centrioles or kinetochores that are most obviously
involved. However, until the formation of the spindle is understood in
molecular terms, it will be difficult to reconcile the variety of observations on spindle formation with the apparent uniformity of its function
in cell division.

Among further questions, answers to which are still being sought,
are: (1) How do the chromosomes attach to the spindle so that the two
chromatids separate one to each pole, rather than both moving to the
same pole? (2) How is the precise alignment of chromosomes at metaphase brought about? Is the notion that the alignment results from an
equal "pull" from both poles valid, or are much subtler mechanisms also
involved? (3) Do special associations exist, during interphase, among
chromatin strands, centrioles, and the nuclear envelope? (For example,
some microscopists claim that portions of chromosomes may be attached to the envelope.) Do such associations account for the behavior
of chromosomes during duplication and early stages of division? (For
example, in some cells, the chromosomes move within the prophase
nucleus along a path paralleling the movement of the centrioles outside the nucleus.) (4) What role, if any, do asters play in cells where
they are present?

4.2.7   ***CHROMOSOME***   In separating at anaphase, chromatids
              MOVEMENT   move at rates of about a micron or two per
                                    minute, (sometimes they are faster). The
movement often involves two components. In most cells, separating
chromatids approach the spindle poles. In many cells, the spindle as
a whole *also* elongates, the ends move apart, and the chromatids separate without actually getting closer to the poles (Fig. IV-16).

When the kinetochore is near the middle of a chromatid, a
characteristic V shape is seen, with the kinetochore in the lead as the
chromosome moves in anaphase. This probably results from the effects
of spindle fibers, which pull on the kinetochores or exert some other
influence leading to kinetochore motion with the rest of the chromosomes pulled behind. Chromosome fragments without kinetochores
can be produced by treating cells with radiation or certain chemicals.
Such fragments do not attach to spindle fibers and do not move.

Many mechanisms have been proposed to explain chromosome
motion. The simplest idea to account for the poleward motion of chromosomes is that the spindle fibers contract. If the fibers were like muscles

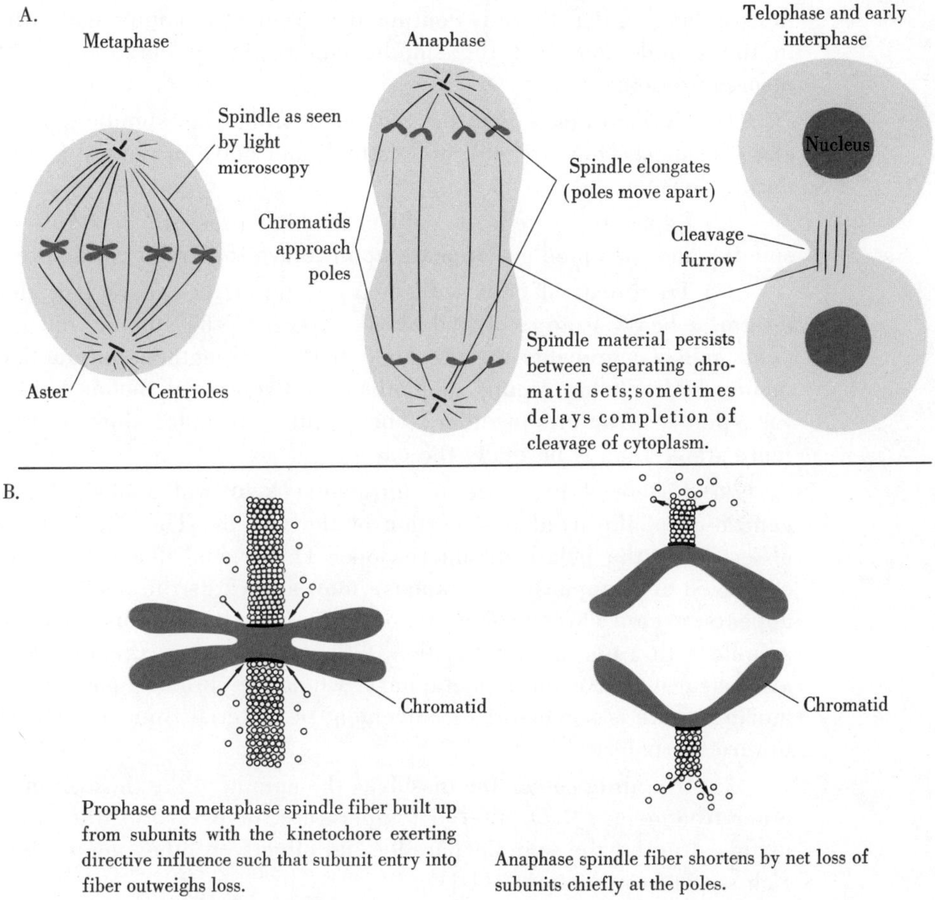

A.

Metaphase   Anaphase   Telophase and early
                                                         interphase

Spindle as seen
by light
microscopy

Chromatids
approach
poles   Spindle elongates
                             (poles move apart)

                             Cleavage
                             furrow

Aster   Centrioles   Spindle material persists
                             between separating chro-
                             matid sets; sometimes
                             delays completion of
                             cleavage of cytoplasm.

Nucleus

B.

Chromatid   Chromatid

Prophase and metaphase spindle fiber built up
from subunits with the kinetochore exerting
directive influence such that subunit entry into   Anaphase spindle fiber shortens by net loss of
fiber outweighs loss.   subunits chiefly at the poles.

Fig. IV-16   *Chromosome movement during cell division.* **A.** *A diagram of a
hypothetical animal cell showing the separation of sister chromatids. The relative contributions to chromosome movements of* spindle elongation *and of*
movement towards the poles *vary in different cell types.* **B.** *A diagram of the
theory developed by Inoue that during cell division spindle fibers grow and
shorten by gain and loss of subunits. The diagrammed fibers are hypothetical
structures presumed to be identical to the microtubule bundles and associated
material found in cells. One key element in the theory being actively investigated
is the nature of the coordinated operation of kinetochores and centrioles in controlling the assembly and disassembly of the fibers.*

or like stretched rubber bands, they should thicken as the chromosomes move toward the pole. Although the spindle fibers shorten (sometimes to only 25 percent or less of their original length), no such thickening of the fibers and no appropriate changes in spindle microtubule
thickness have yet been observed.

An alternative theory has been proposed, based upon observations indicating that there is continued movement of components within the spindle and that the spindle can readily undergo reversible disorganization:

(1) Chromosomes can be detached from the spindle by fine glass needles. They are able to reattach and move in normal fashion during division.

(2) Exposure of cells to cold or to high pressure dissolves the spindle, and the spindle reappears upon return to normal conditions.

(3) Treatment of cells with heavy water, D_2O (where D is deuterium, a heavy isotope of hydrogen), "freezes" the spindle and prevents mitosis; probably the presence of D_2O strengthens interactions among hydrophobic groups and other relatively weak bonds holding the spindle structure together. The spindle structure thus is made more stable than is normally the case.

(4) Localized exposure to ultraviolet light will abolish the organization of the irradiated portion of the spindle. The effect can be observed in the polarizing microscope. The region of a spindle disorganized at metaphase or anaphase moves towards the pole and disappears, even if chromosome motion has not yet begun; this leaves the spindle with a normal organization. Other observations (Section 4.2.9) also suggest that even at metaphase, when the chromosomes are not moving, there is some sort of movement of material from the equator toward the pole.

(5) The drug *colchicine* dissolves the spindle. Like those of many other treatments (D_2O, altered temperature, or pressure), the effects of this drug on the spindle parallels its effects on other microtubule-rich structures (Section 2.11.1).

The key point of the theory that emerges from these considerations (Fig. IV-16) is that a spindle fiber attached to a chromosome is in dynamic equilibrium with subunits that are continually entering and leaving it. During prophase, as the spindle fiber forms, the gain of subunits outweighs the loss, perhaps owing to the influence of the chromosome kinetochores. During anaphase, the situation is reversed and the loss of subunits at the spindle pole outweighs the gain; as a result, the fiber shortens without thickening. In some manner, as the fiber loses subunits, the attached chromosome moves towards the pole (Fig. IV-16). As part of the fiber structure, microtubules might directly participate in bringing about chromosome motion; or other spindle components may be primarily responsible for causing the motion, with microtubules providing a framework that guides the direction of motion of the chromosomes.

The mechanisms of spindle elongation are also elusive. Elongation can be induced in nonliving preparations, as occurs when certain compounds are added to isolated spindles or to cells that have been soaked in glycerol. (This soaking removes soluble components but leaves many structures intact.) ATP is among the effective compounds, but it is not known whether its energy or its negative charge is responsible for the effect; other charged compounds with no high energy bonds also have this effect. The usual models of elongation propose that the spindle "grows" by the addition of material at the center and the consequent outward pushing of the two poles.

4.2.8   ***CYTOPLASMIC***   In general, cells at telophase divide along
                   DIVISION   the plane of the spindle equator (where the
                                      metaphase chromosomes had been located),
and two cells of equal size are produced. The two daughter cells receive similar shares of cytoplasmic organelles. In some algae, where a single plastid that occupies much of the cytoplasm is present, the plastid divides in synchrony with the rest of the cell. In cells with large numbers of mitochondria or plastids, these organelles are often distributed apparently at random in the cytoplasm, so that roughly equal numbers of organelles are contained in the two daughter cells. Sometimes, equal distribution results from a special arrangement of organelles. Thus, lysosomes often tend to cluster near the poles. In the dividing spermatocytes of some insects, the mitochondria group around the spindle to form an aggregate of elongate mitochondria; this is cut in half as the cell divides.

In animal cells, a cleavage furrow moves in perpendicularly to the long axis of the spindle and pinches the cell in two. The furrow is often thought to be the result of the contraction of a ring of peripheral cytoplasm. Some propose that a special system of contractile proteins is responsible. Others point out that special orientation and interaction often occur among molecules at surfaces (see, for example, Section 4.1.1) and that this can generate *surface tension* forces. (Surface tension is the force that, for example, maintains droplets of liquids in more or less rounded forms.) Forces of this type, reflecting localized changes in the organization of the *cortical cytoplasm* just below the cell surface, might participate in furrow formation.

If the mitotic apparatus is experimentally shifted during prophase or metaphase in a fertilized egg cell or tissue culture cell, the plane of the cleavage furrow is shifted accordingly. If the mitotic apparatus is shifted after metaphase, or even removed, it has no effect upon the cleavage furrow. Some interaction between the spindle and the rest of the cell appears to "set" the cleavage plane; once this has occurred, the spindle is no longer required. One view suggests that attachments sometimes found between asters and the cell surface may contribute to setting the cleavage plane.

In plants, cytoplasmic division occurs by the formation of a *cell plate*. Vesicles, many derived from the Golgi apparatus, accumulate in the middle of the cell, flatten, and fuse to establish the new cell boundaries (Fig. IV-17). This begins at the *phragmoplast*, a region of vesicles that accumulate near microtubules that persist at the spindle equator for a time after the chromosomes have separated (Fig. IV-17).

4.2.9   ***NUCLEAR ENVELOPE AND NUCLEOLUS***   Disappearance of the nuclear envelope in prophase results from fragmentation of the envelope into numerous sacs and vesicles. In telophase, the new nuclear envelope forms by fusion of sacs and vesicles that accumulate around the chromosomes. It is not yet known whether the fragments of the "old" envelope persist and contribute to formation of the new one.

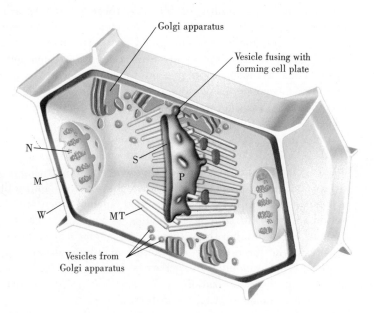

Fig. IV-17 *Schematic representation of a cell of a higher plant as seen at telophase in mitosis. The cell wall is indicated by (W), the plasma membrane by (M) and one of the two daughter nuclei by (N). In the phragmoplast region, a region of membranes and microtubules (MT), a cell plate forms (P) and grows until it separates the cytoplasm into two daughter cells. The cell plate develops as a membrane-delimited structure enclosing a space (S) in which new cell wall will form. The Golgi apparatus contributes many vesicles to the phragmoplast membrane; the vesicle membranes apparently are incorporated into the membrane of the cell plate and the vesicle contents enter the forming cell wall. Modified after M. Ledbetter.*

The nucleolus, as an organized entity, generally disappears in division. However, in at least some cell types, nucleolar components probably are preserved during division and participate in forming a new nucleolus. Microscope and autoradiographic studies suggest that during division, some material from the nucleolus passes to the chromosomes, to the cytoplasm, or to both of them, and that some of this returns to the new nucleolus formed after division. In a few cell types, nucleoli persist as intact bodies throughout the division cycle; sometimes they pass into daughter cells with the chromosomes to which they are attached. In some plant cells, portions of nucleoli, associated with the chromosomes until metaphase, can move to the spindle poles before chromosome movement begins. Apparently, the nucleoli are carried passively by some sort of poleward flow of material within the spindle; thus they are transported independently of the chromosomes.

c h a p t e r **4.3**

GENETIC BASIS OF CELL DIVERSITY

Thus far, discussion has included mechanisms that tend to keep structures and cells *constant*. The present chapter considers the origins of *diversity through genetic mechanisms*, diversity that is the basis of evolution. It is usually not the genes themselves, but rather the genetically determined characteristics of organisms, that are exposed to the environment and thus to natural selection [although there are some special cases of fairly direct interaction of gene and environment, such as enzyme induction (Section 3.2.5)]. Those organisms with characteristics that are advantageous in the particular environment produce more offspring than organisms with fewer advantageous characteristics. They thus contribute a larger number of genes to the next generation of a population. With time, populations undergo evolutionary change.

4.3.1   *MEIOSIS* Mutations in DNA result in altered base sequences which ultimately are translated into proteins with altered amino acid sequences. Sometimes the change of just one amino acid in a protein has profound effects. For example, in *sickle cell anemia*, the human inheritable disease, a single amino acid in the hemoglobin molecule is replaced by another. This results in drastically altered structure and function of red blood cells. (The cells are fragile and under certain conditions, they tend to assume a shrivelled sickle shape, in contrast to the normal disc shape.)

Mutations give rise to *alleles*, which are alternate forms of a gene.

(For example, one allele of an eye color gene may produce blue eyes, whereas brown eyes are produced by another of the several alleles of the same gene.) For the most part, inheritable differences among individuals of a given species are attributable to the fact that different individuals carry different alleles.

All sexually reproducing eucaryotic organisms (that is, many unicellular organisms and almost all multicellular organisms), go through a *zygote* stage, or its equivalent. Usually a single nucleus is formed containing a mixture of alleles from both parents. As outlined earlier in Chapter 3.10, the zygote nucleus may result from the fusion of nuclei carried by distinctive gametes, as in most higher organisms. In many unicellular organisms and in some lower plants, nuclei contributing to the zygote are transferred between two cells without the formation of obviously specialized gametes by processes such as partial and temporary fusion (*conjugation*) of ciliated protozoans.

Meiosis is complementary to zygote formation in the life cycle of organisms, and it accomplishes the segregation of alleles. This results from the fact that the two members of each homologous pair of chromosomes (Section 4.2.1) are separated, by meiosis, into different cells. Usually, the members of a homologous pair both carry the same genes arranged in the same sequence along their lengths; however, the particular alleles present on each homologue may differ. Meiosis involves two rounds of cell division, as outlined in Fig. IV-18. The first meiotic division is preceded by DNA replication during interphase. The division is different from mitosis in two key respects. (1) In the first meiotic prophase, homologues pair gene for gene; they come together and align so that the site (*locus*) of a given gene on one homologue lies next to the same gene locus on the other homologue (Figs. II-17, IV-20, and IV-21); no such pairing occurs in mitotic prophase. (2) The chromosomes remain paired until metaphase. Then, one *chromosome* of each pair goes to each pole; unlike mitosis, the two *chromatids* of each chromosome remain together. Thus the number of chromosomes per nucleus is halved. The *second* meiotic division resembles mitosis in that no pairing occurs and chromatids separate from one another. However,

Fig. IV-18 *Chromosome behavior in meiosis. First meiotic prophase is ►
usually thought of as having 5 major stages: **1**. Leptotene, when the chromosomes have coiled to the point where they are visible as discrete slender threads;
2. Zygotene, when homologues begin visible pairing; **3**. Pachytene, when gene to gene pairing is completed (see Fig. II-17); **4**. Diplotene and **5**. Diakinesis, when the chromosomes coil to reach maximum thickness and then, although they remain associated, relax pairing so that homologues are associated only at some points, the chiasmata (Fig. IV-20). The X and Y chromosomes drawn in the diagram represent the sex-determining chromosomes seen in many organisms (Section 4.3.3).*

A. Overall behavior of chromosomes

At metaphase the chromosomes line up in pairs at the spindles equator.

At mid-prophase of the first of the two meiotic divisions homologous chromosomes are paired; each chromosome is of two chromatids.

At anaphase homologous chromosomes separate to opposite poles while the two chromatids of each chromosome stay together.

Daughter nuclei from the first division contain one chromosome from each homologous pair.

In second division the chromosomes behave as in mitosis. One chromatid of each goes to opposite poles.

Thus, each cell entering meiosis produces four haploid daughters. Each homologous pair segregates independently of the others. For example, A enters the same daughter as B or b with equal frequency. Thus, another (Aa, Bb, XY) cell entering meiosis will produce daughters abY, ABY, aBx and AbY.

B. Crossing over

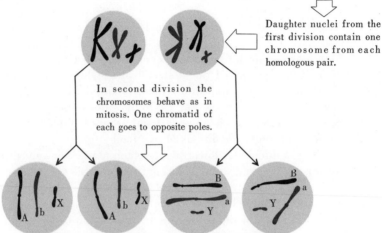

Homologous pair of chromosomes which differ in alleles at 3 sites: M, N, and Q.

Pairing in first meiotic division.

Homologues exchange portions of chromatids.

Chromosomes separate at anaphase of first division.

Second division (chromatids separate).

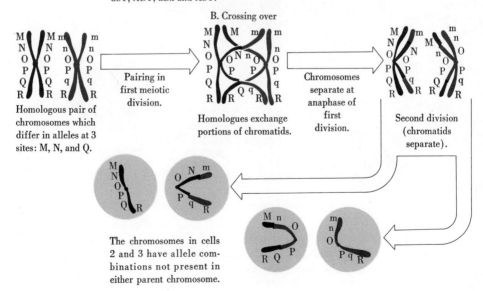

The chromosomes in cells 2 and 3 have allele combinations not present in either parent chromosome.

the second division is not preceded by DNA replication or chromosome duplication; the chromatids that separate are the ones already present during the first division. The net result of the two divisions is four haploid cells (Figs. IV-10 and IV-19), each having half the DNA and half the number of chromosomes (one of each pair) of the parent diploid cell.

The subsequent behavior of the meiotic products varies considerably in different organisms. Many algae and protozoa are normally haploid; the meiotic products each divide by mitosis to establish a new cell line. The members of a line produced in this way are genetically identical, except for mutations occurring subsequent to the meiotic divisions that produced the first cell of the line. Lines differ from one another in alleles, since they have resulted from products of meiosis that differ in alleles (see Fig. IV-18 for an example of such differences). Sexual reproduction is an occasional process; it occurs, for example, in *Paramecia* under conditions of low food supply.

In multicellular lower plants, such as the seaweeds (multicellular algae), the haploid cells resulting from meiosis are released as spores. These divide by mitosis to form a gamete-producing plant that is haploid and multicellular (*gametophyte*). Gametes are produced by differentiation of some of the gametophyte cells. A diploid zygote is formed by

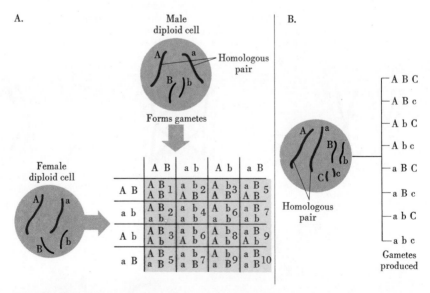

Fig. IV-19 *Some genetic consequences of meiosis.* **A.** *The results of a genetically simple mating. As indicated in the diagram, from the viewpoint under consideration the zygote combination (^{AB}maternal; ^{ab}paternal) can be considered to be the same as (^{AB}paternal; ^{ab}maternal).* **B.** *The gametes that can be produced by an individual whose diploid cells contain 3 pairs of chromosomes.*

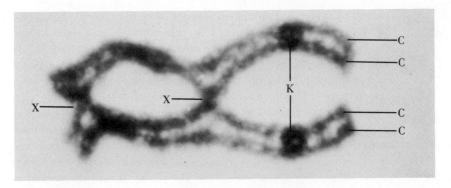

Fig. IV-20 *A pair of homologues during late prophase of the first meiotic division of a salamander spermatocyte. The four chromatids of the two chromosomes are readily seen at (C) and the two kinetochores are indicated by (K). At (X) chiasmata are present; that only one chromatid from each chromosome is involved can be clearly seen in the chiasma at the right. Courtesy of J. Kezer.*

the fusion of gametes. By mitosis, the zygote develops into a diploid multicellular *sporophyte*. In the sporophyte, some cells undergo meiosis to produce the haploid spores. In some species of lower plants, the alternating *sporophyte* and *gametophyte generations* are separate plants. In others, one generation is reduced in size and may occur as a special portion of the more highly developed generation.

In higher plants and virtually all multicellular animals, most cells are diploid, the products of mitotic divisions of the zygote. Each individual thus starts with an essentially unique set of alleles, except for the occasional occurrences, when a single zygote produces two (identical twins) or more organisms. Meiosis occurs in specialized reproductive cells and results in gametes (see Chapter 3.10). In males, all four meiotic products form gametes. In females, three of the cells resulting from meiosis generally degenerate and only one functions in reproduction. The cells resulting from meiosis in animals differentiate directly into sperm and ova. In higher plants, they divide by mitosis to produce microscopic gametophytes that are part of the flower and that form gametes—male nuclei in the pollen and egg cells in the base of the flower.

4.3.2   **THE**   The diversity in detail of gamete and zygote
TRANSMISSION OF   behavior is important for understanding the
HEREDITARY   hereditary patterns of different organisms.
INFORMATION; AN   But the variety of life cycles all involve
EXAMPLE   two constant features: the reduction of
diploid cells to haploid (meiosis), and the
formation of new diploid cells with new combinations of alleles gen-

erated by fusion of haploid nuclei (sexual reproduction). This constancy produces predictable patterns of transmission of chromosomes and genes from one generation to the next, making genetic analysis possible. For example, earlier (Section 2.3.4) it was mentioned that matings of certain individuals of the toad *Xenopus* produce offspring of which one quarter are inviable, due to a chromosomal defect affecting ribosome and nucleolus formation. This pattern of inheritance is explainable by the proposal that the parents in the matings each contain one normal (*N*) and one abnormal (*No*) homologue in the chromosome pair responsible for formation of nucleoli. Since a normal chromosome is present, ribosome synthesis can take place. However, meiosis results in gametes half of which contain the *N*, and half the *No*, homologue. If "m" represents the homologue contributed by the male parent and "f" the homologue contributed by the female parent, then equal numbers of the following types of zygotes will be formed: mNfNo; mNofN; mNfN; mNofNo. Thus, one-quarter of the zygotes are *NN*, one-quarter *NoNo*, and one-half, *NNo*. The *NoNo* category are the ones that die, since they have no normal homologue. Their survival through the early stages of development is based on ribosomes stored in the oocyte before the completion of meiosis. At this time, a normal homologue is still present in the eggs that eventually contain only the *No* homologue, since the oocytes like the other *diploid* cells of the parents are all *NNo*.

This explanation predicts that there should be a class of surviving offspring that contains one *N* and one *No* homologue and another class with two *N* homologues. There are two lines of evidence that this is so. Among the viable offspring of the matings under discussion, some have many cells in which two nucleoli are present (in some cells the two fuse to form a single organelle) while in the cells of other individuals only one nucleolus is ever present. Also, DNA can be extracted from embryos prior to the death of the *NoNo* class and its ability to form molecular hybrids (Section 2.2.3) with purified rRNA determined; the expected three categories of individuals are found: some have essentially no DNA sites that bind rRNA, and, of the rest, one group has twice as many sites as the other. This last finding indicates that the abnormality in *No* chromosomes involves the loss (*deletion*) of the rRNA-producing DNA segments.

It is interesting that the *NN* individuals and the *NNo* individuals both synthesize rRNA at the same rate and in the same amounts, despite the fact that the former individuals presumably have twice as many rRNA genes as the latter. Apparently the mechanism controlling the rates and amounts of synthesis operates to compensate for the differences in chromosomal constitution.

4.3.3   ***GENETIC***   Sexual reproduction and meiosis con-
                   DIVERSITY;   tinually generate new genetic combina-
              CROSSING OVER   tions. In Fig. IV-19 it is shown that a hypo-
                                    thetical organism with 6 chromosomes
(3 pairs) can produce gametes with eight different combinations of
chromosomes. Figure IV-19 diagrams the 10 different zygote combinations that can result from matings of two individuals, each having only
4 chromosomes (2 pairs) with the same arrangement of alleles in both
individuals. These are extremely simple examples. With large numbers
of chromosomes and matings, as is usual between individuals carrying
many different alleles, enormous numbers of combinations are possible.

Even greater diversity of gametes and zygotes results from *crossing
over* between homologous chromosomes. This depends on the reciprocal exchange of portions of chromatids during the first meiotic division
(Figs. IV-18 and IV-20). Crossing over produces characteristic chromosome configurations known as *chiasmata* (Fig. IV-20), which are visible
just before and during the first meiotic metaphase. It is not an occasional phenomenon, but rather an almost invariable feature of the
association of chromosomes in meiosis in most organisms studied.
(There are a few exceptions, such as males of the fly *Drosophila*, in
which pairing occurs, but crossing over is rare.) In any given crossover, the two chromatids usually exchange comparable portions which
may differ in alleles but which carry the same set of gene sites (*gene
loci*). However, when a particular homologous pair is compared in
different cells undergoing meiosis, the chromosome portions involved
in crossing over vary almost at random, so that different sets of loci
are exchanged in different cells.

Sex determination can often (although not invariably) be readily
demonstrated to depend on meiosis. For example, in humans, males
are usually XY and females XX, where X and Y refer to special homologous sex chromosomes which pair at meiosis (Fig. IV-6) although
they differ in genes and in morphology. The segregation of these chromosomes in meiosis in males (Fig. IV-18) results in equal numbers of X
and Y sperm. Since female gametes all are X, half the zygotes will be
XX and half XY. In other organisms, somewhat different situations are
found. For example, in some insects, males have one X and females
two; no Y chromosome is involved. In meiosis in these males, equal
numbers of gametes have either one X or none; all female gametes
have one X.

Meiosis does not occur in procaryotes. The few species of bacteria
that have been adequately studied are haploid in the sense that there
is only a single copy of the chromosome per nuclear region. Although
it is not known how widespread this is, some bacteria are capable of

a form of sexual reproduction (referred to as *genetic recombination*) in which DNA from one cell is transferred to another, recipient cell. This can occur by several mechanisms, including one which involves the fusion of small regions of the cells' surfaces and the passage of the chromosome of one cell into the other cell (see Section 3.2.5 for the involvement of the F episomal factor). The recipient becomes "diploid" by this process, but eventually a single *recombinant* chromosome is produced that is comparable to a crossover product in higher organisms. It contains genes from both of the cells and, normally, haploidy is restored. The number of nuclear regions per bacterial cell may vary under different conditions of growth, but each nuclear region apparently contains a copy of the same chromosome; the multinucleate condition probably represents a feature of cell growth and division rather than one of sexual reproduction.

4.3.4   PROBLEMS OF MEIOSIS

There are still many unanswered questions regarding meiosis. What makes spermatogonia and oogonia, after numerous mitotic divisions, switch their mode of chromosome behavior to meiosis? Many eggs remain in the prophase or metaphase of the first meiotic division until fertilization occurs, which can be weeks or months later; what produces this arrest and how is it released at fertilization? What determines the unequal division of many animal oocytes into a small "polar body," a cell that degenerates, and a much larger cell, the mature egg cell?

Homologous chromosomes pair gene for gene, probably with their kinetochores in some special arrangement. Is there something special about the chemistry or structure of meiotic chromosomes that permits this discriminatory mode of association? Is there any relationship between pairing of chromosomes and base pairing of nucleic acids? Is the *somatic pairing* of homologues seen in the nonmeiotic cells of some insects (such as *Drosophila*) based on mechanisms similar to meiotic pairing? In a few instances, differences have been reported between histones of meiotic and those of mitotic cells, but the significance of this is unknown. In many species, electron microscopy of meiotic stages reveals a special structure, the synaptinemal complex (Fig. IV-21), believed to represent paired chromosomes. Further analysis of this structure may illuminate meiotic mechanisms.

In grasshopper spermatocytes, it is possible to dislodge paired homologues from the meiotic metaphase spindle with a microneedle and to turn the pair around so the chromosome that was about to segregate to one spindle pole now faces the opposite pole. The paired chromo-

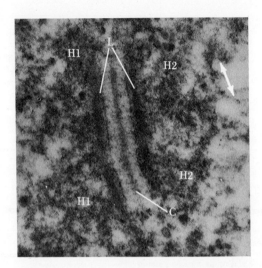

Fig. IV-21   *Portions of paired homologues in the spermatocyte of a rooster at the meiotic stage shown in Fig. II-17. The two closely paired chromosomes are represented by the irregular fibrous regions at H-1 and H-2; the long axes of the chromosomes are indicated by the double-headed arrow. At the surfaces where H-1 and H-2 are associated, a* synaptinemal complex *is seen; this consists of the pair of dense bands (called lateral elements) indicated by L and the single central element (C) found between them. Faintly visible filamentous material runs from the central element to the lateral ones.* × *50,000. Courtesy of M. J. Moses.*

somes reattach to the spindle and separate normally, but they segregate to the poles opposite those they would have moved to without the experimental reorientation. This experiment strongly hints that the chromosomes themselves, rather than some external system, control attachment of homologues to the spindle in the manner necessary for meiotic separation. Presumably the kinetochores are the responsible parts.

The molecular mechanisms of pairing and crossing over still are unknown. Some promising leads have emerged from studies on viruses. Genetic recombination between viral "chromosomes" (DNA molecules) occurs when different viral strains infect the same cell. A single recombinant chromosome is formed that carries hereditary information contributed by both parents; this is comparable to a chromatid resulting from a crossover.

One set of theories (*copy-choice* theories) states that this results from phenomena of DNA replication. During its synthesis a strand of DNA might start to form by alignment of nucleotides along a strand of one parental DNA molecule; after partial completion, it might somehow switch and finish its growth by alignment of nucleotides along a strand of a second parental DNA molecule (Fig. IV-22). For part of its

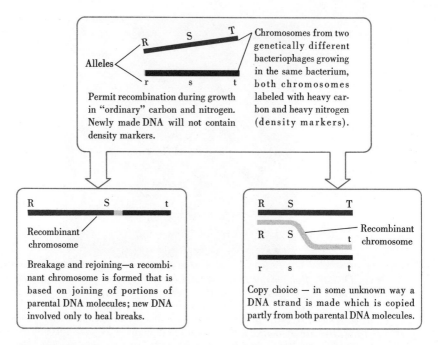

Fig. IV-22 *Alternative hypotheses to explain recombination in bacteriophages. The finding that some of the chromosomes (DNA double helices) that can be shown to be genetically recombinant are essentially as dense as the parent molecules indicates that recombination can take place by breakage and rejoining; extensive DNA replication is not required. However, the experiment is not designed to establish that this is the only mechanism by which recombination is possible. From the work of J. J. Weigle and M. Meselson.*

length, the new strand would contain information specified by the base sequences of one parental DNA, while for the rest of its length, the information would have been specified by the other parental DNA.

An alternative theory is based on breakage and rejoining; two parental DNA molecules might break, and portions of each join with portions of the other. An experiment was designed to determine whether or not this is a mechanism of recombination (Fig. IV-22). Viral DNA can be labeled with *density markers* (such as C^{13} and N^{15}, isotopes of carbon and nitrogen), which permit separation from non-labeled viral material by density gradient centrifugation (see Section 1.2B). The higher the proportion of markers in the DNA, as compared with C^{12} and N^{14} which normally predominate, the greater the DNA density. When recombination takes place between two different strains of viruses, both equally labeled, some of the recombinant chromosomes with genetic information from both parents have DNA essentially as dense as the parental DNA's. This supports a breakage and rejoining mechanism and indicates that extensive synthesis of new DNA is not a necessary part of recombination. Under the conditions of the experiment, newly synthesized DNA would contain no label and the presence of appreciable amounts in the recombinant chromosome would lead to a decrease in density to a level below the parental density.

That breakage and rejoining of chromosomes may also be involved in crossing over in eucaryotes is suggested by use of morphological "markers," homologous pairs in which the two chromosomes differ slightly in morphology. They produce crossover products that are visibly recognizable as combinations of both of the participant chromatids. However, these are observations on complex chromosomes and, while it is possible that DNA molecules break and rejoin as part of crossing over in eucaryotes, this is difficult to study. Since crossing over requires that parts of different chromosomes be close to one another, it seems likely that pairing and crossing over are related processes. However, there is no unequivocal evidence that crossing over must occur *after* the chromosomes are visibly paired in prophase; for example, it could be part of the pairing process. In meiotic cells of some plants, two periods of DNA synthesis have been found. The first is the usual replication period during interphase and it results in doubling of the amount of DNA. The second involves less than 1 percent of the DNA. It occurs during meiotic prophase and thus may possibly participate in pairing or crossing over. If the second period of synthesis is prevented by treatment with an inhibitor of DNA synthesis (*deoxyadenosine*), the chromosomes appear fragmented, normal pairing is prevented, and meiosis is halted. One simple suggestion is that the second period of synthesis is a period when separate or broken DNA molecules are joined as part of pairing or crossing over, but this remains to be proven. Interruption of protein synthesis during meiotic prophase also leads to abnormal chromosome behavior, and it is likely that the chemical events underlying visible pairing and crossing over are complex.

The possibility has been raised several times that crossing over may actually take place long before the obvious pairing stages of meiotic prophase, perhaps even during the period of DNA replication in the interphase preceding meiosis. In this connection, it is sometimes proposed that some kind of pairing takes place during this interphase but that it is not usually seen because the chromosomes are uncoiled and difficult to follow. The evidence for these proposals is indirect and widely disputed, but as long as the mechanisms of crossing over in eucaryotes are unknown, the proposals cannot be completely disregarded. That crossing-over does not *precede* chromosome duplication is indicated by the fact that only one of the two chromatids on each homologue participates in a given exchange.

Exceptions to the general rules of meiosis are known; the exceptions indicate the extent to which even fundamental cellular processes have undergone evolutionary change. During sperm formation in the insect *Sciara*, segregation occurs between *maternal* and *paternal* chromosomes, that is, between the chromosomes derived from those originally contributed by the male and female parents to the zygote that produced the individual under study. At the first meiotic division, the maternal chromosomes move to one pole and their homologues of paternal origin move to the other. Only cells receiving the maternal chromosomes form functional sperm. The paternal chromosomes are not transmitted to the next generation. Accordingly, the patterns of heredity are quite unlike those generally found. In most species, maternal and paternal homologues separate at random to the poles of the first meiotic division, and therefore all gametes usually contain some chromosomes derived from those contributed by both parents. (Crossing over additionally mixes maternal and paternal chromosomal material.) The recognition devices responsible for the unusual behavior of *Sciara* chromosomes are not known.

In the brine shrimp *Artemia*, development is by *parthenogenesis*, the development of an egg without fertilization. The first division of oocyte meiosis is normal, but in the second the chromatids separate without cytoplasmic divisions and a single nucleus eventually forms with both chromatids of each chromosome. Consequently, the egg is diploid and, although there is no contribution of chromosomes by sperm, the egg produces a diploid organism.

4.3.5   ***SOME*** The variation in genetic constitution re-
CONSEQUENCES sulting from mutation, meiosis, and sexual
OF GENETIC reproduction is the raw material upon which
DIVERSITY selection operates in evolution. The pattern
of interaction of gene and environment
differs somewhat in organisms with different types of life cycles. For example, mutations in diploid organisms are less rapidly, or less directly, exposed to selective "testing" by the environment than those of haploid organisms. In a haploid organism, a mutation that results in the production of a defective form of an important enzyme will result in an inviable cell. In diploid organisms, this is not necessarily so, because each gene is represented at least twice in each cell. Thus, the presence of a newly mutated allele on one chromosome may be "masked" (the allele acting as a *recessive*) by the presence on the homologous chromosome of a different (*dominant*) allele that can support the synthesis of adequate amounts of "normal" enzyme. In addition, the presence of two different alleles (*heterozygosity*) in an individual may be more advantageous than the presence of the same

allele on both homologues (*homozygosity*). Also, the extent to which a gene is advantageous or disadvantageous often depends on the environment. Thus, heterozygous individuals with the allele for sickle cell anemia (Section 4.3.1) on one chromosome and an allele for normal hemoglobin on the other homologue may have little or no difficulty (beyond a slight anemia) at low altitudes in temperate climates. However, at high altitudes where the oxygen pressure is low, their red-blood cells will assume the abnormal, shrivelled shape. On the other hand, heterozygosity also confers resistance to forms of malaria, a disease prevalent in the tropics. Individuals homozygous for the sickle cell allele usually die from severe anemia. The net result of factors such as these is that populations of diploid organisms tend to have a much more complex *gene pool* (the total number of genes and alleles in all individuals) than do populations of haploid organisms. The preservation in diploid populations of alleles that may be selectively disadvantageous under some conditions provides a source of evolutionary variability not available to haploids. If the environment changes, these alleles may become advantageous. On the other hand, the large populations and rapid division rates of many haploid unicellular organisms permit rapid evolution by environmental selection among mutants that continually arise at random within all populations.

4.3.6   ***EVOLUTIONARY*** Figure IV-23 is a diagram of the more
   CHANGES IN common chromosomal changes in number
   CHROMOSOMES or morphology that contribute to the differences in karyotype among species.
   Many plant species appear to have evolved, in part, by *polyploidy*, the presence in the cell of extra *sets of chromosomes*. This is related partly to the manner by which polyploidy can overcome sterility of *species hybrids* (the products of matings between organisms of two different species). In most such hybrids, when the two species are only distantly related, the chromosomes contributed by one parent are not homologous to those from the other parent. Normal meiotic pairing cannot occur, and the gametes produced have variable numbers of abnormal combinations of chromosomes; zygotes rarely survive. If, however, the entire chromosome complement of the hybrid is doubled, each chromosome is present twice and can pair with its duplicate; normal meiosis and gamete formation can take place. The doubling may occur naturally by accidental failure of cytoplasmic separation in a dividing cell. Plant breeders induce doubling by the use of agents, such as *colchicine*, which dissolve the spindle. Chromosomes duplicate to form chromatids but, in the absence of a spindle, both chromatids end up in the same cell.

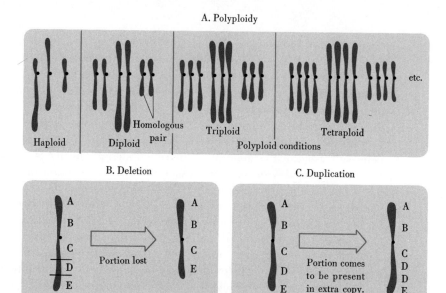

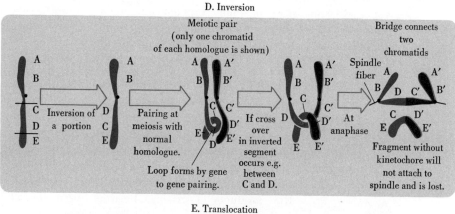

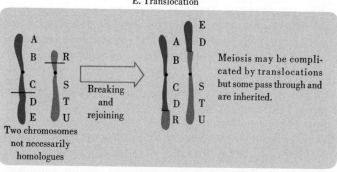

Fig. IV-23 *Chromosome abnormalities. The abnormalities illustrated occur occasionally in nature and may be induced experimentally by certain drugs, irradiation or other means.*

Polyploidy often results in larger cells, with nuclei enlarged in proportion to the number of extra chromosome sets. Some large, polyploid cells are often present along with the diploid cells in tissues (liver, roots and so forth) of multicellular organisms but, beyond the change in size, the significance of this is not clear.

Several human disorders are related to *aneuploidy*, the presence of abnormal numbers of one or several chromosomes of the complement, rather than of entire sets of chromosomes, as in polyploidy. Among these is Down's syndrome (once called "mongolism"), a form of mental retardation. This condition is associated with the presence of three copies of one chromosome, rather than the normal two copies. Thus a total of 47 chromosomes per cell is present, rather than the normal 46.

Radiation and other types of damage as well as abnormalities in crossing over occasionally produce *deletions* and *duplications*. Deletions are usually harmful; for example, defective recessive genes on one homologue may be unmasked (Section 4.3.5) if the corresponding "normal" segment of the other homologue is deleted. If both homologues lack the same regions, functions vital to the cells may be absent; the abnormal nucleolar chromosomes in *Xenopus* (Sections 2.3.4 and 4.3.2) probably have undergone deletion of the DNA segments responsible for rRNA production. Duplications can sometimes result in abnormalities; apparently some sort of balance existing among the genes normally present in two copies is upset by the introduction of additional copies. The fact that some genetic information (DNA nucleotide sequences) normally is present in many duplicate (redundant) copies has already been mentioned; this is true for rRNA sequences in organisms from bacteria to men (See Chaps. 2.3 and 5.1) and may also be true for tRNA and other sequences. Duplications are of interest from another viewpoint. Mutations in the extra region of the chromosome may not dramatically affect the organism, since the cell is still left with two unmutated genes. By successive mutations, the genes in the extra segment can diverge quite widely from their original nucleotide sequences; this can lead to the formation of proteins with new properties, such as the ability to catalyze a reaction for which no enzyme previously existed in the cell. In effect, then, a new gene has arisen from a previously existing gene, while two copies of the "old" gene continue to be present in the cell. (Evidence that this has occurred in evolution will be discussed in Section 4.5.1.) Very similar phenomena probably occur with polyploidy also when "extra" copies of genes come to be present.

Inversions may have a somewhat different significance. Crossing over constantly generates new combinations of alleles of different genes *within* chromosomes. Under some circumstances, it is of selective advantage to the population that particular combinations of alleles present on one chromosome and favorable in a given environment not be disrupted. If a chromosome region becomes inverted, crossing over with a noninverted chromosome is, in effect, suppressed. Close meiotic pairing may be prevented for the region; if pairing and crossing over do take place in the chromosome region, the products are abnormal and usually will not be transmitted by viable offspring to the next generation. As seen in Fig. IV-23, a crossover in the inverted region will produce chromosome fragments and also *bridges* that are connected to the kinetochores of both homologues; these bridges break in meiosis when the homologues separate. Only the chromatids in which no crossover has occurred in the inverted region will be normal, and usually only they will contribute to viable offspring. Thus, the combination of alleles within an inversion tends to be inherited as a block that is not changed from generation to generation by crossing over. Different populations of a single species of the fly *Drosophila*, living at different altitudes or temperatures, show differences in the pattern of inversions on their chromosomes, an evolutionary result of such phenomena.

Since *translocations* (Fig. IV-23) often result in altered lengths of chromosomes, they probably contribute to evolutionary changes in the karyotype. One finding of interest in cells carrying translocations is that the expression of a given gene in terms of cell characteristics may differ depending on its neighbors. Gene *A* may behave differently if placed next to gene *B* than it does when next to *C*; this also is observed with some inversions. Although the basis for these "position effects" is not known, the findings emphasize that genes are not to be considered as totally separate and autonomous agents strung out along the chromosome. The evolution of the chromosome and gene set of a species involves not only specification of the appropriate *kinds* of metabolism, but also the balance and integration of metabolism.

If by chance one of the chromosomal breaks should occur in a nucleolar organizer region, nucleoli may be formed by both chromosomes that receive part of the organizer through translocation. This supports the other evidence that the genetic information in the nucleolar organizer (which includes the nucleotide sequences specifying ribosomal RNA) is present in many similar or identical copies (see Section 2.3.4 and Chapter 5.1). The effect of the translocation is to move some of the copies elsewhere, so that two groups of copies are present rather than one.

4.3.7   ***CYTOPLASMIC***   Cases of *maternal inheritance* are known,
             INHERITANCE   in which characteristics of the offspring are determined solely by the female parent. These cases usually reflect *cytoplasmic inheritance*; that is, trans-

mission of genetic information present in the cytoplasm rather than in the nucleus. Maternal inheritance is based on the fact that in most species of higher organisms, the egg is much larger than the sperm and the sperm contributes relatively little cytoplasm to the zygote.

The mitochondria and plastids are best established as carriers of cytoplasmic genetic information. In yeast cells, the "petite" mutants have mitochondria that are abnormal in structure and function. In some "petite" mutants, the changes arise within the mitochondria themselves; in a few, major alterations in mitochondrial DNA are known to occur. The result of the mitochondrial abnormalities is disruption of aerobic metabolism. The cells use the less efficient anaerobic metabolism; thus, they grow more slowly and form smaller aggregates.

Cytoplasmic mutations are known that affect plastid structure and function. Some result in plants with white portions in which no chlorophyll is made. Offspring of such plants are white or green depending on the site (white portion or green portion) producing the female gametes from which they derive; this is true even with male gametes from normal green plants.

If the cytoplasm of *Euglena*, a unicellular flagellate, is irradiated with ultraviolet light at wavelengths that affect nucleic acids, a colorless "bleached" cell line may result. Irradiation damages the plastid DNA and abolishes the ability of *Euglena* to transmit normal plastids to its offspring. Thus, damage of cytoplasmic DNA results in the irreversible loss of a cytoplasmic function.

Earlier (Section 3.3.3) mention was made of the evidence from protozoa that patterns of basal bodies can be inherited independently of the nucleus. This must also depend on genetic information in the cytoplasm, although the mechanisms are not as clear as they seem to be for mitochondria or plastids.

The occurrence of inheritable mutations or other changes in cytoplasmic organelles does not imply that the organelles duplicate and function in totally autonomous fashion. As outlined in Sections 2.6.3 and 2.7.4, major components of mitochondria and plastids are probably synthesized outside of these organelles. Mutations of nuclear genes can drastically affect mitochondrial and plastid structure and function; for example some yeast "petite" strains result from such mutations. On the other hand, it is becoming increasingly clear that a cell cannot be regarded simply as a product of the nucleus, in which the cytoplasm merely responds to nuclear "commands."

As yet, the evolutionary effects of mutations of genes in cytoplasmic organelles have been little studied. Although such mutations represent a potential source for the variations observed in structure and chemistry, the fact remains that most organelles are basically similar in a vast diversity of organisms. This suggests that the basic features of the organelles were established relatively early in the evolution of present-day organisms; once the organelles had arisen as highly efficient functioning units, most major mutations tended to be deleterious and therefore were not perpetuated.

c h a p t e r **4.4**

DIVERGENCE OF CELLS IN EMBRYONIC DEVELOPMENT

4.4.1   DEVELOPMENT AND DIFFERENTIATION   Figure IV-24 illustrates the major stages in development of eggs of higher animals. The fertilized egg divides by mitosis to produce the *blastula*. As mentioned earlier (Section 4.2.3), the divisions are referred to as *cleavages*. Little cell growth occurs and thus sequential divisions separate the fertilized

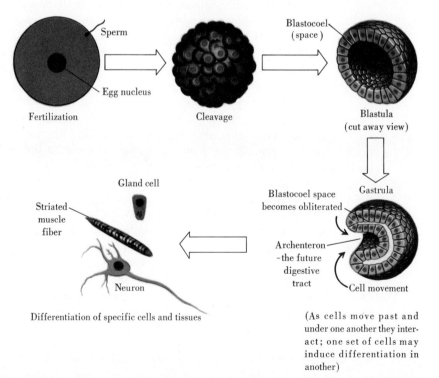

Fig. IV-24   *A schematic outline of early development in some higher animals.*

egg into successively smaller cells. Many eggs have special localized cytoplasmic regions that become included in only a few of the cells resulting from the cleavages. The blastula forms a *gastrula*, by kinds of cell movement that vary according to the species, and during this process, some cells are brought into new spatial relations with others. Subsequently, *differentiation* of cell types occurs, and the embryo increases in size and mass by cell growth and division. The juxta-position of cells in new geometric relations during gastrulation is accompanied by the *induction* of differentiation in groups of cells through interaction with other cells that have come to lie close to or in contact with these groups. Cells acquire characteristics enabling them to "recognize" one another (Section 2.1.6). Additional patterns of cell growth and migration result in the establishment of other specific associations of different cell types; for example, the complex relation-ships among neurons, sensory receptors and muscle cells.

Eventually, embryologists will integrate detailed descriptions of these many processes into a coherent description of molecular changes. Emphasis here is chiefly on the nuclear and chromosomal aspects of differentiation and on the specific activation or repression of genes. Increasing evidence supports the proposition that in different cell types, different genes are switched on or off so that cells come to differ in their structural and chemical characteristics although they possess the same genetic constitution. When different genes are switched on, different specific mRNA's are produced from different DNA templates. Consequently, cell-specific proteins (such as enzymes, secretory ma-terials, or contractile proteins) are synthesized. A point to note is that there must also be some DNA templates that function in virtually all cell types, such as templates for ribosomal or transfer RNAs.

4.4.2 DIFFERENTIA-TION IN EMBRYOS In the few eggs that have been thoroughly studied (sea urchins and amphibia), it appears that protein synthesis in the initial development of the egg depends chiefly on ribosomes and on mRNA that is stored during oogenesis and activated at fertilization (Section 3.10.2). About the time of gastrulation, RNA synthesis and ribosome formation increase dramatically; in many embryos, nucleoli first be-come prominent at gastrulation. The differential synthesis of specific mRNA's in different cells presumably becomes more and more pro-nounced following gastrulation.

Experiments with the sea urchin eggs have shown that cleavages up to the blastula stage can take place in the absence of a nucleus. Differentiation, however, depends on the presence of a nucleus and on the interaction of nucleus and cytoplasm. In some developing amphibian embryos, nuclei can be taken from cells at progressively later stages of development and substituted, by microsurgery, for the nucleus of a zygote that has not yet begun development. The purpose of such experiments is to determine whether the nucleus undergoes irreversible changes during development that are related to the differentiation of cells. Do the offspring of the zygote nuclei retain the ability to give rise to the nuclei of all the varied cells of the organism, or do they change to "specialized" states in which they can no longer support development of many different cell types?

With eggs of the frog *Rana*, nuclei taken from cells up to the blastula stage can support the normal development of eggs. Nuclei taken from most kinds of cells at later stages cannot; with such nuclei, the egg starts to develop but the embryos become abnormal and usually die. This suggests that nuclear changes do accompany differentiation. But interpretation is complicated by the possibility that these changes are not so much alterations in the genetic machinery related to differentiation as they are changes in the response of nuclei to the abnormal experience of transplantation. Perhaps nuclei from later stages are increasingly more easily injured or less able to adapt to the rapidity of early embryonic divisions. That this may be the case is suggested by the observation of abnormalities (for example, the presence of ring-shaped chromosomes) in the nuclei of some transplant embryos. The frequency of these abnormalities is much greater than in ordinary embryos.

If embryos of the toad *Xenopus* are studied, the findings differ significantly from those on *Rana*. The nuclei from advanced stages sometimes support normal development; adult *Xenopus* have been grown from eggs with nuclei transplanted from intestine cells of tadpoles. It was mentioned earlier (Section 4.3.2) that there are some individuals in appropriate strains of *Xenopus* whose nuclei contain only one nucleolus and other individuals whose nuclei contain two nucleoli. The number of nucleoli is a permanent characteristic for the nuclei of each type of individual. Thus, by counting nucleoli after the proper choice of donor and recipient cells for nuclear transplantation, it is possible to demonstrate directly that the nuclei introduced experimentally have given rise to the nuclei of all the cells of the developing individuals. It may be concluded from these experiments that, while nuclei may change in development, the changes are not invariably irreversible. The nuclei of at least some differentiated cells can give rise to all the nuclei of an organism. For these cells, differentiation does not involve the permanent loss of genetic information or the inactivation of a portion of the genetic material in a way that makes reactivation impossible.

Some of the experiments on plant cells mentioned in Chapter 3.11 lead to a similar conclusion.

4.4.3 **INTERACTIONS** In frogs, nuclei from one species, capable **OF NUCLEUS AND** of supporting normal development in eggs **CYTOPLASM** of that species, cannot do so when transplanted into eggs of a different though related species. Some features of nucleo-cytoplasmic interactions are abnormal.

Normally inactive nuclei of chicken red blood cells (birds, unlike mammals, having nucleated red blood cells) can be induced to metabolic activity, such as RNA synthesis, when placed in the cytoplasm of mammalian tissue culture cells. As outlined in Chapter 3.11 this may be done by virus-induced fusion of the red blood cells with HeLa cells, a strain of human tissue-culture cells. The red blood cell nuclei are normally small; their chromatin is densely packed and shows little or no uptake of radioactive RNA precursors. In the fused cells, the chicken nuclei enlarge, the chromatin spreads out, and much RNA synthesis is detectable by autoradiography. Apparently the HeLa cell cytoplasm has "activated" the red blood cell nucleus. DNA synthesis also may be initiated in the red blood cell nuclei; normally, red blood cells do not divide.

These experiments must be interpreted cautiously, since they involve abnormal conditions and drastic manipulation of cells. However, they provide direct evidence that the interaction of nucleus and cytoplasm is a reciprocal process.

Another example of such reciprocity occurs during the development of a number of species of insects. Like those of some other organisms, the eggs of these species contain microscopically distinguishable cytoplasmic regions. One such region, rich in RNA, is at one end of the ovoid egg (the *germinal pole*). In early development, the nuclei undergo mitosis without cytoplasmic division (Section 4.2.3) and they become distributed through the egg cytoplasm. Later, cytoplasmic division separates the egg into uninucleate cells. Only the nuclei that become associated with the germinal pole cytoplasm do *not* undergo an unusual series of mitoses in which a specific group of chromosomes fails to move on the spindle. The chromosomes are eliminated from the nuclei and disintegrate in the cytoplasm. The cells in which the full chromosome complement is retained are those which later produce gametes, the *germ-line* cells. Thus the germ-line cells have a different chromosome complement from the other (*somatic*) cells of the organism. It is not known what mechanisms underlie these striking differences in nuclear behavior that are apparently dependent upon the cytoplasm.

4.4.4   *SALIVARY GLAND CHROMOSOMES*

Some tissues of insects such as *Drosophila*, the fruit fly, and *Chironomus*, the midge, show interesting chromosomal behavior. This has been studied most intensively in the salivary glands. The cells of these glands are extremely large and their nuclei are correspondingly large. Although the cells are non-dividing, the chromosomes are clearly visible. Measurements of the DNA content indicate that the nuclei may contain multiples of the normal amount of DNA as great as 1024 times or more. (This is the amount (2^{10}) that would be expected if the DNA of an ordinary diploid cell replicated 10 times, with all the products of the prior replications taking part in each doubling). Histones and some other nuclear proteins are present in similarly elevated amounts. The number of chromosomes is normal (4 pairs in species of *Drosophila*), and homologous chromosomes are usually paired gene for gene. However, each chromosome consists of a great many parallel fibers (such chromosomes are referred to as *polytene*). The chromosomes are several hundred microns in length and several microns thick, in contrast to the ordinary chromosomes of diploid cells of insects, which are a few microns long and less than a micron thick. The most reasonable interpretation is that the chromosome has undergone uncoiling and has repeatedly replicated without separation of the daughters. It is thought by proponents of the "single-strand" view of chromosome structure (Section 4.2.5) that one fiber of a salivary gland chromosome is comparable to the morphological and functional unit of an ordinary chromosome.

Study of these giant chromosomes has afforded a unique opportunity to correlate cytological, genetic, and developmental information. Figure IV-25 shows a polytene chromosome from *Chironomus*. The striking pattern of transverse bands is evident. Although DNA and protein probably are present throughout the entire length of the chromosome, the bands show a high concentration of DNA and histone; this results from the alignment in the bands of tightly folded or coiled regions (*chromomeres*) of the parallel fibers. The size, appearance, and arrangement of the bands along a given chromosome are different in nonhomologous chromosomes, but identical in the two homologous chromosomes. Several tissues other than the salivary glands also have polytene chromosomes. The banding pattern at a given chromosome is the same in all tissues, and thus there is a parallelism between bands (or more precisely the chromomeres comprising the bands) and genes. Genes also are linearly arranged along chromosomes, and they are the same in the two homologous chromosomes but different in non-homologous chromosomes. Different cells of an individual have identical endowment. By a variety of techniques (including observation of the correlation between altered characteristics of the organism and deletions (Section 4.3.6) of specific chromosome portions identifiable by examination of the salivary gland chromosomes), it is possible to map the chromosomes; that is, to establish the location of specific genes controlling different characteristics (eye color, wing morphology, and many others) at specific bands.

At certain times, some bands show modifications called *puffs* (Fig. IV-25) which result in part from uncoiling of the chromomeres. These puffs are rich in RNA and are seen by autoradiography to incorporate rapidly radioactive precursors of RNA. Of great interest is the fact that different bands form puffs in different tissues; differences

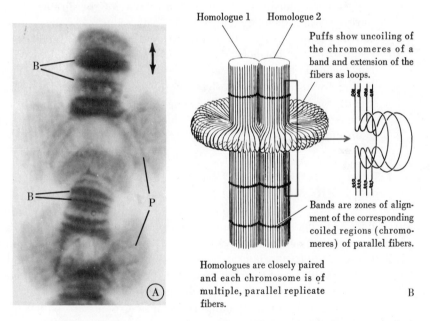

Homologue 1   Homologue 2

Puffs show uncoiling of the chromomeres of a band and extension of the fibers as loops.

Bands are zones of alignment of the corresponding coiled regions (chromomeres) of parallel fibers.

Homologues are closely paired and each chromosome is of multiple, parallel replicate fibers.

B

Fig. IV-25 *Salivary gland chromosomes.* **A.** *Portion of a chromosome from the salivary gland of the insect,* Chironomus. *The double-headed arrow indicates the direction of the long axis of the chromosome. The cross-banding (B) characteristic of these chromosomes is readily visible as are two puffs (P) where the chromosome bands have been altered as indicated in part B of the diagram.* × 1,900. *Courtesy of U. Clever.* **B.** *Schematic representation of a polytene chromosome like the one shown in A. What appears at first glance to be a single thick chromosome is actually two homologues closely paired, gene for gene, and each consisting of many parallel, longitudinally arranged fibers.*

in location of the puffs are seen even among cells of the salivary glands that produce different secretions. Furthermore, specific puffs appear and disappear at specific times of development. From these observations, it has been hypothesized that puffs are sites where particular genes are especially active. Puffs show the expected differences from cell to cell and also the high levels of RNA synthesis one would expect of such gene sites. Polytene chromosome puffing is considered a morphological manifestation of a fundamental molecular and developmental process — the activation of specific genes.

Puffing patterns in salivary gland cells are also influenced by materials originating elsewhere in the body. This is experimentally demonstrated by injection of *ecdyson*, a hormone that plays a key role in control of the developmental cycle of molting (skin shedding) and cocoon building. Injection of the hormone into young larvae induces a puffing pattern that is not normally seen in the salivary chromosomes until later in development — at the time when the organism's own ecdyson normally acts. Interruption of part of the circulatory system during normal development, so that ecdyson reaches only some of the salivary gland cells, will result in puffing in these cells; cells not reached by the hormone do not puff. Thus, intracellular events at the chromosomal level in the cells of an organ (the salivary gland) are controlled by hormones made by cells of another organ (the prothoracic gland). Such cell interactions during development are difficult to study in most organisms because they lack the convenient specializations, giant chromosomes.

4.4.5   **LAMPBRUSH** Another unusual chromosome of great
   CHROMOSOMES convenience for the investigator is found
   in oöcytes of amphibia and a few other animals. During the first meiotic prophase, the oocytes deposit large amounts of yolk and grow enormously. At this time, the chromosomes possess numerous lateral loops, thus giving them their "lampbrush" (or test-tube brush) appearance (Fig. IV-26). Each pair of homologous chromosomes has a characteristic pattern of loops.

When the ends of lampbrush chromosomes are pulled with fine needles, the loops pull out as if they are kinks in a continuous thread. Thus, the loops are not separate structures attached to the chromosomes, but are specialized regions of a continuous structure running the length of the chromosome. Much RNA is present on the loops; its removal by enzymatic (RNAase) treatment does not disrupt the chromosome. On the other hand, DNAase treatment quickly fragments the chromosome into short pieces, indicating that DNA maintains the continuity of the chromosome.

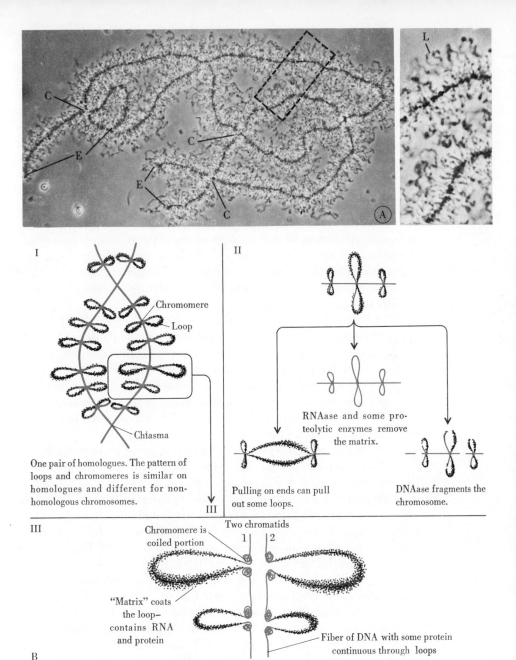

Fig. IV-26 *Lampbrush chromosomes.* **A.** *A pair of homologous chromosomes from an amphibian (Triturus) oöcyte nucleus (first meiotic prophase). The four ends of the two chromosomes are seen at (E) and chiasmata holding the homologues together are seen at (C). Projections (loops) along the entire length of the chromosomes give them a brush-like appearance. The outlined area is enlarged in the photograph at the right to show the lateral loops (L). Approx. ×400; right, ×800. Courtesy of J. Gall.* **B.** *Experiments to reveal the structure of lampbrush chromosomes and the location of key components. The interpretation of the structure is based on experiments of the type shown in II. After the work of Callan, MacGregor, Gall and others.*

Analysis of the rapidity and of other features of the reaction to DNAase treatment has led to the suggestion that each chromatid in the lampbrush chromosomes may be only one DNA molecule thick and that the entire chromatid length is occupied either by one very elongated DNA molecule or by several DNA molecules linked end to end. (The DNA, of course, is not present as a naked molecule; it is associated with histones, other proteins, and RNA which contribute, for example, to the matrix coating of the loops). If this model is correct, the studies of lampbrush chromosomes provide strong support for the concept that chromosomes generally contain a single basic morphological and functional unit equivalent to an elongate DNA molecule (Section 4.2.5). There is however, some dispute as to the details of interpretation of the DNAase-digestion studies.

The lampbrush configuration apparently reflects activation of genes. At the base of each loop, the chromosome is tightly coiled (Fig. IV-26), while in the loop itself the chromosome is uncoiled. Much RNA is concentrated at the loops; by autoradiography, it can be seen that the loops rapidly take up radioactive precursors of RNA. Thus it is probable that they are sites of active RNA production. A similar apparent parallelism between uncoiling and gene activity is encountered in the polytene chromosomes (Section 4.4.4).

At the stage when lampbrush chromosomes are present, the oocyte also shows many nucleoli, sometimes hundreds. At certain times, the nucleoli lie near the nuclear envelope and appear to contribute considerable amounts of material (probably ribosomes or ribosomal precursors) to the cytoplasm; ribosomes are abundant in the cytoplasm. Each nucleolus contains its own DNA, believed to derive from the nucleolar organizers on the chromosome (see Fig. V-1). This has been explained by the proposal that the DNA of the nucleolar organizer makes many copies of itself and that these are released into the nuclear sap where each functions independently to produce a nucleolus. In essence, this leads to a great amplification in the number of sites for specific RNA synthesis. It also results in the presence of much "extra" DNA in the nucleus. Comparable DNA replication at specific sites occurs in some puffs of the polytene chromosomes of insects; this is detected by cytochemical methods for demonstrating DNA or by autoradiography. (These puffs show localized incorporation of DNA precursors.)

The polytene chromosomes of insects are found mainly in cells which are at terminal stages in differentiation; they are destined eventually to degenerate. The chromosomes do not return to "normal" or take part in mitosis. The situation is different with the lampbrush chromosomes. The oocyte nuclei complete meiosis and do return to "normal"; the lampbrush chromosomes change back to "ordinary" chromosomes.

4.4.6   ***CHROMOSOME***   The studies on the giant chromosomes
        CONDENSATION   of insect tissues, and on lampbrush chromo-
        AND ACTIVITY   somes have been of great use in providing
clues to mechanisms of chromosome func-
tion. While the evidence is not complete, the concepts that different
genes are active in different cell types and that some types of gene
activation result in characteristic morphological changes in the chromatin are reasonable proposals with growing experimental backing.
The chromosomes of most cells are much smaller and less suitable
than the polytene or lampbrush chromosomes for detailed morphological study of the chromosomal changes accompanying gene activation. Nevertheless, some suggestive observations have been made.

In the sperm cell, the chromosomes are densely packed, and
little RNA synthesis occurs in the nucleus. The chromosomes of dividing cells are coiled into compact arrays and again there is little
RNA synthesis.

As outlined earlier (Section 2.2.5), in many cells chromosome portions may remain tightly coiled (*heterochromatic*) during interphase.
In female mammals, either one of the X chromosomes may be tightly
coiled as the sex chromatin (see Fig. II-11). In some strains of mice
and cats, the two X chromosomes carry different alleles for coat color.
The genes on one X are active in some cells, and genes on the other X
are active in other cells. This results in characteristic coat patterns
of separate regions of two colors (and also explains why some coat
patterns are found only in females).

In some classes of white blood cells, much of the chromatin is
coiled into dense heterochromatin regions. As studied by autoradiography, these regions show considerably less activity in RNA synthesis
than do the uncoiled euchromatic regions.

From such observations, it is hypothesized that generally, the
tightly coiled ("condensed") regions of the chromosomes are regions
of genetic inactivity and the uncoiled ("extended") regions are regions
of genetic activity. (*Activity* refers to the expression of genetically
controlled characteristics, presumably by the chain of transcription
and translation leading to protein molecules.) Recently, techniques for
isolation of chromosomes have been modified, permitting separation of
smaller chromosomes from larger ones of the same cells; eventually it
may be possible to isolate specific chromosomes. Progress has also been
made toward the separation of condensed and extended chromatin
from a single cell type. Thus it may soon be experimentally feasible to
determine directly the chemical bases for differences among chromosomes and chromosome regions. Another "mapping" technique that
may be of great use is based on molecular hybridization (Fig. II-8).
Initial results suggest that it is possible, for example, to locate sites
of synthesis of some specific RNA's by *autoradiography* of *chromo-*

somes to which specific radioactive RNA's have been hybridized. Radio-activity should appear at chromosomal sites of DNA base sequences complementary to the RNA sequences.

4.4.7   GENE CONTROL   In considering the control of gene expression in bacteria (Section 3.2.5), the "repression" of the activity of specific DNA segments by molecules binding to these segments was described. In the few cases that have been studied, specific proteins appear to be responsible for repression.

Similar mechanisms of repression and derepression probably operate in eucaryotic cells, but the agents responsible have yet to be identified. Histone molecules have been proposed as the molecules that switch genes on and off. There are some suggestive findings: (1) In test-tube experiments, using chromatin isolated from cells or mixtures of purified DNA and proteins, the presence of histones can inhibit greatly several enzyme systems responsible for DNA replication or for RNA transcription from DNA templates; removal of histones bound to the DNA enhances activity of the enzymes; (2) Some sperm have protamines associated with their DNA. Since protamine is an extremely basic protein that might bind to DNA more strongly than histones do, its presence could account, in part, for the metabolic inactivity of sperm nuclei; (3) Histones are closely associated physically with DNA at all stages of the cell life cycle. However, none of these observations establishes the role of histones. The few bacterial repressor proteins so far isolated are of several types, but they do not resemble histones. Regulation of genetic activity requires a high degree of specificity; a molecule that turns a given gene on or off should act specifically on that gene. As yet, histones have not been shown to possess such specificity; there do not appear to be enough different types in a given nucleus to control more than a relatively few genes. There is some evidence that when histones are removed from DNA in the test tube, RNA's can be synthesized with different nucleotide sequences from the RNA's made if histones are present. Perhaps the histones block some DNA regions more effectively than others. On the other hand, histones of different tissues and of different organisms are often quite similar in size, amino acid composition, and other characteristics.

Interesting hypotheses have been put forward to circumvent some of these difficulties. One suggests that the effective agents of gene control are not histones by themselves, but rather that complexes of histones with specific RNA's possess the necessary precise specificity to block DNA sequences selectively. Another proposal is that the binding of histones to specific DNA regions is controlled by mechanisms which modify the groups (chiefly the amino groups, NH_3^+) responsible for the binding. If acetate or other groups are linked to the amino groups of histones, the strength of binding to DNA will be altered; if this is selective, it could provide a mechanism for specific gene control. Evidence for these mechanisms is not yet conclusive. It may be that histones play a structural role in chromosomes and that the observations considered above are unrelated to their main function in the cell. Another possibility is that the control functions of histones are different from the specific repression-derepression mechanisms involved in turning individual genes on and off. One suggestion related to this is that histones control chromosome coiling.

Recently, much attention has focused on hormones as participants in control of gene action. Some, such as the steroid *aldosterone* appear to enter responsive cells and concentrate in the nuclei; probably they bind to specific nuclear proteins. Subsequently, the cell shows altered metabolism resulting, perhaps, from activation of genes. However, the detailed molecular mechanisms for control of gene activity in eucaryotic cells still remain to be described.

chapter **4.5**

DIVERGENCE OF CELLS
DURING EVOLUTION

In the course of evolution, new enzymes and metabolic pathways have arisen; diverse groups of procaryotes, algae, and protozoons have adapted to different environments; and specialized cells within multicellular organisms have assumed distinctive roles. At some point, cellular life must have arisen from precellular life, eucaryotes from procaryotes, diploid cells from haploid cells, and multicellular organisms from unicellular organisms.

Some early cells appear to have been preserved in the fossil record. Objects resembling present-day procaryotes have been identified in rocks formed 2–3 billion years ago (although it is not always possible to determine whether they are truly the remains of cells or merely incidental mineral deposits unrelated to life). It is easier to identify early algae and protozoa because the characteristic cell walls of the algae and the silicon- or calcium-containing exoskeletal structures of some protozoons are well preserved. Information about tissue organization (for example, bones in animals, wood of plants) may be obtained by microscopic examination of fossilized multicellular organisms. Aspects of cell chemistry can be deduced from the analysis of organic deposits such as coal or certain petroleum oils, which are formed by the transformation of living organisms and their products.

Present methods, however, do not permit deductions regarding fine structure or metabolic organization from fossils. Knowledge of cell evolution must, therefore, depend heavily on the comparative study of existing species, on speculation from known features of macromolecules and cells, and on postulated features of primitive environments. For example, nucleic acids are universally found as carriers of genetic information, a similar nucleotide-carried genetic code is universally used, and essentially similar RNA's and ribosomes are used to make protein by all species studied, procaryotes and eucaryotes alike. This suggests that the nucleic acids and present-day mechanisms of replication, transcription, and translation arose very early in the evolution of life. Geological investigations suggest that oxygen became abundant in the earth's atmosphere about 1–2 billion years ago. By then, processes similar to photosynthesis had probably evolved, since such processes are held responsible for the presence of abundant atmospheric oxygen. As the amount of atmospheric oxygen increased, organisms with aerobic metabolism evolved. Similar mechanisms for the anaerobic breakdown of sugar are found in most organisms, from bacteria to higher plants and animals. The mechanisms must have arisen early in the evolution of present-day species, probably in some ancestor common to most present-day organisms. The addition to these anaerobic mechanisms of the aerobic, oxygen-dependent metabolic sequences of sugar breakdown resulted in an enormous increase in the efficiency with which energy could be obtained by cells from their nutrients (see Section 1.3.2). Presumably, this permitted the expenditure of energy for "luxury" items, such as the replication and segregation of very large amounts of DNA and the maintenance of very large, stable multicellular aggregates. As cells evolved more complex structures, the selective advantages in metabolic efficiency, flexibility, or adaptation to different environments must have outweighed the problems of maintenance and reproduction of more complex organization.

4.5.1 SOME FACTS AND SPECULATIONS It is becoming possible to add some detail to the general picture presented in the preceding section. There is a growing body of evidence on the evolution of molecules and metabolic systems, and of speculation on the evolution of organelles.

Comparative studies of the amino acid sequences in the respiratory protein, cytochrome *c*, indicate that it arose early in the evolution of aerobic metabolism. The amino acid sequences of portions of the molecule are identical in the cytochrome *c* of yeast, invertebrates, mammals, and higher plants. Organisms as distantly related as yeast and mammals show the same amino acid sequence in almost half of the molecule (the total number of amino acids in a cytochrome *c* molecule is 104–108 in different organisms). Thus, once cytochrome *c* evolved as part of a highly efficient enzyme system, the great majority of mutations that changed the amino acid sequence of certain portions of the molecule must have been selectively disadvantageous because they adversely affected the functioning of the system; organisms carrying such mutations would be eliminated by natural selection. The portions of the molecule that are similar or identical in all species studied are presumed to include amino acid sequences of particular importance to the respiratory function of the molecule.

However, part of the cytochrome *c* has changed, indicating that not every amino acid in the sequence is fixed in evolution or vital for the functioning of the protein. Under some circumstances, when selective advantages result, new enzymes can evolve from more ancient proteins. Obviously this must be based on changes in the nucleotide sequences of the appropriate genes. This probably can occur with chromosome duplications (Section 4.3.6); mutations in an extra copy of a gene may eventually result in a protein that assumes new functions. That this has occurred in evolution is again indicated by comparative studies of proteins. For example, *hemoglobin*, the iron-containing protein of red blood cells and *myoglobin*, an iron-containing protein found in some muscles and a few other tissues, probably evolved simultaneously along separate lines from a common ancestral protein. Hemoglobin contains four subunits, each a polypeptide chain about the size of a myoglobin molecule (which is a single polypeptide chain) and each having a three-dimensional structure similar to myoglobin. Both proteins bind oxygen and have similar amino acid arrangements in the region of the molecules responsible for oxygen binding and for three-dimensional shape. However, the two proteins have diverged considerably in amino acid sequence, indicating that the evolutionary separation must be of long standing; at some point subsequent to the separation, the evolutionary changes in hemoglobin resulted in the formation of a tetrapartite molecule. The mammalian digestive enzymes, *trypsin* and *chymotrypsin*, catalyze different reactions in the breakdown of proteins; however, they probably have a common ancestor since they show considerable similarity in amino acid sequence. Several different pituitary gland hormones also have similar amino acid sequences. In all these cases, the similarities between distinct proteins can be most simply explained by the proposal that extra copies of genes arose and that this was followed by evolutionary divergence of the original and the extra gene. It has been speculated that the variety of transfer RNA's, each specific for a given amino acid, may have arisen by a similar process, since the few tRNA's that have been studied thoroughly show many similarities in the sequence of nucleotides (apart from the coding nucleotides involved in binding to mRNA). How protein synthesis would take place in the hypothetical organisms in which this evolution started is not known, but it is possible that inefficient mechanisms involving only a few tRNA's in the synthesis of simple proteins might have permitted survival.

Probable sequences of the evolution of organelles are less clear and current ideas are considerably more speculative. Multienzyme complexes might have arisen when mutations resulted in the presence of groups at the surface of two enzymes that promoted binding the two together. A sort of evolutionary self-assembly might have taken place with the selective advantages of increased efficiency leading to the preservation of useful intermolecular associations that arose by chance.

For the next step, the formation of complex organelles such as mitochondria and plastids, an interesting theory is that symbiosis was involved, as may be indicated by some of the facts outlined in the discussion of these organelles. Thus, as one hypothetical sequence suggests, a primitive nonmotile photosynthetic cell containing photosynthetic enzymes and pigments bound to membranes (perhaps resembling present-day photosynthetic procaryotes) might have been phagocytosed by a primitive ameba-like cell; for some reason it was not digested. The combination of motility, photosynthetic capacity and perhaps phagocytic ability would be of mutual advantage. If the two cells multiplied synchronously, the association might be stable; in time it might become necessary for the survival of the partners. (In the evolution of *lichens*, a symbiotic relationship between algae and fungi has apparently evolved to the point where the fungi normally cannot survive without the algae.) The photosynthetic cell might, under these circumstances, evolve into a plastid and lose its ability to live independently. The evidence that symbiosis of this general type might have occurred includes the following: Mitochondria and plastids are partially self-duplicating; they possess nucleic acids and ribosomes which differ from the rest of the cell; and, in some cases, the ribosomes resemble the ribosomes of procaryotes. There are present-day cases of situations in which one cell type lives within another in a symbiosis-like association; thus, while the evidence for origin of organelles through symbiosis is not conclusive the suggestion is not too far-fetched.

An interesting point may be raised with respect to the evolution of peroxisomes. Their biochemistry suggests that these organelles could have served in a primitive respiratory mechanism in which H_2O_2 was involved. However, peroxisome distribution and metabolic capabilities are very variable among different cells, and organisms and many cell types have not yet been found to have any peroxisomes. One explanation for this proposes that the original respiratory function was superceded by the evolution of mitochondrial mechanisms that were more efficient. It is possible that peroxisomes were lost from many cells in subsequent evolution, but that in some cell types portions of their enzymatic machinery (somewhat different portions in different cells) assumed new functions in the increasingly complex cellular metabolism and thus were not eliminated.

The evolution of mitosis, a key mechanism in eucaryotes, must have occurred early, since mitosis occurs in virtually all present-day eucaryotes, despite the fact that the evolutionary separation of groups such as plant cells and animal cells probably is very ancient. The evolutionary significance of mitosis lies in the fact that it accomplishes a regular segregation of large amounts of DNA and thus provides for constancy of complex genetic combinations. The steps in the evolution of the process may be reflected in the presence of alternate division mechanisms in a few eucaryotes. For example, in ciliated protozoa, the micronucleus divides by mitosis, while the macronucleus apparently is pinched in two after duplication of its contents.

At present, the origins of most of the highly diversified cells of present-day organisms remain a matter of conjecture. It is hoped, however, that an understanding of them eventually will be reached, explaining evolutionary steps such as those that have led to the presence in the same organism of pigment cells producing melanin from the amino acid *tyrosine* and gland cells producing the hormone adrenalin, from the same starting material. Study of evolution at the level of whole organisms started from a handful of fossils, a few unusual species, and a mass of anatomical descriptions. By now it has progressed to experimentation on the subtle effects of the environment on the frequency of particular genes in populations. It may be expected that the study of evolution at the cellular and molecular level will progress in similar manner.

FURTHER READING

Molecular and Cellular Aspects of Development. Bell, Ed. (ed.), New York: Harper and Row, 1965. A collection of accounts of important experiments on developing embryos and tissue culture systems.

Cairns, J., "The bacterial chromosome," *Scientific American,* Jan. 1966, 214(1):36.

DuPraw, E. J., *Cell and Molecular Biology.* New York: Academic Press, Inc., 1968. 739 pp. Contains much useful information on chromosomes.

DNA and Chromosomes. DuPraw, E. J. New York: Holt, Rinehart and Winston, 1970. A useful reference source.

Ebert, J. D., *Interacting Systems in Development*, 2d ed. New York: Holt, Rinehart and Winston, Inc., 1965. An introduction to developmental biology.

Gurdon, J. B. "Transplanted nuclei and cell differentiation," *Scientific American*, Dec. 1968, 219(6):24.

Kellenberger, E., "The genetic control of the shape of a virus," *Scientific American*, Dec. 1966, 215(6):32.

Levine, R. P. *Genetics*, 2d ed. New York: Holt, Rinehart and Winston, Inc., 1968. An introduction to genetics.

Prescott, D. M. (ed.), *Advances in Cell Biology, Volume 1*, Appleton-Century-Croft Co., 1970. Useful articles reviewing recent research on the nucleus and other topics.

Sagan, L., "On the origin of mitosing cells," *J. Theoretical Biology*, 1967, 14:225–274. A stimulating speculative account of the evolution of cells.

Sager, R., "Genes outside the chromosome." *Scientific American*, Jan. 1965, 212(1):70.

Swanson, C. P., Merz, T., and Young, W. J. *Cytogenetics*. Englewood Cliffs, N.J.: Prentice-Hall, 1967. A short book on chromosome behavior in heredity and evolution.

Formation and Fate of Cell Organelles. Warren, K. B., (ed.), New York: Academic Press, 1967. A collection of research articles on topics related to self-assembly.

Wolff, S., "Strandedness of chromsomes," *International Review of Cytology* 1969 25:279–295. Summary of "multistrand" view of chromosome structure.

Wolstenholme, G. E., and O'Connor, M. *Principles of Biomolecular Organization*. Boston: Little, Brown & Co., 1966. 491 pp. Good discussions of self-assembly.

Wood, W. B. and R. J. Edgar., "Building a bacterial virus," *Scientific American*, July 1967, 217(1):60.

Zubay, G. L. (ed.), *Papers in Biochemical Genetics*. New York: Holt, Rinehart and Winston, 1968. A collection of key research articles on molecular biology.

part **5**

*TOWARD
A
MOLECULAR
CYTOLOGY*

chapter **5.1**
FROM WILSON TO WATSON

It is now apparent that E. B. Wilson's confidence (see page 5) was well founded when he predicted over forty years ago that many "puzzles of the cell" would be solved as more powerful tools of analysis became available. None has yet been solved completely, but all are yielding to modern methods. Wilson's "puzzles" included the manner by which nuclear genes affect chemical reactions in the cytoplasm; the nature of cell differentiation; the continuity from one cell generation to the next of centrioles, mitochondria, and chloroplasts; and the influence of the cell's complex organization upon the behavior of macromolecules.

In Wilson's day, morphological description was the basic approach; chemical analysis was in its infancy. Today cell biologists have available a wide battery of methods, and qualitative description is now based on a variety of microscopes. Increasingly the problems defined by such descriptions are analyzed by precise quantitative measurements that are based on biochemical and biophysical procedures. Important strides are being made in taking structures apart and then reconstituting them in the test tube. Although Wilson recognized that form and function were inseparable, it has taken the expanded knowledge of biochemical events and their intracellular localizations, coupled with the extension of microscopic observations into the molecular realm, to establish this for all organelles within the cell.

The complex path of progress in cell biology is well illustrated by the history of the study of nucleoli. Clues have come from many directions. Initially, it was noted by light microscopists that nucleoli are found in virtually all eucaryotic cells, but that they are particularly prominent in active cells which synthesize much protein (Chapter 2.3). Later, cytochemists found much RNA in nucleoli and correlated this with the presence of abundant RNA in the cytoplasm; the presumption of transfer of RNA from nucleolus to cytoplasm could be supported by autoradiographic studies (see Fig. I-21). The discovery of genetically distinctive "anucleolate" strains of *Xenopus* (Sections 2.3.4 and 4.3.2) established the fact that the absence of nucleoli is accompanied by the absence of the formation of ribosomes. Molecular biological work explained this by showing that nucleolar RNA gives rise to ribosomal RNA (Section 2.3.4). Large precursor molecules (45 S in HeLa cells and roughly similar size in other cell types) are formed in the nucleolus and are modified then to produce the RNA molecules (18 and 28 S) found in the subunits of the ribosomes (see Fig. III-33).

But nucleoli do not seem to be self-reproducing; they disappear as organized entities in many cells during division. How are they formed? Again, early light microscopy provided an important clue by showing that there is special chromatin associated with nucleoli during interphase and during early stages of cell division. In many organisms, nucleoli are attached to special nucleolar-organizer regions of specific chromosomes (see Figs. II-17 and IV-5); when fusions between nucleoli are taken into account, the number of nucleoli parallels the number of organizer regions. Molecular hybridization studies (Section 2.3.4) supported by other evidence (Section 4.3.2) have now shown that the nucleolar-organizer regions contain the genetic information for ribosomal RNA. They have revealed that this information is present in redundant form; that is, up to several hundred copies of the information, apparently identical, may be present in each organizer. In amphibian oocytes, even more copies of the information are temporarily present under special circumstances of intensive rRNA production; replicates of the nucleolar-organizer region are produced and released from the chromosome, each to make a nucleolus (Section 4.4.5). This special situation has been used to advantage in isolation of the DNA responsible for rRNA synthesis; Figure V-1 is a photograph of DNA and associated structures isolated from one of the extra nucleoli of an amphibian oocyte. It appears that the genes have been "caught in action" as they synthesize large RNA molecules. These are probably the precursor molecules from which rRNA molecules derive. (That the elongate fiber is of DNA is indicated by its disruption when treated with DNAse; the lateral filaments are known to contain RNA since they are rapidly labeled by radioactive RNA precursors, as can be demonstrated by autoradiography.)

The hybridization of 18 and 28 S RNA with purified nuclear DNA is additive rather than competitive; that is, the binding of one of these two types of RNA to DNA does not interfere with subsequent binding of the other. Both can hybridize simultaneously as effectively as they can separately. This indicates that there are separate nucleotide sequences in the organizer for 28 and 18 S RNA. One alternative possibility would have been that 18 and 28 S RNA's shared a common sequence, as might be expected, for example, if 18 S molecules were fragments derived from the larger 28 S molecules. A simple model of arrangements of the repeated base sequences in the organizer suggests that sequences for 28 S RNA alternate with those for 18 S RNA. During transcription, large precursor RNA molecules are made, each containing a region with the sequence of a 28 S molecule and a region corresponding to an 18 S molecule. Subsequently, the precursor is split into fragments that give rise to 28 and 18 S RNA by a series of steps at least some of which take place in the nucleolus.

The functioning of the nucleolus and of the nucleolar organizer are not completely understood. For example, the details of processing of the rRNA precursor molecules are not yet known. (What enzymes fragment them into specific smaller pieces?; what modifications other than fragmentation take place?) The sources of ribosomal proteins and the mechanisms of association with rRNA, the formation of ribosomes from their subunits, the significance of the various structures in the nucleolus, and the controls of rRNA production (Section 4.3.2) are among the many problems that remain to be solved. There is, however, every reason to expect that solutions will be found, so varied and powerful are the techniques now available, and so numerous are the organisms from which to choose material.

Of central importance in elucidating cell metabolism and the genetic machinery that controls metabolism, has been the use of the procaryotic bacteria and of the noncellular viruses. Study of bacteria and viruses has led to revolutionary changes in the way the cell is viewed and in the experimental questions being asked. (These changes are dramatized in *The Molecular Biology of the Gene* (New York: W. A. Benjamin, Inc., 1965) by J. D. Watson who with Francis Crick, first unraveled the structure of DNA.) For example, in Wilson's day and long after, a mutation was an event of unknown nature, affecting a gene of unknown chemistry and resulting in a change detectable in the organism only at a level such as the color of the eye, far removed from the primary effect. Today, largely from work on bacteria and viruses, mutation is explained as a change in base sequence of DNA, resulting in a changed mRNA and protein whose enzymatic or other properties are altered, ultimately producing the visible effect in the organism.

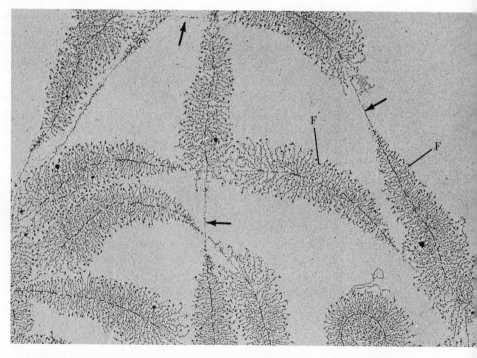

Fig. V-1   *A part of the material isolated from one of the extra nucleoli of a salamander* (Triturus) *oöcyte. Arrows point to the elongate DNA-containing fiber. Lateral filaments (F) protrude from the fiber and are grouped in regions 2-3 microns in length with each region showing about 100 filaments. 2 to 3 microns of DNA contain enough information to specify one of the large (about 45S) rRNA precursor molecules and each group of filaments is thought to represent many such RNA molecules being synthesized simultaneously by one gene. The smaller filaments of a group are considered to be at an earlier stage in synthesis.* × 20,000. *Courtesy of O. L. Miller, Jr. and B. R. Beatty.*

Solution to some of the problems of nucleolar function outlined above may also be advanced by the study of bacteria, even though procaryotes contain no nucleoli. For example, subunits of bacterial ribosomes have been dissociated into RNA and protein and then reconstituted in the test tube; the experiments indicate that the subunits may form from their components by processes resembling self-assembly. At a slightly different level of structure, evidence has accumulated that bacterial ribosomes cyclically dissociate into their subunits and are reconstructed from subunits as part of their normal functioning in protein synthesis; apparently, subunit dissociation takes place as the ribosomes reach the end of the mRNA of a polysome and reassociation is part of incorporation into a new polysome. Further, it is thought likely that association with ribosomes and initiation of protein synthesis takes place at one end of an mRNA molecule while the other end is still being completed and is attached to DNA. Further analysis of the assembly of bacterial ribosomes will undoubtedly provide essential clues to fruitful lines of investigation with eucaryotic cells. A beginning has also been made toward identification of some of the roles of specific ribosomal proteins in the functioning of bacterial ribosomes. And, most of what is known of the properties of mRNA derives from study of bacteria and viruses; processes of information transfer from eucaryote nucleus to cytoplasm are *assumed* to rely chiefly on mRNA but molecular-level study of such processes is really just commencing.

For procaryotes and eucaryotes much is unclear about DNA replication and organization. Mutant bacteria varying in enzymes and features of duplication are being studied to sort out roles of different enzymes in DNA replication and genetic recombination. A striking fact about eucaryotes is that part of their DNA, in addition to nucleolar organizers, seems to contain extensive base sequences that are repeated many thousands of times.

The concepts of molecular biology are already sufficiently developed, and enough information is available about organelles of animal and plant cells, that direct attacks on the most complex cellular functions (such as embryonic differentiation and nervous activity) have begun. Both the usefulness and limitations of simpler systems are becoming apparent. Bacteria switch genes on and off in response to environmental factors such as the presence of specific metabolites. Study of this process is providing insight into molecular mechanisms of gene repression and derepression. Now the task is to determine how such mechanisms fit into the much more complex phenomena of embryonic differentiation. Following fertilization, closely regulated and precise patterns of cell division and cell interaction lead to gene activation and to the production of specific mRNA's. Most study has been concerned with such control of transcription, but mechanisms also operate to control translation, and feedback inhibition and other mechanisms affect the functioning of completed enzymes. All of these controls contribute to determining the particular metabolic pattern of a cell and probably all play important roles in development.

Although the forms and functions of cell organelles require interpretation in molecular terms, organelles constitute higher levels of integration than molecules. A mitochondrion is not a membrane-delimited sac in which respiratory enzymes, DNA, and other molecules are dissolved at random. It is a complex *organized* structure that couples phosphorylation and oxidation, transfers electrons in a highly efficient manner, and carries out other integrated functions. Nuclear DNA in eucaryotes is not a naked template. It is part of an organized chromosome, capable of complex interactions with other chromosomes (for example, in meiotic pairing) and with other organelles (for example, the mitotic spindle).

There is great intellectual excitement and far-reaching practical importance in molecular explanations of cytological events. But there is also profound drama in cell activities at higher levels of integration: the ceaseless beating of heart muscle cells for many decades, the spectacular specializations of structure and function within a single protozoon cell, or the beautifully synchronized and precisely balanced changes in the development of an animal or plant embryo.

One reads Wilson's book, decades after its publication, with astonishment at the variety, complexity, and beauty of cells. One reads Watson's book with enthusiasm for the precision and power with which events at the molecular level can now be described.

c h a p t e r **5.2**

CYTOLOGY AND PATHOLOGY

Even while the concept of the cell as the unit of form and function in higher organisms was being established as a principle of biology, it was extended to the study of diseased tissue. The basis for cell pathology was laid in the mid-nineteenth century by Rudolph Virchow, and since that time pathology, cell biology, and cytology have been interdependent.

With the present focus in cell study upon organelles and molecules, cell pathology is naturally moving in this direction. Explanations of abnormal cell functions, particularly their origins (*pathogenesis*), are being sought in terms of organelles and genes. Tantalizing clues have been reported: for example, changes of endoplasmic reticulum in fatty liver development; of polysomes in fibroblasts during vitamin

C deficiency; of lysosomes in nerve degeneration and regeneration; and of mitochondria in riboflavin deficiency. The disease that is best understood in molecular terms has already been mentioned: in sickle cell anemia, a single amino acid change in a protein, resulting from mutation, leads to grossly abnormal red blood cells (Sections 4.3.1 and 4.3.5). Enzymatic changes in the so-called inborn errors of metabolism have been known for a relatively long time. (These changes include the abnormal enzymes of amino acid metabolism in several rare mental disorders and the absence of an enzyme of pigment metabolism in some forms of albinism.) Recently it has been recognized that, in many of the "storage diseases," (Section 2.8.5) where large quantities of carbohydrates, lipids, or other substances are deposited in the cells of organs, such as the brain and liver, the deposits are often inside abnormal lysosomes. In *Pompe's disease*, a rare, invariably fatal disease in children, large amounts of glycogen are deposited in cells of the heart and skeletal muscle and in lysosomes of liver cells. The latter lack one of the enzymes (normally present in liver lysosomes) that can hydrolyze glycogen. The evidence that exogenous materials may enter the cell by pinocytosis and that these usually reach the lysosomes raised the possibility of supplying the missing hydrolase to these lysosomes. The first attempts at a cure based on intravenous injections of the enzymes did not improve the health of the patient, but the enzyme *was* taken up by the liver and almost all glycogen-filled lysosomes disappeared from the liver cells. Cardiac muscle cells are not active in macromolecule uptake and the glycogen in them is not contained in lysosomes, so heart glycogen was not removed by the injected enzyme; heart abnormalities apparently lead to death in this disease.

The ultimate perspective of cell pathology is to obtain sufficient insight into the basic causes of disease so that cure or prevention is possible. A great task of the moment is the elucidation of the molecular basis of carcinogenesis. Because of extensive biochemical studies, there is more reason for optimism than ever before; of great importance have been the applications of molecular biological techniques to cultured normal and malignant cells and to study of the transformation of normal to malignant cells in culture by addition of oncogenic viruses.

Pathology also contributes to an understanding of normal cells. Often the roles of given cells or organelles can be better understood when a particular function is exaggerated; the abnormal teaches much about the normal. For example, the manner of synthesis of membranes and enzymes of smooth ER in hepatocytes is being studied in animals treated with the drug phenobarbitol, since such treatment leads to the manufacture of large amounts of such ER (Section 2.4.4). Several drugs and carcinogens produce changes in nucleoli that are visible by microcinematography of living cells; at first there is an apparent redistribution of material within the nucleoli and also changes in their size. Subsequently the nucleoli may disintegrate. Electron microscopy of treated cells reveals that the several components of each nucleolus segregate from one another, so that the nucleoli show large separate regions of granules, fibrils, and amorphous material. Of interest is the fact that some of the effective agents have known effects on RNA synthesis (*Actinomycin D*, for example, inhibits the synthesis of RNA). Thus it may be possible to correlate structural and functional changes and to understand better the complexities of nucleolar morphology.

Diseased tissue is often the source material for isolating biologically important substances. It was from pus cells that Miescher, almost 100 years ago, isolated "nuclein," later renamed DNA; pus contains large numbers of white blood cells in easily obtainable form and already partially separated from the anucleate red blood cells.

c h a p t e r **5.3**

EPILOGUE

One of the extraordinary aspects of science is that each generation learns enough to warrant optimism for the future and to arouse excitement in the succeeding generation. The prospects for cytology and cell biology have never been as bright as they are today. Important first steps have been taken that foreshadow test-tube synthesis of living matter. The synthesis of complete viral RNA and DNA was achieved in a mixture containing only initial copies of nucleic acid, enzymes, and small precursor molecules. Test-tube synthesis of proteins from amino acids has been accomplished. The first was the hormone, *insulin,* the second the enzyme, *ribonuclease.* The isolation of specific portions of the genetic apparatus of bacteria (p. 172) and higher organisms (p. 321) and the identification of some of the molecules responsible for control of gene activity (Sect. 3.2.5) also point toward future findings of great interest.

Biology is in the midst of a revolutionary period and experiments inconceivable ten years ago are everyday exercises today. Roughly forty years elapsed between the last edition of Wilson's book and the first edition of Watson's. It is probable that, forty years from today, features of the cell will be described in terms of electrons and atomic nuclei; already, concepts of the new solid-state physics are being applied to electron transport in mitochondria and chloroplasts. Knowledge of the cell changes continuously. The excitement of studying cells derives partly from solving old problems, but each solution brings new questions that require answers.

Index